SPIN AND ISOSPIN
in Particle Physics

SPIN AND ISOSPIN
in Particle Physics

PETER A. CARRUTHERS
Professor of Physics
Cornell University
Ithaca, New York

GORDON AND BREACH SCIENCE PUBLISHERS
New York London Paris

Gordon and Breach, Science Publishers, Inc.
150 Fifth Avenue
New York, N.Y. 10011

Editorial office for the United Kingdom

Gordon and Breach, Science Publishers Ltd.
12 Bloomsbury Way
London W.C.1

Editorial office for France

Gordon & Breach
7–9 rue Emile Dubois
Paris 14e

Library of Congress catalog card number 72-160021. ISBN 0 677 02580 7. Printed in Northern Ireland.

PREFACE

In the past decade experimental studies of collisions of elementary particles have revealed many resonant states of high spin. It is often useful to regard these entities as quasi-stable "particles." The study of the properties of such particles comprises a significant fraction of current research in particle physics. At present there is no widely accepted theoretical understanding of the rich spectrum of strongly interacting particles. At the same time no particular mathematical description of particles of arbitrary spin has dominated theoretical research. (The formalism of helicity amplitudes has proved very useful for the systematic features of scattering amplitudes, but this is a phenomenology rather than a theory.)

In the present volume we describe several approaches to problems involving arbitrary spin and isospin which have been found useful in dealing with "practical" theoretical problems. Our intention is to help graduate students in theoretical physics acquire the technical knowledge about spin and isospin necessary to carry out research. It is also hoped that experimental physicists may find something of value in the book. We have assumed that the reader has mastered the material contained in a typical graduate level course in elementary particle physics, including the rudiments of canonical field theory and Feynman graphs. Much of the material in Chapters 2–6 could easily be included in such a course. Chapters 7–10 deal with selected applications to various problems of current interest. The introductory chapter on the four-dimensional orthogonal group may be omitted entirely if desired. However some readers will find the treatment of the homogeneous Lorentz group simpler after having read Chapter 1. For the sake of clarity we have tried to avoid presenting every idea in its most general and abstract form. A full treatment of our subject would fill a treatise of oppressive size.

The citation of references has become a difficult matter in recent years because of the great volume of literature. The cited references are mainly those consulted during the writing of the manuscript.

Some of the material derives from lecture notes prepared for courses given at Cornell University. Much of the rest is a byproduct of various research projects of the author. Various portions of the manuscript were written at Cornell University, the California Institute of Technology, and the Aspen Center for Physics. The author is grateful to the latter two institutions for their frequent hospitality. The author is indebted to his wife, Lucy, for her careful typing of the manuscript, and to the editors of *Annals of Physics, The Physical Review* and the *Journal of Mathematical Physics* for their permission to reproduce some material from their journals.

P.A.C.

CONTENTS

1

THE FOUR-DIMENSIONAL ORTHOGONAL GROUP *SO*(4)

1.1 INTRODUCTION

The finite dimensional representations of the homogeneous Lorentz group are closely related to the representations of the four-dimensional orthogonal group $SO(4)$. In this chapter we study $SO(4)$ and its representations as an introduction to the physically relevant (but more complicated) Lorentz and Poincaré groups. Before commencing the study of the specific group $SO(4)$ we make some general remarks.

Consider the non-singular linear transformation of the N dimensional vector $x = (x_1, x_2, \ldots, x_N)$:

$$x' = ax \tag{1.1}$$

The components x_i and matrix elements a_{ij} are in general complex, and the group corresponding to the matrix transformations is called $GL(N)$, the general linear group in N dimensions. We shall usually be concerned with subgroups of $GL(N)$ wherein the matrix a is restricted by some condition, such as preservation of some quadratic form. We shall be particularly interested in the four-dimensional orthogonal group, whose transformations preserve the value of $x_1^2 + x_2^2 + x_3^2 + x_4^2$ (x_i real) and the restricted Lorenz group, whose transformations preserve $x_4^2 - x_1^2 - x_2^2 - x_3^2$ (x_i real).

A set of operators $O(a)$ which imitates the composition of linear transformations of (1.1), i.e.

$$O(b)O(a) = O(c) \tag{1.2}$$

when $ba = c$, is said to be a representation of the group defined by the basic transformation (1.1).

A standard method for obtaining representations is to consider changes in functions induced by the transformation $x' = ax$. Define a function $\phi'(x)$ such that its value at $x' = ax$ is the same as ϕ at x

$$\phi'(x') = \phi(x), \quad \text{or} \quad \phi'(x) = \phi(a^{-1}x) \tag{1.3}$$

The operator $O(a)$ associated with the transformation is

$$O(a)\phi(x) \equiv \phi'(x) = \phi(a^{-1}x) \tag{1.4}$$

If $x' = ax$ is regarded as a translation to the right then $O(a)$ applied to $\phi(x)$ gives the value at the point $a^{-1}x$ so that the contour of the function is

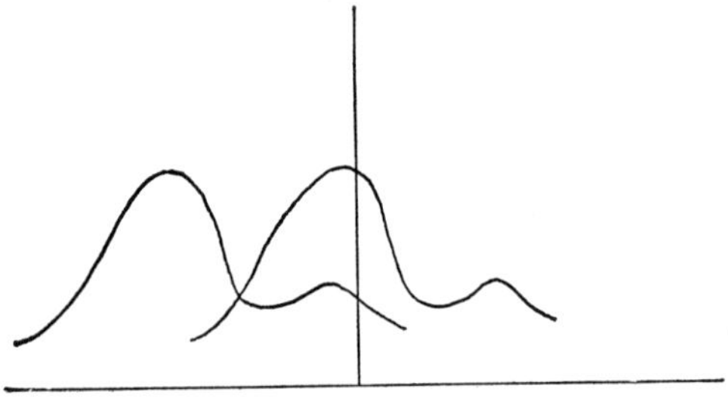

Figure 1.1 The example of a translated function illustrates the basic operation defined in Eq. (1.4).

"shifted" to the right by a. This interpretation is more vivid if one considers translations (Fig. (1.1)).

$$\begin{aligned} \phi'(x) = D(a)\phi(x) &= \phi(x - a) = e^{-a\cdot\partial/\partial x}\phi(x) \\ &= e^{ia\cdot p}\phi(x) \end{aligned} \tag{1.5}$$

$$D(a) = e^{ia\cdot p} \qquad p_k = i\frac{\partial}{\partial x_k}$$

1.2 INFINITESIMAL PROPERTIES OF $SO(4)$; REPRESENTATIONS

The four-dimensional orthogonal group $SO(4)$ is defined by real transformations $x' = ax$ leaving invariant the quantity

$$x^2 = x_1^2 + x_2^2 + x_3^2 + x_4^2 \tag{1.6}$$

The requirement $x_i^2 = (x_i')^2$ gives the orthogonality condition

$$a_{ij}a_{ik} = \delta_{jk} \qquad a^T a = 1 \tag{1.7}$$

where T denotes transpose. As an immediate consequence, we find

$$a^T = a^{-1} \qquad (\det a)^2 = 1 \tag{1.8}$$

We consider only the proper transformations with $\det a = 1$.

To describe infinitesimal transformations we need six real infinitesimal parameters $\delta\omega_{ij}$:

$$\begin{aligned} a_{ij} &= \delta_{ij} + \delta\omega_{ij} \\ \delta\omega_{ij} &= -\delta\omega_{ji} \end{aligned} \tag{1.9}$$

Rather than study the transformation (1.9) directly we consider the operators $O(a)$; since $a^{-1}x - x = \delta x - \delta\omega x$ we can exploit the antisymmetry of $\delta\omega$ to write (here $\partial_i = \partial/\partial x_i$)

$$\begin{aligned} O(a)\phi(x) &= \phi(x) + \delta x_i\, \partial_i\phi(x) \\ &= \left(1 + \frac{i}{2}\,\delta\omega_{ij}M_{ij}\right)\phi(x) \end{aligned} \tag{1.10}$$

where $M_{ij} = -i(x_i\,\partial_j - x_j\,\partial_i)$. M_{ij} is defined in analogy to the angular momentum operators of the three-dimensional rotation group $SO(3)$.

To first order in $\delta\omega$

$$O(1 + \delta\omega) = 1 + \frac{i}{2}\,\delta\omega_{jk}M_{jk} \tag{1.11}$$

Here and elsewhere we use the remarkable result of the theory of Lie groups that the infinitesimal transformations completely determine the local structure of the group.[1] Thus the basic problem is to study the solutions of the Lie algebra of the infinitesimal generators M_{ij}. Using $[x_i, \partial_j] = -\delta_{ij}$, we easily find

$$[M_{mn}, M_{rs}] = i\,\delta_{ms}M_{rn} + i\,\delta_{ns}M_{mr} - i\,\delta_{nr}M_{ms} - \quad \delta_{mr}M_{sn} \tag{1.12}$$

The problem of solving this equation, which is the *defining* equation for representations of the $O(4)$ algebra, is more general than the particular differential operator giving rise to the algebra. The solution to the problem is easily found noting the homomorphism of $O(4)$ to $O(3) \otimes O(3)$ [or $SU(2) \otimes SU(2)$]. This is verified by splitting the algebra into a direct sum of $O(3)$ algebras. Define

$$\begin{array}{lll} L_1 = M_{23} & L_2 = M_{31} & L_3 = M_{12} \\ K_1 = M_{41} & K_2 = M_{42} & K_3 = M_{43} \end{array} \tag{1.13}$$

This gives the algebra

$$\begin{aligned} [L_i, L_j] &= i\varepsilon_{ijk}L_k \\ [L_i, K_j] &= i\varepsilon_{ijk}K_k \\ [K_i, K_j] &= i\varepsilon_{ijk}L_k \end{aligned} \tag{1.14}$$

Inspection of this set of equations suggests that one can decouple the generators by a change of basis:

$$\mathbf{J}^{(1)} = \frac{\mathbf{L} + \mathbf{K}}{2} \qquad \mathbf{J}^{(2)} = \frac{\mathbf{L} - \mathbf{K}}{2} \tag{1.15}$$

$$[J_i^{(\alpha)}, J_j^{(\alpha')}] = i\delta_{\alpha\alpha'}\varepsilon_{ijk}J_k^{(\alpha)}$$

For the particular realization given by Eq. (1.10) we have

$$\begin{aligned} 2J_1^{(1)} &= -i\left[x_2\frac{\partial}{\partial x_3} - x_3\frac{\partial}{\partial x_2} + x_4\frac{\partial}{\partial x_1} - x_1\frac{\partial}{\partial x_4}\right] \\ 2J_2^{(1)} &= -i\left[x_3\frac{\partial}{\partial x_1} - x_1\frac{\partial}{\partial x_3} + x_4\frac{\partial}{\partial x_2} - x_2\frac{\partial}{\partial x_4}\right] \\ 2J_3^{(1)} &= -i\left[x_1\frac{\partial}{\partial x_2} - x_2\frac{\partial}{\partial x_1} + x_4\frac{\partial}{\partial x_3} - x_3\frac{\partial}{\partial x_4}\right] \\ 2J_1^{(2)} &= -i\left[x_2\frac{\partial}{\partial x_3} - x_3\frac{\partial}{\partial x_2} - x_4\frac{\partial}{\partial x_1} + x_1\frac{\partial}{\partial x_4}\right] \\ 2J_2^{(2)} &= -i\left[x_3\frac{\partial}{\partial x_1} - x_1\frac{\partial}{\partial x_3} - x_4\frac{\partial}{\partial x_2} + x_2\frac{\partial}{\partial x_4}\right] \\ 2J_3^{(2)} &= -i\left[x_1\frac{\partial}{\partial x_2} - x_2\frac{\partial}{\partial x_1} - x_4\frac{\partial}{\partial x_3} + x_3\frac{\partial}{\partial x_4}\right] \end{aligned} \tag{1.16}$$

Clearly the particular identification of coordinates made in the definition of the generators is non-unique. This is also true of Eq. (1.13).

Formally the solution to the representation problem is solved by specifying two angular momenta (j_1, j_2). The states within an *IR** (j_1, j_2) are labeled by

* Henceforth we will often use the abbreviation *IR* to denote *irreducible representation.*

$(m_1, m_2) = (J_3^{(1)}, J_3^{(2)})$. There are $(2j_1 + 1)(2j_2 + 1)$ components of each *IR* with basis functions

$$\Psi_\kappa^{j_1 j_2} = \Psi_{j_1 m_1} \Psi_{j_2 m_2} \tag{1.17}$$

The *group* factorization is exhibited by writing the (finite) unitary transformation as

$$O(a) = \exp\left(\frac{i}{2}\,\omega_{ij} M_{jk}\right) \tag{1.18}$$

To express this in terms of **L** and **K** we define

$$\begin{aligned} \boldsymbol{\alpha} &= (\omega_{41}, \omega_{42}, \omega_{43}) \\ \boldsymbol{\beta} &= (\omega_{23}, \omega_{31}, \omega_{12}) \end{aligned} \tag{1.19}$$

$$O = \exp\,(i\boldsymbol{\alpha}\cdot\mathbf{K} + i\boldsymbol{\beta}\cdot\mathbf{L}) \tag{1.20}$$

or,

$$O = \exp\,(i\boldsymbol{\lambda}_1\cdot\mathbf{J}^{(1)})\exp\,(i\boldsymbol{\lambda}_2\cdot\mathbf{J}^{(2)}) \tag{1.21}$$

where $\boldsymbol{\lambda}_1 = \boldsymbol{\alpha} + \boldsymbol{\beta}$, $\boldsymbol{\lambda}_2 = \boldsymbol{\beta} - \boldsymbol{\alpha}$. Define the $O(4)$ representation matrices as usual by*

$$O(a)\Psi_\kappa^{j_1 j_2} = \sum_\lambda \Psi_\lambda^{j_1 j_2} D_{\lambda\kappa}(a) \tag{1.22}$$

Orthogonality gives

$$D_{\lambda\kappa}(a) = \langle \Psi_\lambda^{j_1 j_2}|\, O(a)\, |\Psi_\kappa^{j_1 j_2}\rangle \tag{1.23}$$

We thus obtain

$$\begin{aligned} D_{\lambda\kappa}(a) &= D^{j_1}_{m_1' m_1}(\boldsymbol{\lambda}_1) D^{j_2}_{m_2' m_2}(\boldsymbol{\lambda}_2) \\ \lambda &= (m_1', m_2'); \qquad \kappa = (m_1, m_2) \end{aligned} \tag{1.24}$$

The *IR* may be labeled by the values of the "Casimir invariants"

$$\mathbf{J}^{(1)2} = j_1(j_1 + 1), \qquad \mathbf{J}^{(2)2} = j_2(j_2 + 1) \tag{1.25}$$

or the angular momenta j_1, j_2. $\mathbf{L}\cdot\mathbf{K}$ and $\frac{1}{2}(\mathbf{L}^2 + \mathbf{K}^2)$ are equally good invariants:

$$\begin{aligned} \tfrac{1}{2}(\mathbf{L}^2 + \mathbf{K}^2) &= \mathbf{J}^{(1)2} + \mathbf{J}^{(2)2} = j_1(j_1 + 1) + j_2(j_2 + 1) \\ \mathbf{L}\cdot\mathbf{K} &= \mathbf{J}^{(1)2} - \mathbf{J}^{(2)2} = j_1(j_1 + 1) - j_2(j_2 + 1) \end{aligned} \tag{1.26}$$

---

* The operators (1.11) satisfy (1.2) and easily lead to $D(b)D(a) = D(ba)$ for the representation matrices of (1.22).

Still another form is of use, which is "covariant" and independent of the particular coordinate identification used above.

$$
\begin{aligned}
C_1 &\equiv \tfrac{1}{4}M_{ij}M_{ij} = \tfrac{1}{2}\sum_{i<j} M_{ij}M_{ij} = \tfrac{1}{2}(\mathbf{L}^2 + \mathbf{K}^2) \\
C_1 &= \mathbf{J}^{(1)2} + \mathbf{J}^{(2)2} = j_1(j_1 + 1) + j_2(j_2 + 1) \\
C_2 &\equiv -\tfrac{1}{8}\varepsilon_{ijkl}M_{ij}M_{kl} = -\tfrac{1}{4}M_{ij}M_{ij}^D \\
M_{ij}^D &= \tfrac{1}{2}\varepsilon_{ijkl}M_{kl} \\
\varepsilon_{1234} &= +1
\end{aligned} \tag{1.27}
$$

ε_{ijkl} is the completely antisymmetrical tensor density.* Even (odd) permutations of the indices yield the value $+1(-1)$. M_{ij}^D is the "dual tensor" of the tensor M_{ij}. The components of the dual tensor are

$$
\begin{aligned}
M_{12}^D &= M_{34} \qquad M_{23}^D = M_{14} \qquad M_{31}^D = M_{24} \\
M_{41}^D &= M_{32} \qquad M_{42}^D = M_{13} \qquad M_{43}^D = M_{21}
\end{aligned} \tag{1.28}
$$

The invariant C_2 is now given by

$$
\begin{aligned}
C_2 &= -\tfrac{1}{2}\sum_{i<j} M_{ij}M_{ij}^D = -\tfrac{1}{2}[M_{12}M_{12}^D + \cdots] \\
&= +\tfrac{1}{2}(M_{12}M_{43} + \cdots) \\
C_2 &= +\mathbf{L}\cdot\mathbf{K} = -j_2(j_2 + 1) + j_1(j_1 + 1)
\end{aligned} \tag{1.29}
$$

The various forms are summarized by

$$
\begin{aligned}
C_1 &= \tfrac{1}{2}M_{ij}M_{ij} = \tfrac{1}{2}(\mathbf{L}^2 + \mathbf{K}^2) = \mathbf{J}^{(1)2} + \mathbf{J}^{(2)2} = j_1(j_1 + 1) + j_2(j_2 + 1) \\
C_2 &= -\tfrac{1}{4}M_{ij}M_{ij}^D = \mathbf{L}\cdot\mathbf{K} = -\mathbf{J}^{(2)2} + \mathbf{J}^{(1)2} = -j_2(j_2 + 1) + j_1(j_1 + 1)
\end{aligned} \tag{1.30}
$$

There is another useful set of basis functions which we now describe. Instead of identifying states by (m_1, m_2) we can use (l, m), where l is the value of the "total angular momentum" $\mathbf{L} = (\mathbf{J}^{(1)} + \mathbf{J}^{(2)})$, $(\mathbf{L}^2 = l(l + 1)$, $L_3 = m)$. Taking $j_1 \geqslant j_2$, the allowed values of l are

$$
l = j_1 - j_2, j_1 - j_2 + 1, \ldots, j_1 + j_2 \tag{1.31}
$$

The new basis functions are given by†

$$
\Psi_{lm}^{j_1 j_2} = \sum_{m_1 m_2} C(j_1 j_2 l;\, m_1 m_2 m)\Psi_{j_1 m_1}\Psi_{j_2 m} \tag{1.32}
$$

* ε_{ijkl} is an invariant tensor for $\det a = 1$, since $\det a \cdot \varepsilon_{ijkl} = a_{mi}a_{nj}a_{rk}a_{sl}\varepsilon_{mnrs}$.

† The Clebsch-Gordan coefficients are given in the notation of Rose.[2]

Next we introduce the variables j_0 and n by the definitions

$$
\begin{aligned}
l_{\min} &\equiv j_0 = j_1 - j_2 \qquad & j_1 &= \frac{j_0 + n}{2} \\
l_{\max} &\equiv n = j_1 + j_2 \qquad & j_2 &= \frac{n - j_0}{2}
\end{aligned}
\tag{1.33}
$$

The Casimir invariants C_1 and C_2 now have the values

$$
\begin{aligned}
C_1 &= \tfrac{1}{2}[j_0^2 + n(n + 2)] \\
C_2 &= j_0(n + 1)
\end{aligned}
\tag{1.34}
$$

in an *IR* of $SO(4)$ previously specified by (j_1, j_2).

It will be noticed that the obtained representations are both single- and double-valued, in analogy to the usual treatment of three-dimensional rotations. Though convention is non-uniform, we shall regard the group $SO(3)$ as comprised of all single-valued representations, and $SU(2)$ as including single *and* double-valued representations. Thus the representations of the *algebra* (1.14) of $SO(4)$ is that of the direct product group $SU(2) \times SU(2)$.

1.3 SINGLE-VALUED REPRESENTATIONS IN THE $(j_0\, n)$ BASIS

As in the analogous three-dimensional problem, many features of the representation problem are clarified by the introduction of spherical polar coordinates.

$$
\begin{aligned}
x_4 &= r \cos \psi & x_1 + x_2 + x_3 + x_4 &= r^2 \\
x_3 &= r \sin \psi \cos \theta, & 0 &\leqslant \psi \leqslant \pi \\
x_2 &= r \sin \psi \sin \theta \sin \phi, & 0 &\leqslant \theta \leqslant \pi \\
x_1 &= r \sin \psi \sin \theta \cos \phi, & 0 &\leqslant \phi \leqslant 2\pi
\end{aligned}
\tag{1.35}
$$

The domain of variation of (ψ, θ, ϕ) permits each x_i to go from $-r$ to $+r$. For $\psi = \pi/2$, $x_4 = 0$ and $x_{1,2,3}$ have their usual meaning. Thus θ, ϕ can be regarded as having their normal three-dimensional geometrical significance.

The Jacobian matrix of the transformation is

$$J = \frac{\partial(x_1, x_2, x_3, x_4)}{\partial(r, \psi, \theta, \phi)}$$

$$J = \begin{bmatrix} \sin\psi\sin\theta\cos\varphi & \sin\psi\sin\theta\sin\varphi & \sin\psi\cos\theta & \cos\psi \\ r\cos\psi\sin\theta\cos\varphi & r\cos\psi\sin\theta\sin\varphi & r\cos\psi\cos\theta & -r\sin\psi \\ r\sin\psi\cos\theta\cos\varphi & r\sin\psi\cos\theta\sin\varphi & -r\sin\psi\sin\theta & 0 \\ -r\sin\psi\sin\theta\sin\varphi & r\sin\psi\sin\theta\cos\varphi & 0 & 0 \end{bmatrix} \tag{1.36}$$

The absolute value of the determinant is

$$|\det J| = r^3\sin^2\psi\sin\theta \tag{1.37}$$

The conversion of volume element is given by

$$dx_1\,dx_2\,dx_3\,dx_4 = r^3\,dr\,d\Omega_4, \qquad d\Omega_4 = \sin^2\psi\,d\psi\sin\theta\,d\theta\,d\phi \tag{1.38}$$

in analogy to the three-dimensional relation

$$dx_1\,dx_2\,dx_3 = r^2\,dr\,d\Omega_3 \qquad d\Omega_3 = \sin\theta\,d\theta\,d\phi \tag{1.39}$$

In order to convert differential expressions to polar coordinates,

$$\frac{\partial f}{\partial x_i} = \frac{\partial f}{\partial r}\frac{\partial r}{\partial x_i} + \frac{\partial f}{\partial \psi}\frac{\partial \psi}{\partial x_i} + \frac{\partial f}{\partial \theta}\frac{\partial \theta}{\partial x_i} + \frac{\partial f}{\partial \phi}\frac{\partial \phi}{\partial x_i} \tag{1.40}$$

one needs the inverse matrix

$$[J]^{-1} = \frac{\partial(r, \psi, \theta, \phi)}{\partial(x_1, x_2, x_3, x_4)}$$

$$[J]^{-1} = \begin{bmatrix} \sin\psi\sin\theta\cos\varphi & \dfrac{\cos\psi\sin\theta\cos\varphi}{r} & \dfrac{\cos\theta\cos\varphi}{r\sin\psi} & -\dfrac{\sin\varphi}{r\sin\psi\sin\theta} \\ \sin\psi\sin\theta\sin\varphi & \dfrac{\cos\psi\sin\theta\sin\varphi}{r} & \dfrac{\cos\theta\sin\varphi}{r\sin\psi} & \dfrac{\cos\varphi}{r\sin\psi\sin\theta} \\ \sin\psi\cos\theta & \dfrac{\cos\psi\cos\theta}{r} & -\dfrac{\sin\theta}{r\sin\psi} & 0 \\ \cos\psi & -\dfrac{\sin\psi}{r} & 0 & 0 \end{bmatrix} \tag{1.41}$$

Now we are prepared to explore the eigenvalue problem and its relation to the solutions of the four-dimensional Laplacian equation. The single valued representations of $O(3)$ arise in the solution of Laplace's equation

$$\nabla_3^2\phi = 0, \qquad \nabla_3^2 = \frac{\partial^2}{\partial x_1^2} + \frac{\partial^2}{\partial x_2^2} + \frac{\partial^2}{\partial x_2^3}$$

Because of the invariance of ∇_3^2 under rotations, the rotated function $O_R\phi(x)$ solves the equation when ϕ does. One writes

$$\nabla_3^2 = \frac{1}{r^2}\frac{\partial}{\partial r}\left(r^2\frac{\partial}{\partial r}\right) - \frac{\mathbf{L}_3^2}{r^2} \tag{1.42}$$

where

$$\begin{aligned}\mathbf{L}_3^2 &= -(\mathbf{r}\times\nabla)^2\\ &= -\left[\frac{1}{\sin\theta}\frac{\partial}{\partial\theta}\left(\sin\theta\frac{\partial}{\partial\theta}\right) + \frac{1}{\sin^2\theta}\frac{\partial^2}{\partial\phi^2}\right]\end{aligned} \tag{1.43}$$

Separation of variables leads to the eigenvalue problem

$$\mathbf{L}_3^2\phi = \lambda_3\phi \rightarrow \lambda_3 = l(l+1), \qquad l = 0, 1, 2, \ldots \tag{1.44}$$

This gives a complete set of single-valued functions on the unit sphere (θ, ϕ), the usual spherical harmonics $Y_{lm}(\theta, \phi)$. Define the four-dimensional angular momentum squared (L_4^2) by

$$L_4^2 = \tfrac{1}{2}M_{ij}M_{ij} = 2C_1 \tag{1.45}$$

in analogy to $L_3^2 = \frac{1}{2}M_{ij}M_{ij}$ $(i, j = 1, 2, 3)$. We want to solve the equation

$$L_4^2\phi = \lambda_4\phi \tag{1.46}$$

Actually we know that $\lambda = 2C_1$ will be restricted to a subset of the values $2(j_1(j_1+1) + j_2(j_2+1))$. First we relate L_4^2 to ∇_4^2 and then use an explicit expression for ∇_4^2 to obtain L_4^2 in spherical polar coordinates. Defining p_j to be $-i\partial/\partial x_j$, $\nabla_4^2 = -p_i^2$, and

$$\begin{aligned}M_{ij} &= x_ip_j - x_jp_i \qquad \nabla_4^2 = -p_i^2\\ L_4^2 &= \tfrac{1}{2}(x_ip_j - x_jp_i)^2\\ &= (x_ip_jx_ip_j - x_ip_jx_jp_i)\end{aligned}$$

Now move all p's to the right using $p_jx_i = -i\,\delta_{ji} + x_ip_j$.

$$\begin{aligned}L_4^2 &= x_i^2p_j^2 - i\mathbf{x}\cdot\mathbf{p} - (-4i\mathbf{x}\cdot\mathbf{p} + x_i\cdot(\mathbf{x}\cdot\mathbf{p})p_i)\\ &= -r^2\,\nabla_4^2 + 3i\mathbf{x}\cdot\mathbf{p} - \mathbf{x}\cdot(\mathbf{x}\cdot\mathbf{p})\mathbf{p}\end{aligned}$$

Using the formula $i\mathbf{x} \cdot \mathbf{p} = r\partial/\partial r$ we obtain

$$L_4^2 = -r^2\left[\nabla_4^2 - \frac{1}{r^3}\frac{\partial}{\partial r}\left(r^3\frac{\partial}{\partial r}\right)\right]$$
$$\nabla_4^2 = \frac{1}{r^3}\frac{\partial}{\partial r}\left(r^3\frac{\partial}{\partial r}\right) - \frac{L_4^2}{r^2} \tag{1.47}$$

(cf. Eq. (1.42).) Separation of the equation $\nabla_4^2\phi = 0$ leads to

$$\phi = R\mathscr{Y}$$
$$L_4^2\mathscr{Y} = \lambda_4\mathscr{Y} \tag{1.48}$$
$$\frac{\partial}{\partial r}\left(r^3\frac{\partial R}{\partial r}\right) = \lambda_4 rR$$

The brute force calculation of ∇_4^2 by means of the formula (1.40) is extremely tedious and it is preferable to use the results of tensor calculus. The general Laplacian is a special case of the covariant divergence.[3]

$$T^n_{,n} = \frac{1}{\sqrt{g}}\frac{\partial}{\partial x^n}(\sqrt{g}T^n) \tag{1.49}$$

where $T^n = g^{nr}\,\partial\phi/\partial x^r$ and g_{ij} is the covariant metric tensor, defined by the line element ds^2

$$ds^2 = \sum_{i=1}^{4} dx_i^2 = g_{ij}\,dq^i\,dq^j$$
$$(q^1, q^2, q^3, q^4) = (r, \psi, \theta, \phi) \tag{1.50}$$
$$g_{mn} = \sum_i \frac{\partial x^i}{\partial q^m}\frac{\partial x^i}{\partial q^n}$$

From Eq. (1.36) we easily find*

$$[g_{ij}] = \begin{bmatrix} 1 & 0 & 0 & 0 \\ 0 & r^2 & 0 & 0 \\ 0 & 0 & r^2\sin^2\psi & 0 \\ 0 & 0 & 0 & r^2\sin^2\psi\sin^2\theta \end{bmatrix} \tag{1.51}$$

* The line element is

$$ds^2 = dr^2 + r^2(d\psi^2 + d\theta^2) + r^2\sin^2\psi\sin^2\theta\,d\phi^2$$

$\sqrt{g}$ is given by

$$\sqrt{g} = |\det g_{ij}|^{\frac{1}{2}} = r^3 \sin^2 \psi \sin \theta = |\det J| \tag{1.52}$$

The contravariant metric tensor is

$$[g^{ij}] = [g_{ij}]^{-1} \tag{1.53}$$

The invariant divergence $T^{n}_{,n}$ is by definition

$$\nabla_4^2 \phi = \frac{1}{\sqrt{g}} \frac{\partial}{\partial q^m}\left(\sqrt{g}\, g^{mn} \frac{\partial \phi}{\partial q^n}\right) \tag{1.54}$$

Substitution now leads to

$$\nabla_4^2 = \frac{1}{r^3} \frac{\partial}{\partial r}\left(r^3 \frac{\partial}{\partial r}\right) + \frac{1}{r^2 \sin^2 \psi}\left\{\frac{\partial}{\partial \psi}\left(\sin^2 \psi \frac{\partial}{\partial \psi}\right) + \frac{1}{\sin \theta} \frac{\partial}{\partial \theta}\left(\sin \theta \frac{\partial}{\partial \theta}\right) + \frac{1}{\sin^2 \theta} \frac{\partial^2}{\partial \phi^2}\right\} \tag{1.55}$$

Comparison with Eq. (1.47) gives the desired formula

$$\begin{aligned} L_4^2 &= -\frac{1}{\sin^2 \psi}\left\{\frac{\partial}{\partial \psi}\left(\sin^2 \psi \frac{\partial}{\partial \psi}\right) - L_3^2\right\} \\ &= -\left\{\frac{\partial^2}{\partial \psi^2} + 2 \cot \psi \frac{\partial}{\partial \psi} + \frac{1}{\sin^2 \psi}\left(\frac{\partial^2}{\partial \theta^2} + \cot \theta \frac{\partial}{\partial \theta} + \frac{1}{\sin^2 \theta} \frac{\partial^2}{\partial \phi^2}\right)\right] \end{aligned} \tag{1.56}$$

The structure of L_4^2 suggests that the independent solutions of the four dimensional spherical harmonic equation (1.48) may be written in the form

$$\mathscr{Y} = p_{nl}(\psi) Y_{lm}(\theta, \phi) \tag{1.57}$$

where the symbol n, which denumerates the solutions of the differential equation

$$\frac{d^2 p(\psi)}{d\psi^2} + 2 \cot \psi \frac{dp}{d\psi} - \frac{l(l+1)}{\sin^2 \psi} p + \lambda p = 0 \tag{1.58}$$

turns out to be the integer n introduced previously in Eq. (1.33), provided that we parametrize λ as $\lambda = n(n+2)$. Then it follows that for $n = 0, 1, 2, \ldots$ we have finite solutions of the differential equation. Note that this n is the same as defined earlier as $n = l_{\max}$. For $j_1 = j_2 = \frac{1}{2}n$, $\lambda = n(n+2)$;

also $j_0 = 0$, which implies that $C_2 = 0$ for the solutions of the four-dimensional Laplace equation

$$p''_{nl} + 2 \cot \psi p'_{nl} + \left[n(n+2) - \frac{l(l+1)}{\sin^2 \psi}\right] p_{nl} = 0 \tag{1.58a}$$

This function is not yet in recognizable form. Next make the transformation

$$p_{nl}(\psi) = (\sin \psi)^{-\frac{1}{2}} \chi_{nl}(\psi) \tag{1.59}$$

The equation for χ_{nl} is now

$$\frac{1}{\sin \psi} \frac{d}{d\psi}\left(\sin \psi \frac{d\chi}{d\psi}\right) + \left[n(n+2) + \frac{1}{\sin^2 \psi}\left(\frac{2 - 3\cos^2 \psi}{4} - l(l+1)\right)\right]\chi = 0 \tag{1.60}$$

The further transformation $z = \cos \psi$ gives $\chi(\psi) = p_\nu^\mu(z)$,

$$\frac{d}{dz}\left[(1 - z^2)\frac{dp_\nu^\mu}{dz}\right] + \left[\nu(\nu+1) - \frac{\mu^2}{1 - z^2}\right] p_\nu^\mu = 0 \tag{1.61}$$

where $\nu = n + \frac{1}{2}$.

Thus χ_{nl} is an associated Legendre function $P_\nu^\mu(z)$ with

$$\mu = -(l + \tfrac{1}{2}), \qquad \nu = n + \tfrac{1}{2} \tag{1.62}$$

(The solution $\mu = l + \frac{1}{2}$ is singular at $z^2 = 1$.) In order to establish this one may consult the Bateman project[4], which shows that $p_\nu^\mu(z)$ is $(z^2 - 1)^{\frac{1}{2}} v$, where $v = v(\frac{1}{2} - \frac{1}{2}z)$ is a hypergeometric function with $a = \mu - \nu$, $b = \mu + \nu + 1$, $c = \mu + 1$:

$$\Gamma(1 - \mu) p_\nu^\mu(z) = 2^\mu (z^2 - 1)^{-\frac{1}{2}\mu} F(1 - \mu + \nu, -\mu - \nu; 1 - \mu; \tfrac{1}{2} - \tfrac{1}{2}z) \tag{1.63}$$

Combining (1.62) and (1.63), we find

$$p_{n+\frac{1}{2}}^{-l-\frac{1}{2}}(z) = \frac{2^{-l-\frac{1}{2}}}{\Gamma(l + \frac{3}{2})}(z^2 - 1)^{l+\frac{1}{2}/2} F(l + n + 2; l - n, l + \tfrac{3}{2}; \tfrac{1}{2} - \tfrac{1}{2}z) \tag{1.64}$$

This solution degenerates to a polynomial whenever

$$l - n = -m, \qquad m = 0, 1, 2, \ldots \tag{1.65}$$

Thus when $m = 0, 1, 2, \ldots$ we have polynomial solutions* with $n = 0, 1, 2, \ldots, m$.

---

* Note that $p_{nl}(\psi) \propto (\sin \psi)^l F(l + n + 2, l - n, l + \frac{3}{2}, \frac{1}{2} - \frac{1}{2}\cos \psi)$.

From the general properties of compact groups we know that the irreducible representations are all finite dimensional. Hence the polynomial solutions exhaust the (single valued) representations of $SO(4)$.

The function $P_{n+\frac{1}{2}}^{-l-\frac{1}{2}}$ is essentially the same as the Gegenbauer function $C_{n-l}^{l+\frac{1}{2}}$, as shown on p. 175 of ref. 4.

Further details concerning the representations of the four-dimensional orthogonal group may be found in ref. 5. An interesting application of these representations to the analysis of scattering amplitudes has been given in refs. 6 and 7.

REFERENCES

1. M. Hamermesh, "Group Theory" (Addison-Wesley Publishing Co., Reading, Mass., 1962), Chap. 8.
2. M. E. Rose, "Elementary Theory of Angular Momentum" (John Wiley and Sons, Inc., New York, 1957).
3. J. L. Singh and A. Schild, "Tensor Calculus" (University of Toronto Press, Toronto, 1949).
4. "Higher Transcendental Functions," Vol. I, ed. A. Erdelyi (McGraw-Hill Book Co., New York, 1953), p. 121.
5. J. D. Talman, "Special Functions" (W. A. Benjamin, Inc., New York, 1968), Chap. 10.
6. G. Domokos and P. Suranyi, *Nucl. Phys.* **54,** 529 (1964).
7. D. Z. Freedman and J-M. Wang, *Phys. Rev.* **160,** 1560 (1967).

2

THE HOMOGENEOUS LORENTZ GROUP, I. BASIC IDEAS. IRREDUCIBLE REPRESENTATIONS

2.1 INTRODUCTION

It is assumed that physical quantities are labeled by points in the continuum of space-time coordinates $x^\mu (\mu = 0, 1, 2, 3)$ with the identification $t = x^0$, $x = x^1$, $y = x^2$, $z = x^3$ for the time t and the usual Cartesian space coordinates x, y, z. The usual starting point in the study of the special theory of relativity is to examine the consequences of the invariance of the "interval"

$$s^2 = t^2 - x^2 - y^2 - z^2 \tag{2.1}$$

under (homogeneous) Lorentz transformations

$$x' = \Lambda x \qquad (x'^\mu = \Lambda^\mu_\nu x^\nu) \tag{2.2}$$

where x' is interpreted as the set of coordinates used in a reference frame Σ' moving uniformly with respect to the original (inertial) frame. The matrices Λ describing ordinary Lorentz transformations (excluding space and time inversions) form a group, whose physical content may be regarded as composed of the three pure Lorentz transformations along the coordinate axes and three rotations around the coordinate axes. In the following we denote this group by $\mathscr{L}$.

In classical relativistic physics the basic physical quantities are tensors, and the covariance of physical laws follows automatically from the fact that tensor equations have the same form in every coordinate frame.[1]

In quantum mechanics the situation is somewhat different. In the first place the description of half-integral spin wave functions requires a class of representations of $\mathscr{L}$ (the double valued spinor representations) not envisioned by tensor calculus.* The systematic understanding of this question leads to the correlation of the transformations (2.2) with transformations in a two-dimensional complex spin space

$$\xi' = D(\Lambda)\xi \tag{2.3}$$

where the 2 × 2 matrix D has unit determinant. The matrices $D(\Lambda)$ form a group, called $SL(2, C)$. The spinor calculus is an apparatus for providing equations covariant under $SL(2, C)$ in analogy to the usual tensor calculus, and which includes the latter as a special case. Conceptually the transition from tensor to spinor representations requires only that one admit double valued coordinates in the description of physical systems. However the spinors entering the preceding equations are essentially old-fashioned wave functions, which do not provide a proper description of the state of an elementary particle. This brings us to the second, crucial difference. In quantum field theory the *state* is described by a vector in Hilbert space, while the spinor fields described above are changed to operators (defined on the space time continuum) acting on the vectors of the Hilbert space.[2] Suppose we have two physical states a and b described by vectors† Ψ_a, Ψ_b in Σ and Ψ'_a, Ψ'_b in Σ' where Σ and Σ' are related by the transformation matrix Λ. In order that the transition probability be invariant

$$|\langle\Psi_a \,|\, \Psi_b\rangle| = |\langle\Psi'_a \,|\, \Psi'_b\rangle| \tag{2.4}$$

it is necessary that the state vectors in different frames be related by a *unitary* (or anti-unitary, if time inversions are allowed) transformation

$$\Psi'_a = U(\Lambda)\Psi_a \tag{2.5}$$

Now, the (finite dimensional) representations of $SL(2, C)$ mentioned above in the description of covariant wave equations are non-unitary, as all finite dimensional *IR* of a non-compact group must be.[3] However, one may easily construct unitary representations of $SL(2, C)$. Unfortunately these representations are not the desired ones because the invariants of an *IR* do not directly correspond to the invariant properties of elementary systems (mass and spin).

---

* Historically, these considerations entered physics through Dirac's discovery of his covariant equation describing free particles of spin $\frac{1}{2}$.

† Since the vectors may (apparently) be modified by a constant phase factor without changing the state, one speaks of correspondence between *rays*.

To find the proper description we note that Eq. (2.1) tacitly selects a preferred coordinate, namely the origin of coordinates to which x^μ is referred. If instead of (2.1) we require the preservation of the relative interval

$$s = (t_x - t_y)^2 - (x^1 - y^1)^2 - (x^2 - y^2)^2 - (x^3 - y^3)^2 \tag{2.6}$$

then the most general geometrical transformation leaving (2.6) invariant is obtained by adding a translation a^μ to (2.2)

$$x' = \Lambda x + a \tag{2.7}$$

The ten parameter group thereby obtained is the Poincaré group ($\mathscr{P}$). The group multiplication law is (cf. Chapter 4)

$$(\Lambda_2, a_2) \cdot (\Lambda_1, a_1) = (\Lambda_2\Lambda_1, \Lambda_2 a_1 + a_2) \tag{2.8}$$

The translation group T ($x' = x + a$) and homogeneous Lorentz group $\mathscr{L}$ are clearly subgroups of $\mathscr{P}$.

It turns out that the invariants of $\mathscr{P}$ correspond to the mass and spin for single particle states. It is extremely interesting that the admission of translational degrees of freedom (which permits coupling of translations with 4-dimensional rotations) automatically provides the primitive physical quantities. In addition it is easier to construct the unitary representations for $\mathscr{P}$ than for $\mathscr{L}$.

Nevertheless the representations of $\mathscr{L}$ are of considerable interest. The finite-dimensional representations are needed to describe the transformations of the spin degrees of freedom of field operators. Moreover the unitary representations of $\mathscr{L}$ find increasing use in many problems of current interest. In the present chapter we analyze the representations of $\mathscr{L}$ in some detail.

Following an outline of the main features of the unitary representations we give a few references to the sizeable mathematical literature concerning this problem.

We distinguish covariant and contravariant indices and use the metric tensor $g_{\mu\nu} = \text{diag}\,[1, -1, -1, -1]$. Ordinary space vectors V^i correspond to the 1, 2, 3 components of the contravariant vector V^μ. The invariant scalar product of two four-vectors V^μ and W^μ is

$$V \cdot W = g_{\mu\nu} V^\mu W^\nu = V_\mu W^\mu \tag{2.9}$$

and the (length)^2 of V is $V^2 = V \cdot V$. The operator $\partial_\mu = \partial/\partial x^\mu$ attaches a covariant index to a tensor; e.g. $\partial_\mu \phi$ is a covariant four-vector if ϕ is a scalar.

The invariance of the space-time interval under the transformations (2.2) leads to the following conditions on the parameters Λ^μ_ν

$$\Lambda^\mu_\nu g_{\mu\rho} \Lambda^\rho_\lambda = g_{\lambda\nu} \tag{2.10}$$

This equation differs from the condition (1.7) only in the change of metric. Writing (2.10) in matrix notation $\Lambda^T G \Lambda = G$, we find

$$\det \Lambda = \pm 1 \tag{2.11}$$

Since $\det \Lambda$ is a continuous function of its elements, only $\det \Lambda = 1$ transformations are connected to the identity. (Moreover the set $\{\Lambda^\mu_\nu\}$ with $\det \Lambda = -1$ cannot form a group.) The basic transformations may also be classified according to the sign of Λ_{00}. From (2.10) we see that $(\Lambda^0_0)^2 \geqslant 1$ so that the transformations having $\Lambda^0_0 \geqslant 1$ are disjoint from those with $\Lambda^0_0 \leqslant -1$. Following the notation of Streater and Wightman[2], we separate $\mathscr{L}$ into four pieces:

$$\begin{array}{lll} L^\uparrow_+ & \det \Lambda = 1 & \operatorname{sgn} \Lambda_{00} = + \\ L^\uparrow_- & \det \Lambda = -1 & \operatorname{sgn} \Lambda_{00} = + \\ L^\downarrow_+ & \det \Lambda = 1 & \operatorname{sgn} \Lambda_{00} = - \\ L^\downarrow_- & \det \Lambda = -1 & \operatorname{sgn} \Lambda_{00} = - \end{array} \tag{2.12}$$

Of these transformations, only those of $L^\uparrow_+$ form a group. However the following unions are groups:

$$\begin{array}{ll} L^\uparrow = L^\uparrow_+ \cup L^\uparrow_- & \text{(orthochronous)} \\ L_+ = L^\uparrow_+ \cup L^\downarrow_+ & \text{(proper)} \\ L_0 = L^\uparrow_+ \cup L^\downarrow_- & \text{(orthochorous)} \end{array} \tag{2.13}$$

The transformations $L^\uparrow_-$, $L^\downarrow_+$ and $L^\downarrow_-$ are formed by compounding space, space-time, and time inversions, respectively with the transformations of $L^\uparrow_-$.

Next we review the elementary transformations composing $L^\uparrow_+$. If the primed reference system moves in the positive x^i direction with velocity $v^i = \tanh \alpha_i$, the matrices $\Lambda = L_i$ are

$$L_1 = \begin{bmatrix} \cosh \alpha_1 & -\sinh \alpha_1 & 0 & 0 \\ -\sinh \alpha_1 & \cosh \alpha_1 & 0 & 0 \\ 0 & 0 & 1 & 0 \\ 0 & 0 & 0 & 1 \end{bmatrix} \tag{2.14}$$

etc. If the primed reference system is rotated by angle φ_i about axis i ($i = 1$, 2, 3) the matrices $\Lambda = R_i$ are

$$R_1 = \begin{bmatrix} 1 & 0 & 0 & 0 \\ 0 & 1 & 0 & 0 \\ 0 & 0 & \cos\varphi_1 & \sin\varphi_1 \\ 0 & 0 & -\sin\varphi_1 & \cos\varphi_1 \end{bmatrix} \tag{2.15}$$

etc. The corresponding one-parameter subgroups compound as follows:

$$\begin{aligned} L(\alpha_i)L(\alpha_i') &= L(\alpha_i + \alpha_i') \\ R(\varphi_i)R(\varphi_i') &= R(\varphi_i + \varphi_i') \end{aligned} \tag{2.16}$$

The exponential mapping exhibits the group property quite clearly. Defining the matrices k_i and l_i by*

$$\begin{aligned} k_1 &= -L_1'(0) = \begin{pmatrix} \sigma_x & 0 \\ 0 & 1 \end{pmatrix} \\ l_1 &= -iR_1'(0) = \begin{pmatrix} 1 & 0 \\ 0 & \sigma_y \end{pmatrix} \end{aligned} \tag{2.17}$$

etc., we find that the Lie differential equations

$$\frac{dL_i}{d\alpha_i} = -k_i L_i \qquad \frac{dR_i}{d\varphi_i} = i l_i R_i \tag{2.18}$$

(no sum on i) are solved by

$$L(\alpha_i) = e^{-\alpha_i k_i}; \qquad R_i(\varphi_i) = e^{i\varphi_i l_i} \tag{2.19}$$

The matrices k_i and l_i are Hermitian. As a consequence $L_i^\dagger = L_i$ while $R_i^\dagger = R_i^{-1}$.

2.2 INFINITESIMAL GENERATORS AND THEIR COMMUTATORS

Equations (2.17) and (2.19) exhibit explicitly how the special finite transformations (2.19) are determined by the local transformations. The commutator structure of the infinitesimal operators specifies the group structure

* In these equations the positions of indices do not denote covariant or contravariant transformation properties.

in a well-known way. However the properties of the space in which the operators of a general representation act, and specific properties of those operators, require careful study, as there exist infinitely many spaces besides the four-dimensional one in which representations may exist. Proceeding in analogy to the $SO(4)$ analysis, we denote the infinitesimal transformation by

$$\begin{aligned} \Lambda^\mu_\nu &= g^\mu_\nu + \delta\omega^\mu_\nu \\ \delta\omega_{\mu\nu} &= -\delta\omega_{\nu\mu} \end{aligned} \tag{2.20}$$

As our first example, consider the operators $O(\Lambda)$ acting on the scalar functions $\phi(x)$, as in Eq. (1.4).

$$\begin{aligned} O(\Lambda) &= 1 - \frac{i}{2}\,\delta\omega^{\mu\nu}L_{\mu\nu} \\ L_{\mu\nu} &= i(x_\mu\partial_\nu - x_\nu\partial_\mu) \end{aligned} \tag{2.21}$$

$L_{\mu\nu}$ is Hermitian within the Klein-Gordon scalar product (problem (2.1)). It therefore is interesting to seek Hermitian operators satisfying the commutation relations

$$[M_{\mu\nu}, M_{\rho\sigma}] = i[-g_{\mu\sigma}M_{\rho\nu} + g_{\nu\rho}M_{\mu\sigma} - g_{\nu\sigma}M_{\mu\rho} + g_{\mu\rho}M_{\sigma\nu}] \tag{2.22}$$

This result is easily derived from the explicit form of Eq. (2.21) using the relation $[x_\mu, \partial_\nu] = -g_{\mu\nu}$. We have written $M_{\mu\nu}$ in place of $L_{\mu\nu}$ in equation (2.22) to suggest that there exist representations different from the explicit form (2.21).

The representations corresponding to (2.21) are clearly infinite dimensional. Are such representations irreducible? A basic theorem states that the unitary *IR* of a non-compact group must be infinite-dimensional. Thus in contrast to the finite-dimensional non-unitary representations (2.19), there exist no subspaces of $\{\phi\}$ invariant under (2.22). (See problem 2.2 for a discussion of the analogous problem for the rotation group.)

There also exist important examples of $M_{\mu\nu}$ composed of pieces acting in different spaces and having differing properties under Hermitian conjugation. For example, in textbooks on relativistic quantum mechanics it is shown that the Dirac wave function $\psi'(x')$ in a moving frame at point $x' = \Lambda x$ is related to the original wave function by $S(\Lambda)\psi(x)$. Defining $O(\Lambda)$ by

$$O(\Lambda)\psi'(x) = S(\Lambda)\psi(\Lambda^{-1}x) \tag{2.23}$$

and using the well-known result* for infinitesimal $\delta\omega$

$$S(\Lambda) = 1 - \frac{i}{4}\,\delta\omega^{\mu\nu}\sigma_{\mu\nu}$$
$$\sigma_{\mu\nu} = \frac{i}{2}\,[\gamma_\mu, \gamma_\nu] \tag{2.24}$$

we see that

$$M_{\mu\nu} = L_{\mu\nu} + S_{\mu\nu} \tag{2.25}$$

where $S_{\mu\nu} = \frac{1}{2}\sigma_{\mu\nu}$. $L_{\mu\nu}$ and $S_{\mu\nu}$ separately obey (2.22), but the $S_{\mu\nu}$ (4×4 matrices) are not Hermitian for the Lorentz boosts S^{on} ($n = 1, 2, 3$).

The algebra (2.22) of the Hermitian generators $M_{\mu\nu}$ may be recast in a form similar to (1.14). First note that for all indices distinct,

$$[M_{\mu\nu}, M_{\rho\sigma}] = 0 \tag{2.26}$$

while otherwise we have ($\mu \neq \nu \neq \sigma \neq \mu$)

$$[M^\mu_\nu, M^\mu_\sigma] = i g^{\mu\mu} M_{\sigma\nu} \tag{2.27}$$

Noting that $L_{mn} = -i(x^m\partial/\partial x^n - x^n\partial/\partial x^m)$ (l, m, n cyclic) we define an angular momentum $\mathcal{J}^l = M_{mn}$. We also define a Lorentz boost operator $\mathcal{K}^i$ by $\mathcal{K}^i = M^{oi}$. Then the commutation rules satisfied by the six Hermitian operators $\mathcal{K}^i$, $\mathcal{J}^i$ are

$$[\mathcal{J}^i, \mathcal{J}^j] = i\varepsilon^{ijk}\mathcal{J}^k$$
$$[\mathcal{J}^i, \mathcal{K}^j] = i\varepsilon^{ijk}\mathcal{K}^k \tag{2.28}$$
$$[\mathcal{K}^i, \mathcal{K}^j] = -i\varepsilon^{ijk}\mathcal{J}^k$$

The first equation says that $\mathcal{J}$ is an angular momentum (generator of $SU(2)$), while the second states that $\mathcal{K}$ is a vector operator with respect to this group, in exact analogy to the $SO(4)$ algebra (1.14). However a key difference (the minus sign) appears in the $[\mathcal{K}, \mathcal{K}]$ commutator. For $O(3, 1)$ $\mathcal{K}$ is not an angular momentum, and the representations have structure quite different from those of $SO(4)$. Seemingly trivial differences of signs which arise when comparing $SO(4)$ and $O(3, 1)$ have far reaching consequences. For this reason we have not seen fit to pretend that Minkowski space has the same structure as four-dimensional Euclidean space by introducing judicious factors of i in the usual way.

---

* Conventions with respect to γ matrices are summarized in App. A.

The four dimensional transformation matrices $\Lambda^{\mu}_{\ \nu}$ which define the properties of the group give a representation of (2.22) very different from the differential operators $L_{\mu\nu}$. In order to cast (2.20) in the same form as (2.21) we use the antisymmetry of $\delta\omega_{\mu\nu}$ to write

$$\delta\omega_{\mu\nu} = g_{\mu\alpha}g_{\nu\beta}\,\delta\omega^{\alpha\beta} = \tfrac{1}{2}(g_{\mu\alpha}g_{\nu\beta} - g_{\mu\beta}g_{\nu\alpha})\,\delta\omega^{\alpha\beta} \tag{2.29}$$

allowing the matrix $\delta\omega = [\delta\omega^{\mu}_{\ \nu}]$ by which Λ deviates from unity to be expressed in terms of the matrices $m_{\alpha\beta} = [(m_{\alpha\beta})^{\mu}_{\ \nu}]$

$$\begin{aligned} \delta\omega &= -\frac{i}{2}\,\delta\omega^{\alpha\beta} m_{\alpha\beta} \\ (m_{\alpha\beta})_{\mu\nu} &\equiv i(g_{\mu\alpha}g_{\nu\beta} - g_{\mu\beta}g_{\nu\alpha}) \end{aligned} \tag{2.30}$$

The commutator of the matrices $m_{\mu\nu}$, $m_{\rho\sigma}$ agrees with Eq. (2.22), as shown in problem (2.3). However the matrices comprising the representation are not unitary, in contrast to the operators $L_{\mu\nu}$ of (2.21) acting in the space of scalar functions equipped with the Klein-Gordon inner product.* The matrices $[(m_{\alpha}^{\ \beta})^{\mu}_{\ \nu}]$ belonging to transformations with parameter $\delta\omega^{\alpha}_{\ \beta}$ are

$$\begin{aligned} m_{01} &= -i\begin{bmatrix} 0 & 1 & 0 & 0 \\ 1 & 0 & 0 & 0 \\ 0 & 0 & 0 & 0 \\ 0 & 0 & 0 & 0 \end{bmatrix} & m_{02} &= -i\begin{bmatrix} 0 & 0 & 1 & 0 \\ 0 & 0 & 0 & 0 \\ 1 & 0 & 0 & 0 \\ 0 & 0 & 0 & 0 \end{bmatrix} \\ m_{03} &= -i\begin{bmatrix} 0 & 0 & 0 & 1 \\ 0 & 0 & 0 & 0 \\ 0 & 0 & 0 & 0 \\ 1 & 0 & 0 & 0 \end{bmatrix} & m_{12} &= \begin{bmatrix} 0 & 0 & 0 & 0 \\ 0 & 0 & -i & 0 \\ 0 & i & 0 & 0 \\ 0 & 0 & 0 & 0 \end{bmatrix} \\ m_{23} &= \begin{bmatrix} 0 & 0 & 0 & 0 \\ 0 & 0 & 0 & 0 \\ 0 & 0 & 0 & -i \\ 0 & 0 & i & 0 \end{bmatrix} & m_{31} &= \begin{bmatrix} 0 & 0 & 0 & 0 \\ 0 & 0 & 0 & -i \\ 0 & 0 & 0 & 0 \\ 0 & 0 & i & 0 \end{bmatrix} \end{aligned} \tag{2.31}$$

---

* For positive energy solutions the inner product is positive definite, permitting a standard Hilbert space.

The matrices m^k_l, $k, l = 1, 2, 3$ have (for space components) the same elements as the spin 1 angular momentum matrices in the regular representation $(t^m)^{kl} = -i\varepsilon^{klm}$, as expected since this subgroup corresponds to rotations of the space components of the four vector x^μ.

To find the physical significance of the parameters $\delta\omega^\mu_\nu$ recall that under a rotation of the coordinate system by $\delta\theta$ around axis $\mathbf{n}$, the vector acquires coordinates $\mathbf{x}'$

$$\mathbf{x}' = (1 + i\delta\theta\mathbf{n}\cdot\mathbf{t})\mathbf{x}, \qquad \mathbf{n}^2 = 1 \tag{2.32}$$

where $\mathbf{x}$ is regarded as a three-component vector acted on by the matrix $\mathbf{n}\cdot\mathbf{t}$. For a pure Lorentz transformation with velocity $v_x = \tanh\alpha$ the coordinates of an event in a moving frame appear to be

$$\begin{pmatrix} t' \\ x' \end{pmatrix} = L(\alpha)\begin{pmatrix} t \\ x \end{pmatrix} \tag{2.33}$$

with $L(\alpha)$ given in (2.19). Thus the set $(\delta\omega^2_3, \delta\omega^3_1, \delta\omega^1_2)$ correspond to the vector $\mathbf{n}\,\delta\theta$ and $(\delta\omega^{01}, \delta\omega^{02}, \delta\omega^{03})$ correspond to $\delta\boldsymbol{\alpha}$ for parameters α^2 entering in the pure Lorentz transformations. In terms of the physical quantities we have

$$\Lambda = 1 - \delta\boldsymbol{\alpha}\cdot\mathbf{k} + i\,\delta\theta\mathbf{n}\cdot\mathbf{l} \tag{2.34}$$

where $\mathbf{k}$ and $\mathbf{l}$ are given in Eq. (2.17). Employing the definitions of $\mathscr{K}$ and $\mathscr{J}$ given before Eq. (2.28) and writing the $\delta\omega_{\mu\nu}$ as in (2.34), the operator $O(\Lambda)$ is

$$O(\Lambda) = 1 + i\,\delta\boldsymbol{\alpha}\cdot\mathscr{K} + i\,\delta\theta\hat{n}\cdot\mathscr{J} \tag{2.35}$$

The description given above corresponds to the "passive" interpretation of transformations whereby a fixed system is described from different reference frames.

The examples given above illustrate the fundamental distinction between finite and infinite dimensional representations of the Lorentz group, that the pure Lorentz transformations cannot be represented by unitary operators in the former case. We have seen that the two types can coexist in important practical cases such as the transformation of the Dirac (spin $\frac{1}{2}$) field.

Now we turn to a more systematic development of the representations.

2.3 FINITE DIMENSIONAL REPRESENTATIONS OF THE LORENTZ GROUP

Thus far we have exhibited two examples of operators satisfying the commutator structure (2.22), the differential operators $L_{\mu\nu}$ of Eq. (2.21) and the

4×4 matrices $m_{\mu\nu}$ defined in Eq. (2.30). In Eq. (2.28) a different form was given to the algebra by defining Hermitian operators $\mathscr{K}^j$ and $\mathscr{J}^i$ corresponding to pure Lorentz transformations and pure rotations

$$\begin{aligned} \mathscr{K}^i &= M^{oi} \\ \mathscr{J}^i &= M_{jk} \qquad (ijk \text{ cyclic}) \end{aligned} \tag{2.36}$$

Clearly the same transformation can be carried out for the $m_{\mu\nu}$. Eq. (2.31) shows that $\mathbf{K} \equiv i\mathbf{k}$ is anti-Hermitian, while $l^i = +m_{kl}$ as in (2.36). The commutation relations for the $\mathbf{k}$, $\mathbf{l}$ are, instead of (2.28)

$$\begin{aligned} [l^i, l^j] &= i\varepsilon^{ijk} l^k \\ [l^i, k^j] &= i\varepsilon^{ijk} k^k \\ [k^i, k^j] &= i\varepsilon^{ijk} l^k \end{aligned} \tag{2.37}$$

which coincide exactly with the $O(4)$ commutator structure.

Equation (2.37) indicates the existence of an infinite number of (finite-dimensional) representations having anti-Hermitian $\mathscr{K}$, in addition to the specific example (see Eq. (2.31)) already discovered. The $\mathbf{k}$, $\mathbf{l}$ in question are simply those of Chapter 1. The non-unitary matrices representing $O(3, 1)$ are easily converted to the unitary matrices of $O(4)$ by changing the Lorentz transformation parameter $\boldsymbol{\alpha}$ to the purely imaginary value $-i\boldsymbol{\alpha}$ ($\boldsymbol{\alpha}$ real).

In analogy to Eqs. (1.15)ff we introduce two angular momenta $\mathbf{j}_1$, $\mathbf{j}_2$ by

$$\begin{aligned} \mathbf{j}_1 &= \tfrac{1}{2}(\mathbf{l} + \mathbf{k}) \qquad & \mathbf{l} &= \mathbf{j}_1 + \mathbf{j}_2 \\ \mathbf{j}_2 &= \tfrac{1}{2}(\mathbf{l} - \mathbf{k}) \qquad & \mathbf{k} &= \mathbf{j}_1 - \mathbf{j}_2 \end{aligned} \tag{2.38}$$

Distinguishing the angular momenta by an index α ($\alpha = 1, 2$) and denoting components by i, j, k, we have

$$[j_i^\alpha, j_j^{\alpha'}] = i\varepsilon_{ijk}\, \delta_{\alpha\alpha'}\, j_k^\alpha \tag{2.39}$$

The basis functions are given in (1.17) and the representations may be labeled by (j_1, j_2) or by the invariants $\mathbf{l} \cdot \mathbf{k} = \mathbf{j}_1^2 - \mathbf{j}_2^2$ and $\frac{1}{2}(\mathbf{l}^2 + \mathbf{k}^2) = \mathbf{j}_1^2 + \mathbf{j}_2^2$ in analogy to Eq. (1.30).

The simplest non-trivial representations have $(j_1, j_2) = (\frac{1}{2}, 0)$ and $(0, \frac{1}{2})$ respectively. The corresponding matrices are called $D^{(\frac{1}{2},0)}$ and $D^{(0,\frac{1}{2})}$. We denote the components of the basic spinors ξ and η acted on by these matrices by* ξ_a and $\eta^{\dot b}$:

$$\xi = \begin{pmatrix} \xi_1 \\ \xi_2 \end{pmatrix}, \qquad \eta = \begin{pmatrix} \eta^{\dot 1} \\ \eta^{\dot 2} \end{pmatrix} \tag{2.40}$$

In summary we find the properties

$$\begin{aligned} D^{(\frac{1}{2},0)}: \quad & \mathbf{j}_1 = \tfrac{1}{2}\boldsymbol{\sigma}, \qquad \mathbf{j}_2 = 0 \\ & \mathbf{l} = \mathbf{k} = \tfrac{1}{2}\boldsymbol{\sigma} \\ D^{(\frac{1}{2},0)} &= 1 + \frac{i\,\delta\theta}{2}\mathbf{n}\cdot\boldsymbol{\sigma} - \tfrac{1}{2}\,\delta\boldsymbol{\alpha}\cdot\boldsymbol{\sigma} \\ D^{(0,\frac{1}{2})}: \quad & \mathbf{j}_1 = 0, \qquad \mathbf{j}_2 = \tfrac{1}{2}\boldsymbol{\sigma} \\ & \mathbf{l} = -\mathbf{k} = \tfrac{1}{2}\boldsymbol{\sigma} \\ D^{(0,\frac{1}{2})} &= 1 + \frac{i\,\delta\theta}{2}\mathbf{n}\cdot\boldsymbol{\sigma} + \frac{\delta\boldsymbol{\alpha}}{2}\cdot\boldsymbol{\sigma} \end{aligned} \tag{2.41}$$

Under finite rotations both representations transform as exp $(i\theta n\cdot\boldsymbol{\sigma}/2)$ while pure Lorentz transformations have the form exp $(\mp\boldsymbol{\alpha}\cdot\boldsymbol{\sigma}/2)$.

It is simple to construct the bases for irreducible representations $D^{(j_1,j_2)}$ from the basic spinors ξ and η using the standard $SU(2)$ construction

$$u^{j_1j_2}_{m_1m_2} = \frac{\xi_1^{j_1+m_1}\xi_2^{j_1-m_1}}{\sqrt{(j_1+m_1)!\,(j_1-m_1)!}}\,\frac{\eta^{\dot 1\, j_2+m_2}\eta^{\dot 2\, j_2-m_2}}{\sqrt{(j_2+m_2)!\,(j_2-m_2)!}} \tag{2.42}$$

Corresponding to $\xi' = D^{(\frac{1}{2}\,0)}\xi$, $\eta' = D^{(0,\frac{1}{2})}\eta$, we have

$$u^{j_1j_2}_{m_1'm_2'}(\xi',\eta') = \sum D^{(j_1,j_2)}_{m_1'm_2',m_1m_2}u^{j_1j_2}_{m_1m_2}(\xi,\eta) \tag{2.43}$$

The Clebsch-Gordan series is a direct consequence of its $SU(2)$ analogue

$$\begin{aligned} &D^{(j_1,j_2)} \oplus D^{(k_1,k_2)} \\ &\quad = D^{(j_1+k_1,j_2+k_2)} \oplus D^{(j_1+k_1-1,j_2+k_2)} \oplus \cdots \oplus D^{(|j_1-k_1|,|j_2-k_2|)} \end{aligned} \tag{2.44}$$

If we consider the subgroup of pure rotations we may reduce the representation as follows

$$D^{(j_1,j_2)} \oplus D^{j_1+j_2} \oplus \cdots \oplus D^{|j_1-j_2|} \tag{2.45}$$

where D^j is a standard $SU(2)$ representation matrix.

Now we consider in more detail the properties of the basic representations $D^{(\frac{1}{2},0)}$ and $D^{(0,\frac{1}{2})}$. First consider the nature of the complex conjugate representations. Working with the infinitesimal transformations (2.41) and noting

* The motivation for this notation will be explained in Sec. 3.2.

the relation

$$\sigma_2 \boldsymbol{\sigma}^* \sigma_2 = -\boldsymbol{\sigma} \tag{2.46}$$

in the standard basis

$$\sigma_1 = \begin{pmatrix} 0 & 1 \\ 1 & 0 \end{pmatrix} \qquad \sigma_2 = \begin{pmatrix} 0 & -i \\ i & 0 \end{pmatrix} \qquad \sigma_3 = \begin{pmatrix} 1 & 0 \\ 0 & -1 \end{pmatrix} \tag{2.47}$$

we easily find the relation

$$D^{(\frac{1}{2},0)} = S D^{(0,\frac{1}{2})*} S^{-1} \tag{2.48}$$

where $S = i\sigma_2$. Thus $D^{(\frac{1}{2},0)}$ is equivalent to $D^{(0,\frac{1}{2})*}$. In general one finds (problem 2.4) a matrix C such that

$$C D^{(j_1,j_2)} C^{-1} = D^{(j_2,j_1)*} \tag{2.49}$$

Next we establish a direct connection between the transformations $\xi \to D\xi$ ($D = D^{(\frac{1}{2},0)}$) of spin space and the 4-vector transformations $x \to \Lambda x$. This may be done by studying quadratic forms $\Sigma \xi_i^* A_{ij} \xi_j$, in analogy to the rotation group scalar $\xi^\dagger \xi$ and vector $\xi^\dagger \boldsymbol{\sigma} \xi$. To begin we notice that $\xi^\dagger \xi$ is no longer a scalar under pure Lorentz transformations. Denoting a pure rotation by D_R and a pure Lorentz transformation by D_L

$$\begin{aligned} D_R &= e^{i\theta \mathbf{n}\cdot\boldsymbol{\sigma}/2} = \cos\frac{\theta}{2} + i\mathbf{n}\cdot\boldsymbol{\sigma}\sin\frac{\theta}{2} \\ D_L &= e^{-\boldsymbol{\alpha}\cdot\boldsymbol{\sigma}/2} = \cosh\frac{\alpha}{2} - \hat{\alpha}\cdot\boldsymbol{\sigma}\sinh\frac{\alpha}{2} \end{aligned} \tag{2.50}$$

we note that $\xi^\dagger \xi$ transforms as the time component of a four-vector:

$$\begin{aligned} \xi^\dagger \xi \to \xi^\dagger D_L^2 \xi &= \xi^\dagger e^{-\boldsymbol{\alpha}\cdot\boldsymbol{\sigma}} \xi \\ &= \cosh \alpha \xi^\dagger \xi - \hat{\alpha}\cdot(\xi^\dagger \boldsymbol{\sigma} \xi) \sinh \alpha \end{aligned} \tag{2.51}$$

provided $\xi^\dagger \boldsymbol{\sigma} \xi$ is a vector. The latter property is easily verified†

$$\begin{aligned} \xi^\dagger \boldsymbol{\sigma} \xi &\to \xi^\dagger D_L \boldsymbol{\sigma} D_L \xi \\ &= -\sinh \alpha \xi^\dagger \xi + \hat{\alpha}\cdot(\xi^\dagger \boldsymbol{\sigma} \xi) \cosh \alpha \end{aligned} \tag{2.52}$$

These results show that $(\xi^\dagger \xi,\ \xi^\dagger \boldsymbol{\sigma} \xi)$ transform exactly as a four vector under the 2×2 matrix transformation $\xi \to D\xi$. Thus we may define a "4-vector"

† We remind the reader of the identities

$$\{\sigma_i, \sigma_j\} \equiv \sigma_i \sigma_j + \sigma_j \sigma_i = 2\,\delta_{ij}$$

$$\sigma_i \sigma_j = \delta_{ij} + i\varepsilon_{ijk}\sigma_k.$$

Pauli matrix σ^μ

$$\sigma^\mu = (I, \boldsymbol{\sigma}) \tag{2.53}$$

from which we may form a four vector

$$V^\mu \equiv \xi^\dagger \sigma^\mu \xi \tag{2.54}$$

such that $V \to \Lambda V$ when $\xi \to D\xi$. This correlation of Minkowski transformations in space time with spin space transformations lies at the root of the whole subject. The analogy is the same as that between space rotations and $SU(2)$ transformations in a similar spin space. The transformations $\xi \to D\xi$ define a matrix group of linear transformations in a complex two-dimensional space having det $D = 1$. This group, the special (S) linear (L) group operating in a two (2) dimensional complex (C) space is generally referred to as $SL(2, C)$. To explore this question in more detail we show that the matrix $D = \exp(i\theta\mathbf{n} \cdot \boldsymbol{\sigma}/2 - \boldsymbol{\alpha} \cdot \boldsymbol{\sigma}/2)$ has the aforementioned property. Consider the transformation (α, β, γ, δ complex numbers)

$$\begin{pmatrix} \xi_1' \\ \xi_2' \end{pmatrix} = \begin{pmatrix} \alpha & \beta \\ \gamma & \delta \end{pmatrix} \begin{pmatrix} \xi_1 \\ \xi_2 \end{pmatrix} \tag{2.55}$$

subject to the condition

$$\det \begin{pmatrix} \alpha & \beta \\ \gamma & \gamma \end{pmatrix} = \alpha\delta - \beta\gamma = 1 \tag{2.56}$$

Because of (2.56) the transformation (2.55) is specified by six real numbers. To show that D is such a matrix note that any 2×2 matrix X can be written in the form

$$X = AI + i\mathbf{B} \cdot \boldsymbol{\sigma} \tag{2.57}$$

where A is a complex number and $\mathbf{B}$ a complex three-dimensional "vector." The condition (2.56) is easily seen to be

$$\det X = A^2 + \mathbf{B}^2 = 1 \tag{2.58}$$

Making the definitions $B = (\mathbf{B}^2)^{\frac{1}{2}}$ and $\hat{B} = \mathbf{B}/B$ we have $A^2 + B^2 = 1$, so that it is possible to parametrize A and B by

$$A = \cos(\phi/2), \qquad B = \sin(\phi/2) \tag{2.59}$$

with ϕ complex so that

$$X = \cos\frac{\phi}{2} + i\hat{B} \cdot \boldsymbol{\sigma} \sin\frac{\phi}{2} = \exp(i\phi\hat{B} \cdot \boldsymbol{\sigma}/2) \tag{2.60}$$

where ϕ and B are complex. Thus we may identify (2.60) with

$$D = \exp(i\theta\mathbf{n}\cdot\boldsymbol{\sigma}/2 - \boldsymbol{\alpha}\cdot\boldsymbol{\sigma}/2)$$

provided we make the identification

$$\phi\hat{B} = \mathbf{n}\theta + i\boldsymbol{\alpha} \tag{2.61}$$

The construction (2.54) provides the fundamental connection between transformations in spin space and Minkowski space, which relation is developed fully in Sec. 3.3. Another interpretation of Eq. (2.54) is that the 4-vector matrix σ^μ acts as the set of Clebsch-Gordan coefficients which reduce $D^{(\frac{1}{2},0)*} \otimes D^{(\frac{1}{2},0)}$ to transform as a four vector. Since $D^{(\frac{1}{2},0)*} \sim D^{(0,\frac{1}{2})}$ and $D^{(0,\frac{1}{2})} \otimes D^{(\frac{1}{2},0)} = D^{(\frac{1}{2},\frac{1}{2})}$, we find the important result that the usual four vector is equivalent to the representation $D^{(\frac{1}{2},\frac{1}{2})}$. Generalizing Eq. (2.54) slightly, so that

$$\begin{gathered} V^\mu = \xi_a^\dagger\sigma^\mu\xi_b \qquad V^{\mu'} = \Lambda^\mu_\nu V^\nu \\ V'^\mu = \xi_a'^\dagger\sigma^\mu\xi_b' \qquad \xi_c' = D\xi_c \\ \xi_a^\dagger D^\dagger\sigma^\mu D\xi_b = \Lambda^\mu_\nu\xi_a^\dagger\sigma^\nu\xi_b \end{gathered} \tag{2.62}$$

Since ξ_ν are arbitrary we have the transformation law for σ^μ:

$$D^\dagger\sigma^\mu D = \Lambda^\mu_\nu\sigma^\nu; \qquad D \equiv D^{(\frac{1}{2},0)} \tag{2.63}$$

(The preceding analysis only considered special transformations, i.e. pure rotations or pure Lorentz transformations.)

Under a pure Lorentz transformation a momentum four vector $p_0^\mu = (m, \mathbf{0})$ transforms to $p^\mu = m(\cosh\zeta, -\hat{p}\sinh\zeta)$ while the spinors ξ and η are acted on by $D^{(\frac{1}{2},0)} = \exp(-\zeta\hat{p}\cdot\boldsymbol{\sigma}/2)$ and $D^{(0,\frac{1}{2})} = \exp(\zeta\hat{p}\cdot\boldsymbol{\sigma}/2$ respectively. Expanding exponentials as in (2.50), we find the useful formulas

$$\begin{aligned} [D^{(0,\frac{1}{2})}]^2 &= \cosh\zeta + \hat{p}\cdot\boldsymbol{\sigma}\sinh\zeta = \frac{p^\mu\sigma_\mu}{m} \\ [D^{(\frac{1}{2},0)}]^2 &= \cosh\zeta - \hat{p}\cdot\boldsymbol{\sigma}\sinh\zeta = \frac{p^\mu\tilde{\sigma}_\mu}{m} \end{aligned} \tag{2.64}$$

where $\tilde{\sigma}^\mu$ is defined as the set $(I, -\boldsymbol{\sigma})$ in analogy to $\sigma^\mu = (I, \boldsymbol{\sigma})$. The properties of $\tilde{\sigma}^\mu$ will be studied in Chapter III. In Chapter V we shall derive generalizations of Eq. (2.64) for the representations $D^{(j,0)}$ and $D^{(0,j)}$.

If we adjoin space inversions to the proper Lorentz transformations the previous representations are no longer representations in general. Define space inversion by

$$I_s\mathbf{x} = -\mathbf{x} \qquad I_s t = t \tag{2.65}$$

For either definition of generators Eq. (2.21) or (2.29), $\mathbf{l}$ is a pseudo vector and $\mathbf{k}$ a vector. In general we define

$$I_s\mathbf{l} = \mathbf{l}I_s \qquad I_s\mathbf{k} = -\mathbf{k}I_s \tag{2.66}$$

so that

$$\begin{aligned} I_s\mathbf{j}^{(1)} &= I_s\tfrac{1}{2}(\mathbf{l} + \mathbf{k}) = \tfrac{1}{2}(\mathbf{l} - \mathbf{k})I_s \\ &= \mathbf{j}^{(2)}I_s \\ I_s\mathbf{j}^{(2)} &= j^{(1)}I_s \end{aligned} \tag{2.67}$$

Hence $I_s\Psi(j_1m_1, j_2m_2)$ does not belong to $D^{(j_1\,j_2)}$ unless $j_1 = j_2$. In particular we have

$$\begin{aligned} (\mathbf{j}^{(1)})^2 I_s\Psi(j_1m_1, j_2m_2) &= j_2(j_2 + 1)I_s\Psi(j_1m_1, j_2m_2) \\ j_3^{(1)}I_s\Psi(j_1m_1, j_2m_2) &= m_2I_s\Psi(j_1m_1, j_2m_2) \\ (\mathbf{j}^{(2)})^2 I_s\Psi(j_1m_1, j_2m_2) &= j_1(j_1 + 1)I_s\Psi(j_1m_1, j_2m_2) \\ j_3^{(2)}I_s\Psi(j_1m_1, j_2m_2) &= m_1I_s\Psi(j_1m_1, j_2m_2) \end{aligned} \tag{2.68}$$

Thus we may define

$$I_s\Psi(j_1m_1, j_2m_2) = e^{ia}\Psi(j_2m_2, j_1m_1) \tag{2.69}$$

where e^{ia} is an arbitrary phase factor, which we choose to be unity. Thus if we want to have an *IR* for operations including space inversion as well as proper linear transformations we have to expand the space. The representation irreducible under these operations is

$$D^{(j_1,j_2)} \oplus D^{(j_2,j_1)} \tag{2.70}$$

The dimension of the representation is $2(2j_1 + 1)(2j_2 + 1)$. This is the reason the Dirac spinor has four components instead of two. We can form a spin $\frac{1}{2}$ wave equation with a two-component spinor, but it is not covariant under the parity operation. Later we shall find that the Dirac spinor $\psi(x)$ transforms as $D^{(\frac{1}{2},0)} \oplus D^{(0,\frac{1}{2})}$.

The material of this section and of much of Chapter 3 has been treated several times in various textbooks and monographs. In references 4–7 we list some of these works.

2.4 UNITARY REPRESENTATIONS OF THE LORENTZ GROUP

In the present section we investigate the unitary irreducible representations of the homogeneous Lorentz group. In contrast to the finite-dimensional representations, the unitary representations are necessarily infinite-dimensional. Since only unitary representations have physical significance in transforming quantum mechanical state vectors, we are led to consider the infinite-dimensional representations in problems of quantum mechanical significance.

Curiously enough, the invariant quantities characterizing the unitary irreducible representations do not correspond simply to the natural physical quantities of elementary particle states, the mass and spin. Only when we include translations together with homogeneous Lorentz transformations do we obtain representations appropriate to asymptotic particle states. Even so, the homogeneous Lorentz group is interesting and constitutes a significant subgroup of the Poincaré group. The latter is studied in Chapter 4.

As much as possible we pattern our treatment of the homogeneous Lorentz group after that of the four-dimensional orthogonal group treated in Chapter 1. This development follows closely that given in ref. 8. For a full mathematical description of the representations of the homogeneous Lorentz group, the reader is referred to refs. 9–12.

In Eq. (2.21) we defined a set of differential operators which were Hermitian within the Klein-Gordon scalar product. These operators could be factored into subsets obeying the algebra of Eq. (2.28). The second of Eqs. (2.28) implies that the operator $\mathbf{K}$ is a vector operator, that is, it induces changes of the angular momentum by ± 1 or 0 when acting on the basis defined by the full set of angular momentum states diagonalizing $\mathbf{J}^2$ and J_3. Once this fact is taken into account, the only new information is contained in the final equation of (2.28). In order to separate the geometrical information contained in the statement that the boost operators $\mathbf{K}$ constitute vector operators, we employ the Wigner-Eckart theorem.

In order to employ standard conventions we define the spherical components of the vector according to

$$\begin{aligned} K_{\pm 1} &\equiv \mp(K^1 \pm iK^2)/\sqrt{2} \\ K_0 &\equiv K^3 \end{aligned} \tag{2.71}$$

These equations obey the Hermiticity condition

$$K_\mu^\dagger = (-1)^\mu K_{-\mu} \tag{2.72}$$

The operator K_μ increases the component of J_3 by μ in the standard manner. The matrix elements of K_μ between two angular momentum states is, according to the Wigner-Eckart theorem, given by

$$(j'm'|\, K_\mu \,|jm) = C(j1j';\, m\mu m')\frac{(j'\|\, K \,\|j)}{\sqrt{2j'+1}} \tag{2.73}$$

If we employ standard symmetry relations of the Clebsch-Gordan coefficients, we obtain the reality condition

$$(j'\|\, K \,\|j)^* = (-1)^{j'-j}(j\|\, K \,\|j') \tag{2.74}$$

for the reduced matrix elements.

We shall now determine the reduced matrix elements of **K**. For a given j there are only three, corresponding to the values $j' = j$, or j' differing by only one unit from j.

The basic commutator is now

$$[K_{+1}, K_{-1}] = J_3 \tag{2.75}$$

If we sandwich this equation between angular momentum eigenstates Φ_{jm}, we obtain ($j \geqslant 1$)

$$2j+1 = \frac{|(j+1\|\, K \,\|j)|^2}{j+1} - \frac{|(j\|\, K \,\|j-1)|^2}{j} - \frac{|(j\|\, K \,\|j)|^2}{j(j+1)} \tag{2.76}$$

In deriving this equation we have used the following symmetry relations and explicit forms for the Clebsch-Gordan coefficients:

$$C(j1j';\, m, \pm 1, m\pm 1) = (-1)^{j'-j+1}\left(\frac{2j'+1}{2j+1}\right)^{\frac{1}{2}} C(j'1j;\, m\pm 1, \pm 1, m)$$

$$C^2(j+1, 1, j;\, m\pm 1, \mp 1, m) = \frac{(j+2\pm m)(j+1\mp m)}{2(j+1)(2j+3)} \tag{2.77}$$

In order to study Eq. (2.76) it is useful to define the quantities $a(j)$ and $b(j)$ as follows

$$a(j) \equiv |(j\,\|K\|\,j-1)|^2/j$$
$$b(j) \equiv |(j\,\|K\|\,j)|^2/j(j+1) \tag{2.78}$$

The $a(j)$ obey the recursion relation

$$a(j+1) - a(j) = b(j) + 2j + 1 \tag{2.79}$$

It should be noted that a and b are positive by definition.

In the case of the four-dimensional orthogonal group it was noted that one can form two invariant quantities from the generators of the Lie algebra. A similar construction is valid for the present group. The quantities are given by

$$\begin{aligned} F &= \tfrac{1}{4} M_{\mu\nu} M^{\mu\nu} \\ G &= \tfrac{1}{4} M_{\mu\nu} M^{\mu\nu D} \\ M^D_{\mu\nu} &= \tfrac{1}{2} \epsilon_{\mu\nu\rho\sigma} M^{\rho\sigma} \end{aligned} \tag{2.80}$$

If we were to replace the Lorentz generators $M_{\mu\nu}$ by the electromagnetic field tensor, we would obtain the familiar statement that the quantities $\mathbf{E}^2 - \mathbf{H}^2$ ($\mathbf{E}$ and $\mathbf{H}$ respectively being the electric and magnetic field vectors) and the pseudo-scalar quantity $\mathbf{E} \cdot \mathbf{H}$ are invariants.

From the commutation relations obeyed by the generators $M_{\mu\nu}$ we see that F and G commute with the $M_{\mu\nu}$ and hence are Lorentz invariants:

$$[F, M_{\mu\nu}] = [G, M_{\mu\nu}] = 0 \tag{2.81}$$

F and G are constant within an irreducible representation, according to Schur's Lemma.

In order to relate the Casimir operators F and G to the reduced matrix elements of the boost operators $\mathbf{K}$, we expand them in terms of L and K as defined following Eq. (2.27).

$$\begin{aligned} F &= \tfrac{1}{2}(\mathbf{L}^2 - \mathbf{K}^2) \\ G &= \mathbf{L} \cdot \mathbf{K} \end{aligned} \tag{2.82}$$

It will be noted that this breakup is a non-covariant breakup; the operators L and K do not have tensorial properties under Lorentz transformations. If we rewrite $\mathbf{K}^2$ and $\mathbf{L} \cdot \mathbf{K}$ in terms of spherical basis vectors, as in Eq. (2.72),

$$\begin{aligned} \mathbf{K} \cdot \mathbf{K} &= \sum_\mu K_\mu K_\mu^\dagger \\ \mathbf{L} \cdot \mathbf{K} &= \sum_\mu L_\mu K_\mu^\dagger \end{aligned} \tag{2.83}$$

we can easily relate the diagonal matrix elements to those of the reduced matrix elements:

$$2f = j(j+1) - [ja(j) + (j+1)a(j+1) + j(j+1)b(j)]/(2j+1)$$
$$g = \left(\frac{j(j+1)}{2j+1}\right)^{\frac{1}{2}}(j\| K \| j); \tag{2.84}$$
$$f \equiv (jm| F |jm)$$
$$g \equiv (jm| G |jm)$$

It will soon become evident that the Casimir operators F and G are not completely independent. Substituting the last of Eqs. (2.84) into Eq. (2.79) we find

$$a(j+1) - a(j) = (2j+1)\left(1 + \frac{g^2}{j^2(j+1)^2}\right) \tag{2.85}$$

This relation is valid for all j greater than or equal to one. The solution of this recursion relation is easily seen to be given by the following relation, valid for $j \geqslant 1$:

$$a(j) = c + j^2 - g^2/j^2 \tag{2.86}$$

The constant c is not undetermined but can be expressed in terms of the quantity f. In order to see this one simply substitutes $a(j)$ into Eq. (2.82). We thus obtain the following solution for the reduced matrix element connecting the states $j - 1$ and j, given by

$$a(j) \equiv \frac{|(j\| K \| j-1)|^2}{j} = -2f - 1 + j^2 - g^2/j^2 \tag{2.87}$$

The quantities $a(j)$ are all positive and, according to Eq. (2.85), increase as j increases. However Eq. (2.87) indicates that some correlation must exist among f, g, and j in order that $a(j)$ actually be positive. From this equation we see that the only nontrivial possibilities involve $a(j)$ vanishing for all j smaller than a critical value j_0. Equation (2.87) indicates that for sufficiently large j, $a(j)$ is positive for any fixed values of f and g. However as we decrease j, the right hand side may turn negative. The only way to avoid this is to have f and g correlated so that at j_0, $a(j_0)$ vanishes. It is convenient to define a real parameter λ such that

$$g = j_0\lambda \tag{2.88}$$

In that case the condition $a(j_0) = 0$ determines f in terms of j_0 and λ, by

$$f = \tfrac{1}{2}(j_0^2 - \lambda^2 - 1) \tag{2.89}$$

The similarity of these formulas to the corresponding ones for the four-dimensional orthogonal group (Eq. 1.34) should be noted.

It should now be noted that apart from questions of phases, we have determined the matrix elements of the boost operators **K** in terms of the eigenvalues of the two Casimir operators, or equivalently the parameters j_0 and λ.

$$
\begin{aligned}
(j\| K \| j) &= j_0\lambda\left(\frac{2j+1}{j(j+1)}\right)^{\frac{1}{2}} \\
|(j\| K \| j-1)| &= \left[\frac{(j^2-j_0^2)(j^2+\lambda^2)}{j}\right]^{\frac{1}{2}}
\end{aligned}
\tag{2.90}
$$

j_0 is a typical angular momentum quantum number, but the only constraint on the parameter λ is that it be real. We shall actually find that there is a range of purely imaginary λ which gives rise to a special class of unitary irreducible representations.

Equations (2.90) relate the reduced matrix elements of **K** to the convenient parameters j_0 and λ. The action of the boost operators **K** in a spherical basis on the angular momentum states Φ_{jm}, which span the Hilbert space of the Lorentz group, is given by

$$
\begin{aligned}
K_\mu\Phi_{jm} &= C(j1j+1;\, m\mu m+\mu)\,\frac{(j+1\| K \| j)}{\sqrt{2j+3}}\,\Phi_{j+1,m+\mu} \\
&\quad + C(j1j;\, m\mu m+\mu)\,\frac{(j\| K \| j)}{\sqrt{2j+1}}\,\Phi_{j,m+\mu} \\
&\quad + C(j1j-1;\, m\mu m+\mu)\,\frac{(j-1\| K \| j)}{\sqrt{2j-1}}\,\Phi_{j-1,m+\mu}
\end{aligned}
\tag{2.91}
$$

It is now necessary to make a definite phase choice. For a given value of the angular momentum j, the components of the multiplet Φ_{jm}, m running from $-j$ to $+j$, are connected by the usual Condon-Shortley phase convention. However the boost operators **K** may connect states with distinct j. Clearly the basis states $\{\Phi_{jm}\}$ with distinct j can be rephased arbitrarily, as indicated by

$$\Phi_{jm} \rightarrow \eta_j\Phi_{jm} \tag{2.92}$$

It is useful to determine the phase η_j by requiring positivity for the reduced matrix element

$$(j\,\|K\|\,j-1) = |(j\,\|K\|\,j-1)| \tag{2.93}$$

With this convention we notice the antisymmetry of the reduced matrix elements of **K** between states of differing j.

We now exhibit explicitly the action of the boost operators on the angular momentum basis.

$$
\begin{aligned}
K_\mu \Phi_{jm} = {} & \left(\frac{[(j+1)^2 - j_0^2][(j+1)^2 + \lambda^2]}{(j+1)(2j+3)}\right)^{\frac{1}{2}} C(j1j+1;\, m\mu m+\mu)\Phi_{j+1,m+\mu} \\
& + \frac{j_0 \lambda}{[j(j+1)]^{\frac{1}{2}}} C(j1j;\, m\mu m+\mu)\Phi_{jm+\mu} \\
& - \left(\frac{(j^2 - j_0^2)(j^2 + \lambda^2)}{j(2j-1)}\right)^{\frac{1}{2}} C(j1j-1;\, m\mu m+\mu)\Phi_{j-1,m+\mu}
\end{aligned}
\tag{2.94}
$$

Introducing explicit forms for the Clebsch-Gordan coefficients, we obtain the final result for the matrix elements of the boost operators.

$$
\begin{aligned}
K^3 \Phi_{jm} = {} & \frac{1}{j+1}\left(\frac{[(j+1)^2 - j_0^2][(j+1)^2 + \lambda^2][(j+1)^2 - m^2]}{(2j+1)(2j+3)}\right)^{\frac{1}{2}} \Phi_{j+1,m} \\
& + \frac{m j_0 \lambda}{j(j+1)} \Phi_{jm} + \frac{1}{j}\left(\frac{(j^2 - j_0^2)(j^2 + \lambda^2)(j^2 - m^2)}{(2j+1)(2j-1)}\right)^{\frac{1}{2}} \Phi_{j-1,m}
\end{aligned}
$$

$$
\begin{aligned}
(K^1 \pm iK^2)\Phi_{jm} & \\
= \mp \frac{1}{j+1} & \\
\times & \left(\frac{[(j+1)^2 - j_0^2][(j+1)^2 + \lambda^2](j \pm m + 2)(j \pm m + 1)}{(2j+1)(2j+3)}\right)^{\frac{1}{2}} \Phi_{j+1,m\pm 1} \\
+ & \frac{j_0 \lambda}{j(j+1)} [(j \pm m + 1)(j \mp m)]\, \Phi_{j,m\pm 1} \\
\pm & \frac{1}{j}\left(\frac{(j^2 - j_0^2)(j^2 + \lambda^2)(j \mp m - 1)(j \mp m)}{(2j-1)(2j+1)}\right)^{\frac{1}{2}} \Phi_{j-1,m\pm 1}
\end{aligned}
\tag{2.95}
$$

It will be noticed that the matrix elements of the raising operators $K_\pm$, defined in the usual manner, as well as those of K^3, are real. The matrix elements of the Cartesian boost operators K^i are in addition Hermitian. This may be verified for the so-called *principal series*, which is characterized by real λ and j_0 equal to either the series $\frac{1}{2}$, $\frac{3}{2}$, $\frac{5}{2}$, etc., or 0, 1, 2, etc.

$$j_0 = \begin{cases} \frac{1}{2}, \frac{3}{2}, \frac{5}{2}, \ldots \\ 0, 1, 2, \ldots \end{cases} \tag{2.96}$$

$$-\infty < \lambda < \infty$$

Having obtained the representation matrices, it is of interest to consider the possibility of their continuation to complex values of the parameter λ. The first such case leads to the so-called complementary series. We notice that unless $j_0 = 0$, the boost matrices become complex when λ becomes complex. In this case one cannot have unitary representations. However when j_0 vanishes, one can have real boost matrices K^3 and $K_\pm$ for a range of purely imaginary λ. If we let $\lambda = i\kappa$, where κ is real, we obtain the following equations:

$$K^3\Phi_{jm} = \left(\frac{[(j+1)^2 - \kappa^2][(j+1)^2 - m^2]}{(2j+1)(2j+3)}\right)^{\frac{1}{2}}\Phi_{j+1,m} \times \left(\frac{(j^2 - \kappa^2)(j^2 - m^2)}{(2j+1)(2j-1)}\right)^{\frac{1}{2}}\Phi_{j-1,m} \tag{2.97}$$

$$(K^1 \pm iK^2)\Phi_{jm} = \mp\left(\frac{[(j+1)^2 - \kappa^2][j \pm m + 2][j \pm m + 1]}{(2j+1)(2j+3)}\right)^{\frac{1}{2}}\Phi_{j+1,m\pm 1} \pm \left(\frac{(j^2 - \kappa^2)(j \mp m - 1)(j \mp m)}{(2j+1)(2j-1)}\right)^{\frac{1}{2}}\Phi_{j-1,m\pm 1} \tag{2.98}$$

If we consider the action of the boost operators on the lowest state Φ_{00}, we see that κ must lie between 0 and 1 in order to obtain unitary representations.

$$\begin{aligned} K^3\Phi_{00} &= [(1-\kappa^2)/3]^{\frac{1}{2}}\Phi_{1,0} \\ (K^1 \pm iK^2)\Phi_{00} &= \mp[2(1-\kappa^2)/3]^{\frac{1}{2}}\Phi_{1,\pm 1} \end{aligned} \tag{2.99}$$

The complementary series has not yet been found useful in physical applications. The principal series by itself comprises a complete set of representation matrices which are oscillating and bounded.

Another important special case makes contact with the finite-dimensional non-unitary representations treated in Sec. 2.3, and also thereby with the representations of $O(4)$. It will be noticed that when $\lambda = i(n+1)$, the Casimir operators for the homogeneous Lorentz group coincide with those of the four-dimensional orthogonal group, where n is an integer or half-integer. For this particular value of λ, the representation contains no states having j greater than n because of the factor $[(j+1)^2 - (n+1)^2]^{\frac{1}{2}}$, which implies that the reduced matrix element of $\mathbf{K}$ between n and $n+1$ vanishes. Hence the boost matrices K act within the set of angular momentum states having j between j_0 and n. Note that since j is less than or equal to n, the matrices K_3

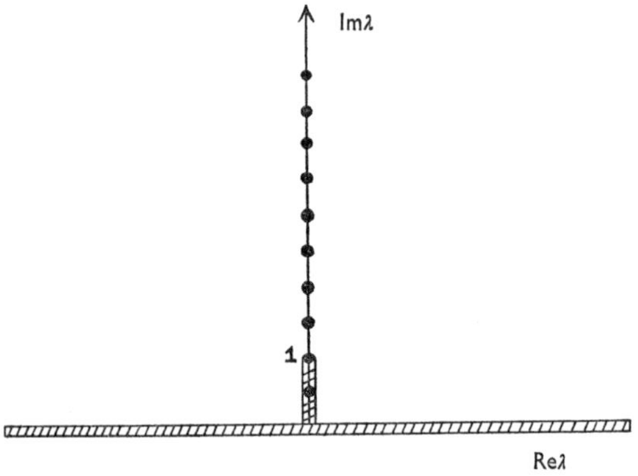

Figure 2.1 The irreducible representations of the homogeneous Lorentz group are characterized by the value of the parameter λ occurring in Eq. (2.96) in addition to j_0. The finite-dimensional non-unitary representations correspond to the special $\lambda = in/2$, n = positive integer.

and $K_\pm$ are purely imaginary. We define the sense of the square root by

$$\begin{aligned} [(j+1)^2 - (n+1)^2]^{\frac{1}{2}} &= i[(n+1)^2 - (j+1)^2]^{\frac{1}{2}} \\ [j^2 - (n+1)^2]^{\frac{1}{2}} &= i[(n+1)^2 - j^2]^{\frac{1}{2}} \end{aligned} \tag{2.100}$$

If we write $\mathbf{K} = i\mathbf{k}$ where $\mathbf{k}^\dagger = \mathbf{k}$, we recover the commutation relations: $[k_i, k_j] = i\epsilon_{ijk}k_k$ as in the case of $SO(4)$. An interesting exercise is to derive the $SO(4)$ representations in the coupled basis directly from the method used here for the homogeneous Lorentz group.

The various cases are conveniently represented in the (j_0, λ) plane as in Fig. 2.1. The unitary representations correspond to the full real λ-axis, as well as the strip from zero to one along the positive imaginary λ-axis. The special points where the imaginary part of λ is $\frac{1}{2}$, 1, $\frac{3}{2}$, etc. correspond to the finite-dimensional non-unitary representations which are isomorphic to the $SO(4)$ representations.

It is also instructive to represent these various cases according to the

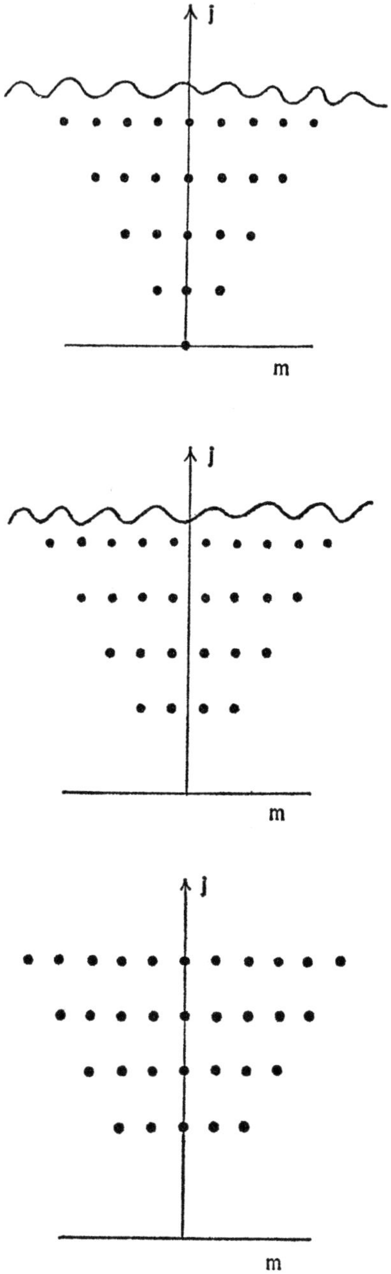

Figure 2.2 The functions Φ_{jm} comprising the bases for various representations are identified by their eigenvalues (j, m). The complementary series has integral j only, down to $j_0 = 0$, as indicated in Fig. 2.2a. The minimum $2j_0$ can be any positive integer for the principal series. In either case the values of j increase indefinitely $(j \geqslant j_0)$. The finite dimensional non-unitary representations have states Φ_{jm} confined to a finite domain as indicated in Fig. 2.2c.

eigenvalues j and m, as indicated in Fig. 2.2 for the principal series, and complementary series, and the finite-dimensional non-unitary representations.

Further information about the representations may be found in refs. 9–12.

PROBLEMS

1. Consider all $\phi(x)$ satisfying the Klein-Gordon equation

$$K_x\phi = 0, \qquad K_x = \Box + m^2, \qquad \Box = \partial_\mu \partial^\mu$$

Show that $L_{\mu\nu}$ is Hermitian within the (invariant) scalar product

$$(\phi, \psi) = i \int d^3x \, \phi^*(\mathbf{x}, t) \overleftrightarrow{\partial_0} \psi(\mathbf{x}, t)$$

where $\overleftrightarrow{\partial_\mu}$ is defined by

$$A \overleftrightarrow{\partial_\mu} B = A \cdot \partial_\mu B - \partial_\mu A \cdot B$$

2. For three-dimensional rotations by angle θ (right-handed) about the axis $\mathbf{n}(\mathbf{n}^2 = 1)$ the rotation operator is $O = \exp(i\theta\mathbf{n} \cdot \mathbf{L})$ where $L = -i\mathbf{r} \times \nabla$. Show that O is infinite dimensional. Show that O is reducible to a direct sum of an infinite number of finite dimensional IR of dimension $2l + 1$, $l = 0, 1, 2, \ldots$.
3. Show that the matrices $m_{\alpha\beta}$ defined in (2.30) $m_{\alpha\beta} = [(m_{\alpha\beta})^\mu{}_\nu]$ satisfy the Lie algebra of Eq. (2.22).
4. Using the basis functions (2.42) prove the equivalence of $D^{(j_1, j_2)}$ and $D^{(j_2, j_1)*}$. Give an explicit expression for the transformation matrix S in Eq. (2.49).
5. Prove Eq. (2.55).
6. Prove (2.57) for a general Lorentz transformation described by $D = \exp(i\theta\mathbf{n} \cdot \boldsymbol{\sigma}/2 - \boldsymbol{\alpha} \cdot \boldsymbol{\sigma}/2)$. (One way is to decompose a general proper Lorentz transformation into the product of a rotation and a pure Lorentz transformation.) Prove that under a rotation $\mathbf{x} \to R\mathbf{x}$ ($\xi \to D_R\xi$, $D_R = \exp(i\theta\mathbf{n} \cdot \boldsymbol{\sigma}/2)$) the components of the "vector" $\xi^\dagger\boldsymbol{\sigma}\xi$ transform as an ordinary vector:

$$\mathbf{V} = \xi^\dagger\boldsymbol{\sigma}\xi \to \xi^\dagger D_R^{-1}\boldsymbol{\sigma}D_R\xi = R\xi^\dagger\boldsymbol{\sigma}\xi$$

$$R\mathbf{V} = \mathbf{V}\cos\theta + \mathbf{V} \times \mathbf{n}\sin\theta + \mathbf{n}(\mathbf{n} \cdot \mathbf{V})(1 - \cos\theta)$$

e.g. for $\mathbf{n} = \hat{z}$

$$V_x' = V_x\cos\theta + V_y\sin\theta$$

$$V_y' = V_y\cos\theta - V_x\sin\theta$$

$$V_z' = V_z$$

REFERENCES

1. W. Rindler, "Special Relativity" (Oliver and Boyd, Edinburgh, 1960).
2. R. F. Streater and A. S. Wightman, "PCT, Spin and Statistics, and All That" (W. A. Benjamin, Inc., New York, 1964).

3. R. Hermann, "Lie Groups for Physicists" (W. A. Benjamin, Inc., New York, 1966).
4. E. M. Corson, "Introduction to Tensors, Spinors and Relativistic Wave Equations" (Blackie and Son, London, 1953).
5. H. Umezawa, "Quantum Field Theory" (North-Holland Publishing Co., Amsterdam, 1956).
6. P. Roman, "Theory of Elementary Particles" (North-Holland Publishing Co., Amsterdam, 1960).
7. W. L. Bade and H. Jehle, *Rev. Mod. Phys.* **25,** 714 (1953).
8. H. Joos, *Fortschritte d. Physik* **10,** 65 (1962).
9. I. M. Gel'fand, M. I. Graev, N. Ya. Vilenkin, "Generalized Functions," Vol. V.
10. M. A. Naimark, "Linear Representations of the Lorentz Group" (The Macmillan Co., New York, 1964).
11. I. M. Gel'fand, R. A. Minlos and Z. Ya. Shapiro, "Representations of the Rotation and Lorentz and their Applications" (The Macmillan Co., New York, 1963).
12. S. Ström, *Arkiv for Fysik* **29,** 467 (1965); *ibid* **33,** 465 (1966).

3

THE HOMOGENEOUS LORENTZ GROUP, II. SPINOR CALCULUS AND RELATIVISTIC WAVE EQUATIONS

3.1 INTRODUCTION

In the preceding chapter the basic spinors ξ and η (Eq. 2.40) were simply pairs of complex numbers acted on by the representations $D^{(\frac{1}{2},0)}$, $D^{(0,\frac{1}{2})}$ respectively. The existence of the space-time continuum enters only when one makes the appropriate identification of the $SL(2, C)$ parameters with the Lorentz transformations and regards ξ and η as spinor fields. The foregoing representations and their generalizations (Eq. 2.42) are finite dimensional and cannot include the type of transformation typified by Eq. (2.21). On the other hand the prototype unitary transformation (2.21) $\phi'(x) = O(\Lambda)\phi(x)$ was defined for a single field $\phi(x)$ rather than a set transforming among themselves.

The Dirac spinor $\psi(x)$, transforming simultaneously on its four spinor indices and its argument via Eq. (2.23) ($\psi_\alpha'(x') = S_{\alpha\beta}\psi_\beta(x)$) provides an example of the generalization encompassing both special cases mentioned in the first paragraph. In addition this relation has a natural physical interpretation when ψ is regarded as the wave function. The components of the spinor ψ' at the point x' in the moving reference frame Σ' (x' corresponds to the same physical event as x in Σ) are related to the components $\psi(x)$ by a finite (4×4) matrix S. The equivalent form (2.23) suggests a different mathematical interpretation of the unitary part, i.e. that it results from a comparison of the transformed function *at the same point* with the original function at that point. The connection of this comparison with the angular momentum operator for three dimensional rotations is well known.

Consider a spinor u transforming according to the representation $D^{(j_1, j_2)}$. We may imagine associating this spinor with a certain point x, and the transformed spinor u' with the point $x' = x$. This generalization (by *definition*) preserves the group property and also lends itself to a transparent physical interpretation. A set of objects transforming according to the rule*

$$u'_\alpha(x') = \sum_\beta D^{(j_1, j_2)}_{\alpha\beta} u_\beta(x) \tag{3.1}$$

is said to comprise an irreducible spinor field (which we shall often call a spinor, for simplicity). In order to give explicit meaning to the notation, one should consult Eq. (2.43).

As long as we remember to associate the transformed spinor with the transformed coordinates, only the finite (non-unitary) representations $D^{(j_1, j_2)}$ appear. The need for writing the transformation in a form like (2.23) only becomes apparent when we go over to quantum field theory, where the Lorentz transformation operators act on the Hilbert space but do not change the coordinate x.

The introduction of spinor fields transforming according to (3.1) is very similar to the way tensor fields are defined in tensor calculus.† (We shall see that the latter subject is a special case of the spinor calculus.) The subject matter in question has a large and somewhat chaotic literature. Much of the older literature is summarized in refs. 4–7 of Chapter 2.

3.2 SPINOR TRANSFORMATION RULES; DOTTED SPINORS

In order to make a systematic development of spinor calculus it is essential to develop the analogue of the covariant-contravariant distinction between transformations occurring in tensor calculus. In the latter formalism only real quantities normally occur.‡ For a complete description of spinor transformation properties it is also useful to contrast a transformation and its complex conjugate.

---

* Note that Eq. (3.1) can be used to *define* what $u'(x')$ is for all x.

† In conventional tensor calculus one often works with reducible representations.

‡ In quantum field theory tensor fields are not real in the usual sense but the *transformation* matrices themselves are real (or can be chosen real in some basis).

In analogy to the metric tensor of ordinary tensor calculus, we seek a metric spinor ϵ^{ab} such that the quadratic form

$$\epsilon^{ab}\psi_a\xi_b = \text{invariant} \tag{3.2}$$

where ψ_a and ξ_b transform as $\psi \to D\psi$, $\xi \to D\xi$, $D = D^{(\frac{1}{2},0)}$. As usual, like indices are summed over the appropriate range when one is up and the other down. In detail the *covariant* spinor ξ_a transforms as (see Eq. (2.40))

$$\xi'_a = D_a^b\xi_b \tag{3.3}$$

In addition we found the spinor η transforming as $D^{(0,\frac{1}{2})}$. The representations are connected by the relations

$$\begin{aligned} [D^{(\frac{1}{2},0)}]^\dagger &= [D^{(0,\frac{1}{2})}]^{-1} \\ i\sigma_2 D^{(\frac{1}{2},0)}(i\sigma_2)^{-1} &= D^{(0,\frac{1}{2})} \end{aligned} \tag{3.4}$$

Thus we have ($D \equiv D^{(\frac{1}{2},0)}$)

$$\xi' = D\xi, \qquad \eta'^\dagger = \eta^\dagger D^{-1} \tag{3.5}$$

under $SL(2, C)$ so that

$$(\eta, \xi) \equiv \eta^\dagger\xi = \eta'^\dagger\xi' \tag{3.6}$$

is invariant and qualifies as an inner product. Although (3.6) is different from (3.2), the second of relations (3.4) shows how to relate $\eta^\dagger$ to an object transforming as ξ and hence gives the form of the metric spinor ϵ. The spinor $\zeta = i\sigma_2\eta$ transforms as $D^{(\frac{1}{2},0)*}$ by (3.4).

$$\zeta' = D^*\zeta, \qquad \zeta = i\sigma_2\eta \tag{3.7}$$

i.e. the complex conjugate spinor ζ^* transforms as ξ, and as ψ assumed in Eq. (3.2). Identifying† ψ with ζ^* we are led to the metric spinor

$$[\epsilon^{ab}] = i\sigma_2 = \begin{bmatrix} 0 & 1 \\ -1 & 0 \end{bmatrix} \tag{3.8}$$

The overall phase of ϵ is conventional and chosen for simplicity.

In order to verify that ϵ^{ab} has useful properties like those of the metric tensor we note that Eq. (3.2), which says that $\epsilon^{ab}\psi_a\xi_b$ is invariant suggests defining the contravariant spinor ξ^a by

$$\xi^a = \epsilon^{ab}\xi_b \tag{3.9}$$

---

† As a particular case we note that $\xi^\dagger i\sigma_2\xi$ is invariant.

Then the invariant takes a familiar form $\psi_a\xi^a$.† The spinor $[\xi^a]$ transforms as $D^{(0,\frac{1}{2})}$ according to Eq. (3.4).

Equation (3.9) is clearly invertible ($[\xi_b] = -i\sigma_2[\xi^a]$ so that the lowering of an index may be written as

$$\xi_a = \epsilon_{ba}\xi^b \tag{3.10}$$

where the symbol ϵ_{ab} is numerically the same as ϵ^{ab} in Eq. (3.7).

Our notation suggests that ϵ_{ab} and ϵ^{ab} are to be regarded as second rank spinors transforming on their indices as well as numerical constants. The consistency of these requirements is easily demonstrated

$$\begin{aligned} \epsilon_{a'b'} &= D^a_{a'} D^b_{b'} \epsilon_{ab} = D^1_{a'} D^2_{b'} - D^2_{a'} D^1_{b'} \\ \epsilon_{1'1'} &= \epsilon_{2'2'} = 0 \\ \epsilon_{1'2'} &= -\epsilon_{2'1'} = \det D = 1 \end{aligned} \tag{3.11}$$

The consistency of $[\epsilon^{ab}] = [\epsilon_{ab}]$, i.e. $\epsilon_{ab} = \epsilon_{ac}\epsilon_{bd}\epsilon^{cd}$ is equivalent to the matrix identity $i\sigma_2 = i\sigma_2(i\sigma_2)(i\sigma_2)^{-1}$. The identity

$$\epsilon_{ab}\epsilon^{ac} = \delta^c_b \tag{3.12}$$

is equivalent to $(i\sigma_2)^2 = -1$. The Kronecker delta δ^c_b is also an invariant spinor of mixed (co- and contravariant) character.

Thus far we have considered spinors ξ_a transforming as $D^{(\frac{1}{2},0)}$, and ψ^a transforming as $D^{(0,\frac{1}{2})*} = i\sigma_2 D^{(\frac{1}{2},0)}(i\sigma_2)^{-1}$. These representations are equivalent but not identical. The metric spinors ϵ_{ab} and ϵ^{ab} allow us to change from one basis to another (raising and lowering indices). In addition the metric spinor is a set of Clebsch-Gordan coefficients for extracting that part of $D^{(\frac{1}{2},0)} \otimes D^{(\frac{1}{2},0)}$ transforming as $D^{(0,0)}$ (i.e. a scalar).

Next consider the transformation of the complex conjugate spinor ξ^* whose components we denote by a dot $\xi_{\dot{a}} \equiv (\xi_a)^*$. Similarly writing $D^{\dot{b}}_{\dot{a}} = (D^b_a)^*$ we have

$$\xi'_{\dot{a}} = D^{\dot{b}}_{\dot{a}}\xi_{\dot{b}} \tag{3.13}$$

More generally when a spinor possesses a dotted subscript, that index transforms as (3.13), i.e. via D^*. When the same symbol occurs with dotted

† The scalar product $\psi^a\xi_a$ is in detail

$$\psi^a\xi_a = \psi^1\xi_1 + \psi_2\xi_2 = \psi_1\xi_2 - \psi_2\xi_1$$

Thus if $\psi = \xi$, $\xi^a\xi_a = 0$. Some care is needed when raising and lowering indices since

$$\psi^a\xi_a = -\psi_a\xi^a.$$

and undotted indices, it will be understood to be related by the appropriate operations of conjugation and raising and lowering.

Comparison with Eq. (3.2) indicates that

$$\epsilon^{\dot{a}\dot{b}}\psi_{\dot{a}}\xi_{\dot{b}} = \text{invariant} \tag{3.14}$$

where $[\epsilon^{\dot{a}\dot{b}}] = [\epsilon^{ab}]$ is an invariant contravariant second rank spinor. We may raise and lower dotted indices by rules obtained by conjugating Eqs. (3.9) and (3.10)

$$\xi^{\dot{a}} = \epsilon^{\dot{a}\dot{b}}\xi_{\dot{b}} \qquad \xi_{\dot{a}} = \epsilon_{\dot{b}\dot{a}}\xi^{\dot{b}} \tag{3.15}$$

By definition the quantity $[\xi_{\dot{a}}]$ transforms as D^*. In contrast, $[\xi^{\dot{a}}]$ transforms as $D^{(0,\frac{1}{2})}$. This is easily shown as follows

$$\begin{aligned} \xi'^{\dot{a}} &= \epsilon^{\dot{a}\dot{b}}\xi'_{\dot{b}} = \epsilon^{\dot{a}\dot{b}}D_{\dot{b}}^{\dot{c}}\xi_{\dot{c}} \\ &= \epsilon^{\dot{a}\dot{b}}D_{\dot{b}}^{\dot{c}}\epsilon_{\dot{d}\dot{c}}\xi^{\dot{d}} \end{aligned} \tag{3.16}$$

i.e.

$$[\xi^{\dot{a}}]' = i\sigma_2 D^*(i\sigma_2)^{-1}[\xi^{\dot{a}}] = D^{(0,\frac{1}{2})}[\xi^{\dot{a}}]$$

In summary we have defined four distinct (though not independent) types of transformation laws. We have dotted and undotted indices for both upper (contravariant) and lower (covariant) indices. The matrices effecting the transformations are summarized in Eq. (3.17):

$$\begin{aligned} &\xi_a; \quad [D^b{}_a] = D = D^{(\frac{1}{2},0)} \\ &\xi_{\dot{a}}; \quad [D^{\dot{b}}{}_{\dot{a}}] = D^* \\ &\xi^a; \quad [D^a{}_b] = [(D^{-1})^a{}_b] = (D^{-1})^T \\ &\xi^{\dot{a}}; \quad [D^{\dot{a}}{}_{\dot{b}}] = [(D^{-1})_{\dot{b}}{}^{\dot{a}}] = (D^\dagger)^{-1} = D^{(0,\frac{1}{2})} \end{aligned} \tag{3.17}$$

From the discussion of Sec. (2.3) one might regard spinors transforming as ξ_a and $\xi^{\dot{a}}$ ($D^{(\frac{1}{2},0)}$ and $D^{(0,\frac{1}{2})}$) as independent* and the spinors like $\xi_{\dot{a}}$ and ξ^a as useful auxiliary objects derived from the former by complex conjugation.

A general mixed spinor $\xi^{a_1\cdots a_k, \dot{a}_1\cdots\dot{a}_l}_{m_1\cdots m_r, \dot{m}_1\cdots\dot{m}_s}$ is a set of objects transforming exactly as the appropriate product of individual spinors (3.17)

$$\xi^{a_1}\xi^{a_2}\cdots\xi^{a_k}\xi^{\dot{a}_1}\cdots\xi^{\dot{a}_l}\xi_{m_1}\cdots\xi_{m_r}\xi_{\dot{m}_1}\cdots\xi_{\dot{m}_s}$$

(The individual ξ may be completely independent; only their transformation

* i.e. independent under $SL(2, C)$ transformations, which of course do not include the operation of complex conjugation.

properties matter). Contraction of upper and lower indices of like type (i.e. both dotted or both undotted) has the same effect as in tensor calculus, and the $SL(2, C)$ transformations preserve symmetry relations among upper or lower indices (dotted and undotted taken separately).

The mixed spinor introduced above is not identical to the irreducible spinor defined in Eq. (3.1) since in general it is not irreducible and in addition the equivalent representations $D^{(\frac{1}{2},0)}$ and $D^{(0,\frac{1}{2})*}$ have been distinguished.

3.3 THE SPIN TENSOR AND THE CONNECTION BETWEEN FOUR-VECTORS AND SPIN SPACE

Equation (2.54) shows how to construct a four-vector $V^\mu = \xi^\dagger \sigma^\mu \xi$ from the (constant) "4-vector" spin matrix $\sigma^\mu = (I, \boldsymbol{\sigma})$. We now examine this connection in more detail, utilizing the results of the preceding section. Since the components $\xi_{\dot{a}}$ comprise $\xi^\dagger$, we are led to write

$$\xi^\dagger \sigma^\mu \xi = \xi_{\dot{a}} \sigma^{\mu \dot{a} b} \xi_b \tag{3.18}$$

The notation suggests that the matrix σ^μ be considered to represent a mixed spinor. Since our previous interpretation required that σ^μ be constant, this would be impossible were it not possible to simultaneously transform on the index μ. By cancelling the effect of the spinor transformations and the Lorentz transformation we can maintain the constancy of σ^μ even though the indices transform as indicated by the notation. We show that this is implicit in Eq. (2.63).

Using the results (3.17) we find the transformed $\sigma'^{\mu \dot{a} b}$

$$\sigma'^{\mu \dot{a} b} = \Lambda^\mu_\nu [(D^\dagger)^{-1}]^{\dot{a}}{}_{\dot{c}} [(D^{-1})^T]^b{}_d \sigma^{\nu \dot{c} d} \tag{3.19}$$

which has the matrix form

$$\sigma'^\mu = \Lambda^\mu_\nu (D^\dagger)^{-1} \sigma^\nu D \tag{3.20}$$

the last equality following from Eq. (2.63).

The "spin-tensor" $\sigma^{\mu \dot{a} b}$ is so named because of its mixed transformation properties. Since it is important in applications we give variants of the spin-tensor obtained by changing indices. Lowering both spinor indices in

$$\sigma^\mu = [\sigma^{\mu \dot{a} b}] = (I, \boldsymbol{\sigma}) \tag{3.21}$$

we get

$$[\sigma^\mu_{\dot{a} b}] = i\sigma_2 (I, \boldsymbol{\sigma})(i\sigma_2)^{-1} = (I, -\boldsymbol{\sigma}^*) \tag{3.22}$$

Lowering the index μ and taking the complex conjugate we obtain

$$[\sigma_{\mu a\dot{b}}] = [\sigma_{\mu\dot{a}b}]^* = (I, \boldsymbol{\sigma}) \tag{3.23}$$

Since $\sigma_{\mu a\dot{b}}$ and $\sigma_{\mu\dot{a}b}$ are numerically equal but have distinct spinor character, we denote the matrix $[\sigma_{\mu a\dot{b}}]$ by $\tilde{\sigma}_\mu$

$$\tilde{\sigma}_\mu = [\sigma^\mu_{a\dot{b}}] = (I, \boldsymbol{\sigma}) \tag{3.24}$$

Writing $\tilde{\sigma}_\mu = \sigma^\mu$ indicates numerical equality of components but *not* identical transformation properties.

The transformation of $\tilde{\sigma}_\mu$ follows from

$$\sigma'^\mu_{a\dot{b}} = \Lambda^\mu_\nu D_a{}^c D_{\dot{b}}{}^{\dot{d}} \sigma^\nu_{c\dot{d}} \tag{3.25}$$

which in matrix notation reads

$$D\tilde{\sigma}_\mu D^\dagger = \Lambda^\nu_\mu \tilde{\sigma}_\nu \tag{3.26}$$

Equation (3.18) shows how $\sigma^{\mu\dot{a}b}$ is used to convert the special tensor $\xi_{\dot{a}}\xi_b$ to a four-vector V^μ by contraction. The inverse operation can also be studied. To the four-vector x^μ we associate a spinor $X_{a\dot{b}}$ by

$$X_{a\dot{b}} \equiv x_\mu \sigma^\mu_{a\dot{b}} \tag{3.27}$$

In matrix notation we have

$$\begin{aligned} X \equiv [X_{a\dot{b}}] &= x^\mu \tilde{\sigma}_\mu = tI + \boldsymbol{\sigma}\cdot\mathbf{x} \\ &= \begin{pmatrix} t+z & x-iy \\ x+iy & t-z \end{pmatrix} \end{aligned} \tag{3.28}$$

From the transformation law for $\tilde{\sigma}$ (Eq. (3.26)) we find

$$DXD^\dagger = x^\mu \Lambda^\nu_\mu \tilde{\sigma}_\nu = x'^\nu \tilde{\sigma}_\nu \equiv X' \tag{3.29}$$

The transformation in question is a Lorentz transformation since

$$\begin{gathered} \det X' = |\det D|^2 \det X = \det X; \\ t'^2 - x'^2 - y'^2 - z'^2 = t^2 - x^2 - y^2 - z^2 \end{gathered} \tag{3.30}$$

i.e. $x'^2 = x^2$.

Equation (3.27) can be inverted using the identity*

$$\sigma_{\mu a\dot{b}}\sigma_{\nu}^{a\dot{b}} = 2g_{\mu\nu} \tag{3.31}$$

to give

$$x_{\mu} = \tfrac{1}{2}X_{a\dot{b}}\sigma_{\mu}^{a\dot{b}} \tag{3.32}$$

By means of (3.31) we can write the scalar product $x_{\mu}y^{\mu}$ as

$$x \cdot y = \tfrac{1}{2}X_{a\dot{b}}Y^{a\dot{b}} \tag{3.33}$$

It is useful to express the 4-gradient $\partial_{\mu} = \partial/\partial x^{\mu}$ as a mixed spinor

$$\partial_{a\dot{b}} = \sigma_{a\dot{b}}^{\mu}\partial_{\mu} \tag{3.34}$$

since the spinor character of the gradient has to be known to write spinor differential equations of known $SL(2, C)$ transformation property. As a special case of (3.33) we note

$$\Box = \partial_{\mu}\partial^{\mu} = \tfrac{1}{2}\partial_{a\dot{b}}\partial^{a\dot{b}} \tag{3.35}$$

Another useful identity follows from the consistency of Eqs. (3.27) and (3.32):

$$\sigma_{\mu a\dot{b}}\sigma^{\mu c\dot{d}} = 2\delta_{a}^{c}\delta_{\dot{b}}^{\dot{d}} \tag{3.36}$$

From the properties of the Pauli spin matrices we find

$$\sigma_{\mu}\tilde{\sigma}_{\nu} + \tilde{\sigma}_{\nu}\sigma_{\mu} = 2g_{\mu\nu} \tag{3.37}$$

Restoring all the spinor labels this is

$$\sigma_{\mu}^{a\dot{b}}\sigma_{\nu\dot{b}c} + \sigma_{\nu}^{a\dot{b}}\sigma_{\mu\dot{b}c} = 2g_{\mu\nu}\delta_{c}^{a} \tag{3.38}$$

Finally we give an explicit formula expressing the Lorentz transformation matrix Λ_{ν}^{μ} in terms of the $SL(2, C)$ matrix D. Multiplying Eq. (3.26) by σ_{μ}, taking the trace and using $Tr(\sigma_{\mu}\tilde{\sigma}_{\nu}) = 2g_{\mu\nu}$, we find

$$\Lambda_{\nu}^{\mu} = \tfrac{1}{2}\,\mathrm{Tr}\,(\sigma^{\mu}D\tilde{\sigma}_{\nu}D^{\dagger}) \tag{3.39}$$

Note that numerically $\tilde{\sigma}_{\nu} = \sigma^{\nu}$; the form (3.39) preserves the covariant notation. In spinor notation Eq. (3.39) is

$$\Lambda_{\nu}^{\mu} = \tfrac{1}{2}\sigma^{\mu a\dot{b}}D_{a}^{\ c}D_{\dot{b}}^{\ \dot{d}}\sigma_{\nu c\dot{d}} \tag{3.40}$$

---

* In order to prove (3.31) we use the Hermiticity of $\sigma_{\mu a\dot{b}}$ to write $\sigma_{\mu\dot{a}b} = \sigma_{\mu b\dot{a}}$

$$\begin{aligned} \sigma_{\mu\dot{a}b}\sigma_{\nu}^{\dot{a}b} &= \sigma_{\mu b\dot{a}}\sigma_{\nu}^{\dot{a}b} = \mathrm{Tr}\,(\tilde{\sigma}_{\mu}\sigma_{\nu}) \\ &= 2;\ \mu = \nu = 0 \\ &= -2\delta_{mn};\ m, n = (1, 2, 3) \\ &= 0;\ \mu \neq \nu \end{aligned}$$

This result is easily derived beginning with Eq. (3.32) for x'^μ. Equation (3.39) shows how to calculate Λ^μ_ν from D. The matrices $\pm D$ clearly correspond to the same Λ.

3.4 LINEAR WAVE EQUATIONS COVARIANT UNDER $SL(2, C)$

The results of the preceding section provide a simple systematic way of writing equations covariant under $SL(2, C)$ transformations. One only has to be sure that the indices (upper and lower, dotted and undotted) balance on either side of an equality. The basic ingredients of a theory are a set of spinor fields from which one constructs "equations of motion". A typical equation of motion is a partial differential equation (or equivalent integral equation) by means of which the causal evolution of the system may be computed. Historically the first such equations were linear in the field, since the latter was originally identified with the state vector, which must obey the (linear) superposition principle. Such equations remain relevant since they describe free particles and hence are the subject of direct experimental interest. In the succeeding stage of development the physical interpretation was changed so that the state vector was no longer the spinor field; the latter became an operator whose equation of motion was generally non-linear. Nevertheless most such fields are believed to obey linear (free particle) equations of motion when the components of a state are widely separated in space-time.

In the preceding section we learned how to write the gradient as a mixed spinor (Eq. (3.34)). The following simple example illustrates the general method and leads to the physically important Dirac equation for spin $\frac{1}{2}$ particles. If we operate on the spinor field $\xi_b(x)$ with $i\partial^{\dot{a}b}$ we obtain an object which transforms as a spinor $\chi^{\dot{a}}$. Thus we could write

$$i\partial^{\dot{a}b}\xi_b = m\chi^{\dot{a}} \tag{3.41}$$

where m has dimension (length)$^{-1}$ and will turn out to be the mass parameter. Later we shall study the special cases $m = 0$ and $\chi \propto \xi$. Equation (3.41) is indeed covariant provided the fields ξ, χ transform as

$$\begin{aligned} \xi'_b(x') &= D^a{}_b\xi_a(x) \\ \chi'^{\dot{d}}(x') &= D^{\dot{d}}{}_{\dot{c}}\chi^{\dot{c}}(x) \end{aligned} \tag{3.42}$$

The covariance is easily demonstrated explicitly (using (3.17))

$$\begin{aligned}\partial'^{\dot{d}b}\xi'_b(x') &= D^{\dot{d}}{}_{\dot{c}}D^{b}{}_{a}D_b{}^{f}\,\partial^{\dot{c}a}\xi_f(x)\\ &= D^{\dot{d}}{}_{\dot{c}}\,\partial^{\dot{c}b}\xi_b(x) = D^{\dot{d}}{}_{\dot{c}}\,\chi^{\dot{c}}(x)\\ &= \chi'^{\dot{d}}(x')\end{aligned} \tag{3.43}$$

Having understood the transformation properties it is useful to write the above in more streamlined form. If we write σ_μ as the (constant) matrix formed from $\sigma_\mu^{\dot{a}b}$, $\xi = \text{col}\ \xi_b$, $\chi = \text{col}\ \chi^{\dot{a}}$, then Eq. (3.41) becomes

$$i\sigma^\mu\,\partial_\mu\xi = m\chi \tag{3.44}$$

Under a Lorentz transformation the various quantities transform as $\sigma'_\mu = \sigma_\mu$, $\partial'_\mu = \Lambda_\mu^{\ \nu}\,\partial_\nu$, $\xi' = D\xi$, $\chi' = (D^\dagger)^{-1}\chi$. To demonstrate covariance we note that the transformed Eq. (3.44) $i\sigma^\mu\,\partial'_\mu\xi' = m\chi'$ reduces to (3.44) if Eq. (3.20) is used.

Equation (3.44) is translation invariant: if $\xi(x)$, $\chi(x)$ are solutions, so are $\xi(x + a)$, $\chi(x + a)$. Thus we have solutions of the form $\exp(\pm ip\cdot x)$, Choosing the conventional positive energy solution $\xi(x) = \xi(p)e^{-ip\cdot x}$. $\chi(x) = \chi(p)e^{-ip\cdot x}$ we find

$$\sigma\cdot p\xi(p) = m\chi(p) \tag{3.45}$$

From this equation we can compute $\chi(p)$ from $\xi(p)$. To invert (3.45) we note that according to the identity $\sigma\cdot p\tilde{\sigma}\cdot p = p^2$ and

$$p^2\xi(p) = m\tilde{\sigma}\cdot p\chi(p) \tag{3.46}$$

In order to satisfy the basic postulate that the four components p^μ represent the physical energy momentum four-vector, which satisfies $p^2 = m^2$, we have to put an additional constraint on the functions ξ and χ. We further require that $\xi(x)$, $\chi(x)$ obey the free Klein-Gordon equation

$$\begin{aligned}K_x\xi &= K_x\chi = 0,\\ K_x &\equiv \partial^2 + m^2\end{aligned} \tag{3.47}$$

Equation (3.47) restricts the solutions of (3.44) to those obeying the mass shell condition $p^2 = m^2$.

Our results may be put in more symmetrical form by noting that Eqs. (3.41) and (3.47) are equivalent to the pair of coupled equations

$$\begin{aligned}i\partial^{\dot{a}b}\xi_b &= m\chi^{\dot{a}}\\ i\partial_{\dot{a}b}\chi^{\dot{a}} &= m\xi_b\end{aligned} \tag{3.48}$$

or in different notation

$$i\sigma^\mu \partial_\mu \xi = m\chi$$
$$i\tilde{\sigma}^\mu \partial_\mu \chi = m\xi \tag{3.49}$$

For example if we use the second of Eqs. (3.48) to eliminate ξ_b in the first line we obtain

$$-\partial^{\dot{a}b} \partial_{\dot{c}b} \chi^{\dot{c}} = m^2 \chi^{\dot{a}} \tag{3.50}$$

The left hand side is simply $-\partial^2 \chi^{\dot{a}}$, using the identity $\partial^{\dot{a}b} \partial_{\dot{c}b} = \delta^{\dot{a}}_{\dot{c}} \partial^2$ which follows when $\partial^\mu \partial^\nu$ is contracted into Eq. (3.38).

We now show that Eqs. (3.48) (or (3.49)) are completely equivalent to the celebrated Dirac equation. The latter is therefore the simplest non-trivial first order linear differential spinor equation subject to the Klein-Gordon condition (3.47). Equations (3.49) may be written as a single matrix equation if we introduce the four-component object ψ and the set of 4×4 matrices γ^μ by

$$\psi(x) = \begin{pmatrix} \xi(x) \\ \chi(x) \end{pmatrix} \qquad \gamma^\mu = \begin{pmatrix} 0 & \tilde{\sigma}^\mu \\ \sigma^\mu & 0 \end{pmatrix} \tag{3.51}$$

$$i\gamma^\mu \partial_\mu \psi(x) = m\psi(x) \tag{3.52}$$

The "gamma matrices" γ^μ obey the anticommunication rules

$$\gamma^\mu \gamma^\nu + \gamma^\nu \gamma^\mu = 2g^{\mu\nu} I \tag{3.53}$$

as a consequence of (3.37)*. Under proper Lorentz transformations ψ transforms† as the direct sum $D^{(\frac{1}{2},0)} \oplus D^{(\frac{1}{2},0)}$:

$$\psi = \begin{pmatrix} \xi_1 \\ \xi_2 \\ \chi_{\dot{1}} \\ \chi_{\dot{2}} \end{pmatrix} = \begin{pmatrix} \xi \\ \chi \end{pmatrix} \to \begin{pmatrix} D\xi \\ (D^\dagger)^{-1}\chi \end{pmatrix} \tag{3.54}$$

Thus the four-component "spinor" ψ gives a basis for a reducible representation of $SL(2, C)$. However the representation is irreducible if we also require parity covariance. This situation was already noted in the discussion following Eq. (2.65). It is useful to reexamine the question to study the x-dependence of the spinors.

---

* γ_μ is defined to be $g_{\mu\nu}\gamma^\nu$, as suggested by the notation.

† Note that for pure rotations ξ and χ transform identically.

If we change coordinate systems so that $x'^{\mu} = g^{\mu\mu}x^{\mu} = (t, -\mathbf{x})$ the form of the Dirac equation should be $i\gamma^{\mu}\,\partial'_{\mu}\psi'(x') = m\psi'(x')$. Since $\gamma^{\mu}\,\partial'_{\mu} = \gamma^{\mu T}\,\partial_{\mu} = \gamma_0\gamma^{\mu}\gamma_0\,\partial_{\mu}$, we see that $\psi'(x') = \eta_p\gamma_0\psi(x)$ is the transformation which preserves the form of the equation under space inversion. Here η_p is a unimodular phase factor. In terms of the two component spinors we have a simple interchange of upper and lower components.

$$\left.\begin{aligned}\xi'(x') &= \eta_p\chi(x)\\ \chi'(x') &= \eta_p\xi(x)\end{aligned}\right\}\psi'(x') = \eta_p\gamma_0\psi(x) \tag{3.55}$$

The free Dirac equation does not naturally lead to the proper transformation under time inversion. This problem is considered in a later section.

It is useful to consider two special cases.

(1) Weyl equation. If we set $m = 0$, the equations for ξ and χ decouple and we obtain

$$\partial^{\dot{a}b}\xi_b = 0 \quad \text{or} \quad \sigma^{\mu}\,\partial_{\mu}\xi = 0 \tag{3.56a}$$

$$\partial_{\dot{a}b}\chi = 0 \quad \text{or} \quad \tilde{\sigma}^{\mu}\,\partial_{\mu}\chi = 0 \tag{3.56b}$$

Writing these equations in detail gives

$$\left(\frac{\partial}{\partial t} + \boldsymbol{\sigma}\cdot\nabla\right)\xi = 0 \tag{3.57a}$$

$$\left(\frac{\partial}{\partial t} - \boldsymbol{\sigma}\cdot\nabla\right)\chi = 0 \tag{3.57b}$$

For plane wave solutions $\xi = \xi(p)e^{-ip\cdot x}$, $\chi = \chi(p)e^{-ip\cdot x}$ these equations reduce to

$$(E - \mathbf{p}\cdot\boldsymbol{\sigma})\xi(p) = 0 \quad \text{or} \quad \boldsymbol{\sigma}\cdot\hat{p}\xi = \xi \tag{3.58a}$$

$$(E + \mathbf{p}\cdot\boldsymbol{\sigma})\xi(p) = 0 \quad \text{or} \quad \boldsymbol{\sigma}\cdot\hat{p}\chi = -\chi \tag{3.58b}$$

Thus $\xi(p)$ describes massless particles with right-handed polarization, while $\chi(p)$ describes massless particles of left-handed polarization.

It will be noted that the equations for ξ and χ, taken separately, are not parity-covariant. These equations are useful in describing neutrinos.

(2) Majorana Field. The Dirac equation is consistent if χ is not distinct from ξ in the sense $\chi^{\dot{a}} = \xi^{\dot{a}}$, i.e. $\chi = i\sigma_2\xi^*$. In this case the Dirac equation is

$$i\,\partial^{\dot{a}b}\xi_b = m\xi^{\dot{a}} \tag{3.59}$$

and the Klein-Gordon condition is automatically satisfied. This situation coincides with the case in which a fermion is its own antiparticle. In Chapter 6 we shall study in more detail the properties of the "charge-conjugate" spinor ($|\eta_c| = 1$)

$$\psi_c = \eta_c C\bar{\psi}^T = \eta_c C_1 \psi^* \tag{3.60}$$

where the matrix C_1 can be defined to be $-i\gamma_2$. The quantity $C_1\psi^*$ is exactly ψ when $\chi = i\sigma_2\xi^*$, so the field is self conjugate with charge parity $+1$. (If we had chosen $\chi = -i\sigma_2\xi^*$ then the charge parity would have been -1.)

The particular set of matrices γ^μ in Eq. (3.51) are especially useful in describing massless (or high-energy) particles since the upper and lower components ξ and χ decouple when $m = 0$ or $p \to \infty$. Explicitly these γ's are

$$\gamma^0 = \begin{pmatrix} O & I \\ I & O \end{pmatrix}, \qquad \gamma^k = \begin{pmatrix} O & -\sigma_k \\ \sigma_k & O \end{pmatrix} \tag{3.61}$$

In addition the (Hermitian) matrix γ_5 is useful

$$\gamma_5 = -i\gamma_0\gamma_1\gamma_2\gamma_3 = i\gamma^0\gamma^1\gamma^2\gamma^3 = \begin{pmatrix} I & O \\ O & -I \end{pmatrix} \tag{3.62}$$

The operators $a_\pm = \frac{1}{2}(1 \pm \gamma_5)$ project ψ into the upper and lower components

$$a_+\psi = \begin{pmatrix} \xi \\ O \end{pmatrix}, \qquad a_-\psi = \begin{pmatrix} O \\ \chi \end{pmatrix} \tag{3.63}$$

The basic structure of the Dirac equation and the commutation rules (3.53) are unchanged under $\psi \to S\psi$, $\gamma_\mu \to S\gamma_\mu S^{-1}$. For formal purposes it is often convenient to express theoretical results in a basis-independent manner. However many practical calculations are best performed using a particular set of γ-matrices.

By a simple extension of the preceding analysis we derive the Dirac-Fierz-Pauli equations for general spin. We begin by introducing a spinor ξ transforming as $D^{(j_1, j_1')}$:

$$\xi^{\dot{a}_1\cdots\dot{a}_l}_{bb_1\cdots b_k}; \qquad j_1 = \tfrac{1}{2}(k+1), \qquad j_1' = \tfrac{1}{2}l \tag{3.64}$$

There is complete symmetry in the upper and in the lower indices. Note that an upper dotted index transforms as $D^{(0,\frac{1}{2})}$, a lower undotted one as $D^{(\frac{1}{2},0)}$.

This spinor does not correspond to a unique spin, since under the rotation subgroup

$$D^{(j_1,j_1')} \supset D^{(j_1+j_1')} \oplus D^{(j_1+j_1'-1)} \oplus \cdots \oplus D^{|j_1-j_1'|} \tag{3.65}$$

Hence the desired spinors (having a unique spin) must be restricted by supplementary conditions.

As for spin $\frac{1}{2}$, parity covariance leads to the introduction of a second spinor χ transforming as $D^{(j_2,j_2')}$

$$\chi^{\dot{a}\dot{a}_1\cdots\dot{a}_l}_{b_1\cdots b_k}; \qquad j_2 = \tfrac{1}{2}k, \qquad j_2' = \tfrac{1}{2}(l+1) \tag{3.66}$$

These quantities differ from Dirac case only by the addition of l dotted contravariant indices and k undotted covariant indices:

$$\begin{aligned} D^{(\frac{1}{2}0)} &\to D^{(\frac{1}{2}k+\frac{1}{2},\frac{1}{2}l)} \\ D^{(0\frac{1}{2})} &\to D^{(\frac{1}{2}k,\frac{1}{2}+\frac{1}{2}l)} \end{aligned} \tag{3.67}$$

The suggested equations of motion[1,2]

$$\begin{aligned} i\,\partial^{\dot{a}b}\xi^{\dot{a}_1\cdots\dot{a}_l}_{bb_1\cdots b_k} &= m\chi^{\dot{a}\dot{a}_1\cdots\dot{a}_l}_{b_1\cdots b_k} \\ i\,\partial_{\dot{a}b}\chi^{\dot{a}\dot{a}_1\cdots\dot{a}_l}_{b_1\cdots b_k} &= m\xi^{\dot{a}_1\cdots\dot{a}_l}_{bb_1\cdots b_k} \end{aligned} \tag{3.68}$$

are a natural generalization of the spin $\frac{1}{2}$ case. As before the spinors ξ and χ obey the Klein-Gordon equation. The equations of motion (3.68) provide the supplementary condition which picks out the highest spin from Eq. (3.65).

To consider the spin carried by the fields, go to the rest frame for a free particle, writing $\xi(x) = \xi(p)e^{-ipx}$, etc. In the rest frame, $i\,\partial^{\dot{a}b} \to (p^0)^{\dot{a}b} = \delta^{\dot{a}b}m$. Hence we find

$$\begin{aligned} \xi^{\dot{a}_1\cdots\dot{a}_l}_{ab_1\cdots b_k}(0) &= \chi^{\dot{a}\dot{a}_1\cdots\dot{a}_l}_{b_1\cdots b_k}(0) \quad \text{(equation of motion)} \\ &= \chi^{\dot{a}_1\dot{a}\cdots\dot{a}_l}_{b_1\cdots b_k}(0) \quad \text{(symmetry)} \\ &= \xi^{\dot{a}\cdots\dot{a}_l}_{a_1b_1\cdots b_k}(0) \quad \text{(equation of motion)} \end{aligned} \tag{3.69}$$

Therefore ξ is symmetrical under $a_1 \leftrightarrow a$, i.e. under interchange of *upper* and *lower* indices. Thus in the rest frame we have total symmetry in *all* indices.

Thus we have a totally symmetric spinor of $k + l + 1 = 2s$ components, which corresponds to spin s

$$s = \tfrac{1}{2}(k + l + 1) \tag{3.70}$$

in the rest frame. Clearly there are as many spin s equations as there are ways to add up k and l to get $2s - 1$.

The preceding points are well-illustrated by a close study of spin 1.

(A) $l = 1$, $k = 0$ (B) $l = 0$, $k = 1$: both give $s = 1$ Case (A) involves $\xi^{\dot{a}}_{b}$ and $\chi^{\dot{a}\dot{a}_1}$; case (B) involves ξ_{bb_1}, and $\chi^{\dot{a}}_{b}$. The corresponding coupled equations of motion are

$$\begin{aligned} i\,\partial^{\dot{a}b}\xi^{\dot{a}_1}_{b} &= m\chi^{\dot{a}\dot{a}_1} \\ i\,\partial_{\dot{a}b}\chi^{\dot{a}\dot{a}_1} &= m\xi^{\dot{a}_1}_{b} \end{aligned} \tag{3.71}$$

$$\begin{aligned} i\,\partial^{\dot{a}b}\xi_{bb_1} &= m\chi^{\dot{a}}_{b_1} \\ i\,\partial_{\dot{a}b}\chi^{\dot{a}}_{b_1} &= m\xi_{bb_1} \end{aligned} \tag{3.72}$$

We already know that the equations of motion lead to spin 1 only. An invariant expression of this fact follows (consider Eqs. (3.71)). From the assumed symmetry of χ we find

$$\partial^{\dot{a}b}\xi^{\dot{a}_1}_{b} - \partial^{\dot{a}_1 b}\xi^{\dot{a}}_{b} = 0 \tag{3.73}$$

The non-trivial case $\dot{a} = 1$, $\dot{a}_1 = 2$ is equivalent to

$$\partial^{\dot{a}b}\xi_{\dot{a}b} = 0 \tag{3.74}$$

which in turn implies $\chi^{\dot{a}}_{\dot{a}} = 0$.

We can make contact with the usual vector description by introducing the field V_μ by $\xi_{\dot{a}b} = \sigma_{\mu_{\dot{a}b}}V^\mu$, i.e.

$$[\xi^{\dot{a}b}] = \begin{bmatrix} V^0 + V^3 & V^1 - iV^2 \\ V + iV^2 & V^0 - V^3 \end{bmatrix} \tag{3.75}$$

Now $\partial_\mu V^\mu = 0$ expresses the subsidiary condition (3.74). In addition V_μ obeys the Klein-Gordon equation

$$K_x V_\mu(x) = 0 \tag{3.76}$$

The system of equations (B) can be analyzed in a similar manner. For free fields cases (A) and (B) are equivalent. But differences occur when an electromagnetic field is present.

Next we consider the case of spin $\frac{3}{2}$, which leads to the Rarita-Schwinger equation. We take $k = l = 1$ in Eq. (3.68), so that

$$\begin{aligned} i\,\partial^{\dot{a}b}\xi^{\dot{a}_1}_{bb_1} &= m\chi^{\dot{a}\dot{a}_1}_{b_1} \\ i\,\partial_{\dot{a}b}\chi^{\dot{a}\dot{a}_1}_{b_1} &= m\xi^{\dot{a}_1}_{bb_1} \end{aligned} \tag{3.77}$$

These equations are like the $s = 1$ case with an extra spinor index (b_1). We can convert a free dotted and undotted index to a vectorial one by defining

$$\begin{aligned} \xi_{b\mu} &= \sigma_{\mu\dot{a}_1}^{b_1}\xi_{bb_1}^{\dot{a}_1} \\ \chi_{\mu}^{\dot{a}} &= \sigma_{\mu}^{b_1}{}_{\dot{a}_1}\chi_{b_1}^{\dot{a}\dot{a}_1} \end{aligned} \tag{3.78}$$

This gives

$$\begin{aligned} i\,\partial^{\dot{a}b}\xi_{b\mu} &= m\chi_{\mu}^{\dot{a}} \\ i\,\partial_{\dot{a}b}\chi_{\mu}^{\dot{a}} &= m\xi_{b\mu} \end{aligned} \tag{3.79}$$

or equivalently

$$\begin{aligned} i\sigma\cdot\partial\xi_{\mu} &= m\chi_{\mu} \\ i\tilde{\sigma}\cdot\partial\chi_{\mu} &= m\xi_{\mu} \end{aligned} \tag{3.80}$$

Just as for spin 1 we can derive the subsidiary condition

$$\partial^{\mu}\xi_{\mu} = 0 \tag{3.81}$$

(which then implies $\partial^{\mu}\chi_{\mu} = 0$). The Dirac form may be obtained by defining a "four-vector spinor"

$$\psi_{\mu} = \begin{pmatrix} \xi_{\mu} \\ \chi_{\mu} \end{pmatrix} \tag{3.82}$$

This obeys a Dirac equation

$$i\gamma^{\mu}\,\partial_{\mu}\psi_{\nu}(x) = m\psi_{\nu}(x) \tag{3.83}$$

and the subsidiary condition

$$\partial^{\mu}\psi_{\mu} = 0 \tag{3.84}$$

From these equations we can also prove the result

$$\gamma^{\mu}\psi_{\mu} = 0$$

Fermions of spin $s = k + \frac{1}{2}$ may be treated by a simple extension of the preceding case. Add k upper dotted indices and k undotted lower indices to the Dirac spinor ($k = l$ in Eq. (3.68)). Define

$$\begin{aligned} \xi_{b\mu_1\cdots\mu_k} &= \prod_{j=1}^{k}\sigma_{\mu_j\dot{a}_j}^{b_j}\xi_{bb_1\cdots b_k}^{\dot{a}_1\cdots\dot{a}_k} \\ \chi_{\mu_1\cdots\mu_k}^{\dot{a}} &= \prod_{j=1}^{k}\sigma_{\mu_j\dot{a}_j}^{b_j}\chi_{b_1\cdots b_k}^{\dot{a}\dot{a}_1\cdots\dot{a}_k} \end{aligned} \tag{3.85}$$

The μ's can be permuted without change of ξ or χ, owing to the symmetry in the spinor indices.

Next we note that the spinor

$$\psi_{\{\mu\}} = \begin{pmatrix} \xi_{\{\mu\}} \\ \chi_{\{\mu\}} \end{pmatrix} \tag{3.86}$$

satisfies the equations

$$\begin{aligned} i\gamma^{\mu}\,\partial_{\mu}\psi_{\{\mu\}} &= m\psi_{\{\mu\}} \\ \partial^{\mu_i}\psi_{\{\mu\}} &= 0 \end{aligned} \tag{3.87}$$

for any i and that $\gamma^{\mu_i}\psi_{\{\mu\}} = 0$ for any i as a consequence.

In a similar way, bosons of spin s are described by a symmetric tensor

$$V_{\mu_1\cdots\mu_j\cdots\mu_s} = V_{\mu_j\cdots\mu_1\cdots\mu_s} = \cdots \tag{3.88}$$

subject to the conditions

$$\begin{aligned} \partial^{\mu_i}V_{\{\mu\}} &= 0 \\ K_x V_{\{\mu\}} &= 0 \end{aligned} \tag{3.89}$$

In the rest frame the components with any index $= 0$ vanish. Thus $V_{(\mu)}$ is the symmetrized tensor in s vectors, i.e. has spin s.

In the presence of interactions many difficulties arise. In fact it is hard to believe that fundamental fields of high spin (>1) exist, if ordinary field theory is valid. Thus one may prefer a different approach to free high spin fields in which there are no redundant components at all. However the generalized Rarita-Schwinger wave functions are often useful in applications because of their familiar vector-spinor character. A more explicit discussion of these questions is given in Chapter 6.

3.5 RARITA-SCHWINGER WAVE FUNCTIONS FOR HIGH SPIN

For practical applications it is frequently useful to have explicit forms for the Rarita-Schwinger wave functions in momentum space. It is also useful to know how such functions behave under complex conjugation and inversions, and further properties such as completeness relations. For Fermions we have "positive-energy" wave functions $u_{\mu}(p, \lambda)$ and "negative-energy" wave functions $v_{\mu}(p, \lambda)$ satisfying

$$\begin{aligned}(\not{p} - m)u_{\mu}(p, \lambda) &= 0 \\ p^{\mu_i}u_{\mu}(p, \lambda) = \gamma^{\mu_i}u_{\mu}(p, \lambda) &= 0 \\ (\not{p} + m)v_{\mu}(p, \lambda) &= 0 \\ p^{\mu_i}v_{\mu}(p, \lambda) = \gamma^{\mu_i}v_{\mu}(p, \lambda) &= 0\end{aligned} \tag{3.90}$$

For Bosons we have wave functions $e_{\mu}(p, \lambda)$ obeying

$$p^{\mu_i}e_{\mu}(p, \lambda) = 0 \tag{3.91}$$

When referring to all such wave functions we shall use the symbol $\chi_{\mu}(p, \lambda)$. χ_{μ} is symmetric in the four vector indices $\boldsymbol{\mu} = \mu_1 \cdots \mu_n$. The label λ characterizes the spin state. We shall use helicity (i.e. component of J along the direction of motion) to label the spin state.* The wave functions for arbitrary spin are easily constructed using recursion relations. First we describe some useful properties of spin $\frac{1}{2}$ and spin 1 wave functions in the helicity basis. We use a real metric with $g_{00} = 1 = -g_{ii}$ $(i = 1, 2, 3)$ and γ matrices

$$\gamma_0 = \begin{pmatrix} I & 0 \\ 0 & -I \end{pmatrix} \qquad \gamma^k = \begin{pmatrix} 0 & \sigma_k \\ -\sigma_k & 0 \end{pmatrix} \tag{3.92}$$
$$\{\gamma_{\mu}, \gamma_{\nu}\} = 2g_{\mu\nu} \qquad \gamma_0\gamma_{\mu}^{\dagger}\gamma_0 = \gamma_0$$

where σ_k are the usual Pauli spin matrices.

The spin state of a wave function usually refers to a fixed direction (the famous z axis) or to the direction of momentum (helicity basis). Although we generally use the latter description, it is useful to review both for the spin $\frac{1}{2}$ case. Let χ_s $(s = \pm\frac{1}{2})$ be the usual real two-component spinor quantized along the z axis $(\sigma_z\chi_s = 2s\chi_s)$. (We write $\chi_s(\hat{z})$ whenever a chance for confusion with helicity spinors occurs.) Then the Dirac wave function $u(p, s)$ for a particle of momentum $\mathbf{p}$ and "spin" s is

$$u(p, s) = \begin{pmatrix} \cosh \frac{1}{2}\zeta\chi_s \\ \boldsymbol{\sigma} \cdot \hat{p} \sinh \frac{1}{2}\zeta\chi_s \end{pmatrix} \tag{3.93}$$

where the parameter ζ is related to energy and momentum by $E = m \cosh \zeta$,

---

* More information about helicity states is given in Chapter IV. The discussion of this section is derived from refs. 4–8.

$p = m \sinh \zeta$. The antiparticle wave function is defined by

$$v(p, s) = C_1\bar{u}^T(p, s) = (-1)^{\frac{1}{2}-s}\begin{pmatrix} \boldsymbol{\sigma} \cdot \hat{p} \sinh \frac{1}{2}\zeta\chi_{-s} \\ \cosh \frac{1}{2}\zeta\chi_{-s} \end{pmatrix} \tag{3.94}$$

where the phase of the charge conjugation matrix $C_1 = i\gamma_0\gamma_2$ has been chosen in a natural way ($v(p, \frac{1}{2}) \to$ col. $(0, \chi_{-\frac{1}{2}})$ when $\mathbf{p} \to 0$).

Inspection reveals the inversion formulas

$$\begin{aligned} u(\mathbf{p}, s) &= Pu(-\mathbf{p}, s) \\ v(\mathbf{p}, s) &= -Pv(-\mathbf{p}, s) \end{aligned} \tag{3.95}$$

where P is the matrix γ_0. To invert both $\mathbf{p}$ and s, we use the time-reversal matrix $T = \gamma_3\gamma_1$ (chosen real) and the formula

$$\chi_{-s} = (-1)^{\frac{1}{2}+s} i\sigma_2\chi_s \tag{3.96}$$

to obtain

$$w^*(\mathbf{p}, s) = (-1)^{\frac{1}{2}+s} Tw(-\mathbf{p}, -s) \tag{3.97}$$

where w stands for either u or v. For simplicity of writing we write the four-momentum $p = (p_0, \mathbf{p})$, even though the quantity in question only depends on the three-vector $\mathbf{p}$.

When space and time inversions are considered, we exhibit the three-vector explicitly.

To discuss the analogous properties of Dirac helicity spinors, it is useful to summarize the corresponding properties of the two-component helicity spinors. The latter are defined by a rotation in the usual way, by rotation of a state with helicity λ in the $\hat{z}$ direction. Let $\hat{p}$ and the rotation axis $\hat{n}$ be given by

$$\begin{aligned} \hat{p} &= (\sin\theta\cos\phi, \sin\theta\sin\phi, \cos\theta) \\ \hat{n} &= (-\sin\phi, \cos\phi, 0) \end{aligned} \tag{3.98}$$

The state $\chi_\lambda(\hat{p})$ is then given by

$$\chi_\lambda(\hat{p}) = \exp(-i\hat{n} \cdot \boldsymbol{\sigma}\theta/2)\chi_\lambda(\hat{z}) \tag{3.99}$$

The state $\chi_\lambda(-\hat{p})$ is conventionally reached by the rotation described by the angles $(\pi - \theta, \phi + \pi)$ to avoid any confusion due to the double-valueness of $SU(2)$ transformations.

The following three formulas are useful (only two are independent):

$$\begin{aligned}\chi_\lambda(\hat{p}) &= (-1)^{\frac{1}{2}-\lambda}e^{2i\lambda\phi}\chi_{-\lambda}(-\hat{p})\\ \chi_\lambda(\hat{p}) &= (-1)^{\frac{1}{2}-\lambda}i\sigma_2\chi^*_{-\lambda}(\hat{p})\\ \chi_\lambda(\hat{p}) &= -e^{2i\lambda\phi}i\sigma_2\chi^*_\lambda(-\hat{p})\end{aligned}\tag{3.100}$$

The helicity spinor $u(p, \lambda)$ is given by

$$\begin{aligned}u(p, \lambda) &= \exp(-i\mathbf{n}\cdot\boldsymbol{\sigma}\theta/2)u(p = p_z, s = \lambda)\\ &= \begin{pmatrix}\cosh\tfrac{1}{2}\zeta\chi_\lambda(\hat{p})\\ 2\lambda\sinh\tfrac{1}{2}\zeta\chi_\lambda(\hat{p})\end{pmatrix}\end{aligned}\tag{3.101}$$

The antiparticle helicity spinor is

$$v(p, \lambda) = C_1\bar{u}^T(p, \lambda) = (-1)^{\frac{1}{2}-\lambda}\begin{pmatrix}-2\lambda\sinh\tfrac{1}{2}\zeta\chi_{-\lambda}(\hat{p})\\ \cosh\tfrac{1}{2}\zeta\chi_{-\lambda}(\hat{p})\end{pmatrix}\tag{3.102}$$

(2λ is the same as $(-1)^{\frac{1}{2}-\lambda}$ for spin $\frac{1}{2}$.)

The formulas corresponding to Eqs. (3.95) and (3.97) are

$$\begin{aligned}u(\mathbf{p}, \lambda) &= (-1)^{\frac{1}{2}-\lambda}e^{2i\lambda\phi}Pu(-\mathbf{p}, -\lambda)\\ v(\mathbf{p}, \lambda) &= -(-1)^{\frac{1}{2}-\lambda}e^{-2i\lambda\phi}Pv(-\mathbf{p}, -\lambda)\\ u^*(\mathbf{p}, \lambda) &= e^{-2i\lambda\phi}Tu(-\mathbf{p}, \lambda)\\ v^*(\mathbf{p}, \lambda) &= e^{2i\lambda\phi}Tv(-\mathbf{p}, \lambda)\end{aligned}\tag{3.103}$$

These results are somewhat more complicated than the "old-fashioned" ones, but are more useful because of the physical significance of the helicity states.

The vector helicity wave functions $e^\mu(p, \lambda)$ are obtained by rotation of the wave functions polarized along the z axis:

$$\begin{aligned}e^\mu(p_z, \pm 1) &= \pm(0, 1, \pm i, 0)/2^{\frac{1}{2}}\\ e^\mu(p_z, 0) &= (p, 0, 0, E)/m\end{aligned}\tag{3.104}$$

For arbitrary directions, one obtains by rotation

$$\begin{aligned}2^{\frac{1}{2}}e^\mu(p, \pm 1) = \mp[0,\ &1 - (1 - \cos\theta)e^{\pm i\phi}\cos\phi,\\ &\pm i(1 \pm i(1 - \cos\theta)e^{\pm i\phi}\sin\phi),\ -\sin\theta e^{\pm i\phi}]\\ e^\mu(p, 0) = (p,\ &E\sin\theta\cos\phi,\ E\sin\theta\sin\phi,\ E\cos\theta)/m\end{aligned}\tag{3.105}$$

The polarization vectors e_μ obey the orthogonality-completeness relations

$$\begin{aligned} p \cdot e(p, \lambda) &= 0 \\ e^*_\mu(p, \lambda) e^\mu(p, \lambda') &= -\delta_{\lambda\lambda'} \\ \sum_\lambda e_\mu(p, \lambda) e^*_\nu(p, \lambda) &= -g_{\mu\nu} + \frac{p_\mu p_\nu}{m^2} \end{aligned} \tag{3.106}$$

A short calculation leads to the relations

$$\begin{aligned} e_\mu(\mathbf{p}, \lambda) &= (-1)^{1-\lambda} e^{2i\lambda\phi}(-g_{\mu\mu}) e_\mu(-\mathbf{p}, -\lambda) \\ e^*_\mu(\mathbf{p}, \lambda) &= e^{-2i\lambda\phi} g_{\mu\mu} e_\mu(-\mathbf{p}, \lambda) \\ e^*_\mu(\mathbf{p}, \lambda) &= (-1)^\lambda e_\mu(-\mathbf{p}, -\lambda) \end{aligned} \tag{3.107}$$

(A quicker way to derive these results is to identify $e_\mu(2p, 2\lambda)$ with $\bar{v}(p, \lambda) \times \gamma_\mu u(p, \lambda)$ and then to use Eqs. (3.102) to reach (3.107).)

The Rarita-Schwinger helicity wave functions for higher spin are easily found by combining spin-$\frac{1}{2}$ and spin-1 helicity wave functions using ordinary Clebsch-Gordan coefficients. For meson wave functions we have the recursion formula relating spin S and spin $S + 1$:

$$e^{s+1}_{\mu_1\mu_2\cdots\mu_{s+1}}(p, \lambda) = \sum_{\lambda_1\lambda_2} C(s1s+1; \lambda_1\lambda_2\lambda) e^s_{\mu_1\cdots\mu_s}(p, \lambda_1) e_{\mu_{s+1}}(p, \lambda_2) \tag{3.108}$$

For Fermions, we have the spinor-tensor ($S = k + \frac{1}{2}$)

$$\begin{aligned} u^{s+1}_{\mu_1\cdots\mu_{k+1}}(p, \lambda) &= \sum_{\lambda_1\lambda_2} C(s1s+1; \lambda_1\lambda_2\lambda) u^s_{\mu_1\cdots\mu_k}(p, \lambda_1) e_{\mu_{k+1}}(p, \lambda_2) \\ v^{s+1}_{\mu_1\cdots\mu_{k+1}}(p, \lambda) &\equiv C_1 \bar{u}^{s+1,T}_{\mu_1\cdots\mu_{k+1}}(p, \lambda) \\ &= \sum_{\lambda_1\lambda_2} C(s1s+1; \lambda_1\lambda_2\lambda) v^s_{\mu_1\cdots\mu_k}(p, \lambda) e^*_{\mu_{k+1}}(p, \lambda) \end{aligned} \tag{3.109}$$

For spin $\frac{3}{2}$, the ordinary Dirac spinors $u(p, \lambda)$ or $v(p, \lambda)$ appear on the right-hand side of Eqs. (3.109).

For notational reasons we write $\chi_\mu(p, \lambda)$ for either the meson wave function $e^s_{\mu_1\cdots\mu_k}(p, \lambda)$ or the Fermion wave function $u^s_{\mu_1\cdots\mu_k}(p, \lambda)$ and $\chi^c_{\boldsymbol{\mu}}$ for the generalized conjugate wave function $e^{s*}_{\mu_1\cdots\mu_s}$ for mesons, and $C_1\chi^{s,T}_{\mu_1\cdots\mu_k} = v^s_{\mu_1\cdots\mu_k}$ for Fermions. The vector $\boldsymbol{\mu}$ stands for the set of four-vector indices μ_i, in which $\chi_{\boldsymbol{\mu}}$ is totally symmetric and divergenceless. ($p^{\mu_i}\chi_{\boldsymbol{\mu}}(p, \lambda)$ vanishes; for Fermions $\gamma^{\mu_i}\chi_{\boldsymbol{\mu}}(p, \lambda)$ also vanishes.)

The orthogonality conditions are easily found by use of the recursion formulas:

$$\begin{aligned} e_\mu^{s*}(p,\lambda)e^{s\mu}(p,\lambda') &= (-1)^s\delta_{\lambda\lambda'} \\ \bar{u}_\mu^s(p,\lambda)u^{s\mu}(p,\lambda') &= (-1)^{s-\frac{1}{2}}\,\delta_{\lambda\lambda'} \\ \bar{v}_\mu^s(p,\lambda)v^{s\mu}(p,\lambda) &= -(-1)^{s-\frac{1}{2}}\,\delta_{\lambda\lambda'} \end{aligned} \tag{3.110}$$

Symmetry relations involving the change of sign of p or λ follow easily from the recursion relations, using induction and Eqs. (3.103) and (3.107) for spin $\frac{1}{2}$:

$$\begin{aligned} e_\mu^s(\mathbf{p},\lambda) &= (-1)^{s-\lambda}e^{2i\lambda\phi}\prod(-g_{\mu\mu})e_\mu^s(-\mathbf{p},-\lambda) \\ e_\mu^{s*}(\mathbf{p},\lambda) &= e^{-2i\lambda\phi}\prod(g_{\mu\mu})e_\mu^s(-\mathbf{p},\lambda) \\ u_\mu^s(\mathbf{p},\lambda) &= (-1)^{s-\lambda}e^{2i\lambda\phi}\prod(-g_{\mu\mu})Pu_\mu^s(-\mathbf{p},-\lambda) \\ u_\mu^{s*}(\mathbf{p},\lambda) &= e^{-2i\lambda\phi}\prod(g_{\mu\mu})Tu_\mu^s(-\mathbf{p},\lambda) \\ v_\mu^s(\mathbf{p},\lambda) &= -(-1)^{s-\lambda}e^{-2i\lambda\phi}\prod(-g_{\mu\mu})Pv_\mu^s(-\mathbf{p},-\lambda) \\ v_\mu^{s*}(\mathbf{p},\lambda) &= e^{2i\lambda\phi}\prod(g_{\mu\mu})Tv_\mu^s(-\mathbf{p},\lambda) \end{aligned} \tag{3.111}$$

Again P and T are the real matrices γ_0 and $\gamma_3\gamma_1$, respectively. The products $\Pi(\pm g_{\mu\mu})$ run over S or $S - \frac{1}{2}$ factors $g_{\mu_i\mu_i}$.

All the preceding inversion formulas are summarized by

$$\begin{aligned} \chi_\mu^s(\mathbf{p},\lambda) &= (-1)^{s-\lambda}e^{2i\lambda\phi}\prod(-g_{\mu\mu})P\chi_\mu(-\mathbf{p},-\lambda) \\ \chi_\mu^{s*}(\mathbf{p},\lambda) &= e^{-2i\lambda\phi}\prod(g_{\mu\mu})T\chi_\mu(-\mathbf{p},\lambda) \\ \chi_\mu^{sc}(p,\lambda) &= C_1\bar{\chi}_\mu^{sT}(p,\lambda) \end{aligned} \tag{3.112}$$

provided we replace all matrices by unity for mesons.

In order to write the field commutation relations in a neat form, it is useful to have projection operators, expressed in terms of the helicity wave functions. For spin $\frac{1}{2}$ we have the well-known projection operators

$$\begin{aligned} \Lambda_+(p) &= \sum_\lambda u(p,\lambda)\bar{u}(p,\lambda) = \frac{m+\not{p}}{2m} \\ \Lambda_-(p) &= -\sum_\lambda v(p,\lambda)\bar{v}(p,\lambda) = \frac{m-\not{p}}{2m} \end{aligned} \tag{3.113}$$

For spin 1, we define a second-rank tensor projection operator

$$\begin{aligned} M_{\mu\nu}(p) &= -\sum_\lambda e_\mu(p,\lambda)e_\nu^*(p,\lambda) \\ &= g_{\mu\nu} - p_\mu p_\nu/m^2 \\ M_{\mu\nu}(p)M_\lambda^\nu(p) &= M_{\mu\lambda}(p) \end{aligned} \tag{3.114}$$

For spin-S mesons, the latter generalizes to

$$M^s_{\mu\nu}(p) = (-1)^s \sum_\lambda e^s_\mu(p, \lambda) e^{s*}_\nu(p, \lambda)$$
$$M^s_{\mu\nu}(p) M^{s\nu}_\lambda = M^s_{\mu\nu} \tag{3.115}$$

For our purposes we only need to note that it is a real tensor constructed from an appropriate number of $g_{\mu\nu}$ and p_μ factors.

From Eqs. (3.111) we find the symmetry relations

$$M^s_{\mu\nu}(\mathbf{p}) = \prod (-g_{\mu\mu}) \prod (-g_{\nu\nu}) M^s_{\mu\nu}(-\mathbf{p})$$
$$M^{s*}_{\mu\nu}(\mathbf{p}) = \prod (g_{\mu\mu}) \prod (g_{\nu\nu}) M^s_{\mu\nu}(-\mathbf{p}) \tag{3.116}$$

Since the number of μ_i equals the number of ν_i, the minus signs cancel in the first of (3.116), proving the reality of $M^s_{\mu\nu}$:

$$M^s_{\mu\nu}(p) = M^{s*}_{\mu\nu}(p) \tag{3.117}$$

In addition, the first equation indicates that, when $\mathbf{p}$ changes sign, $M_{\mu\nu}$ transforms as an ordinary vector in each of its indices, thereby excluding pseudotensors from the construction of $M^s_{\mu\nu}(p)$.

For baryons of spin S, we write the projection operators as

$$\mathscr{P}^s_{+\mu\nu}(p) = (-1)^{s-\frac{1}{2}} \sum_\lambda u^s_\mu(p, \lambda) \bar{u}^s_\nu(p, \lambda)$$
$$\mathscr{P}^s_{-\mu\nu}(p) = -(-1)^{s-\frac{1}{2}} \sum_\lambda v^s_\mu(p, \lambda) \bar{v}^s_\mu(p, \lambda) \tag{3.118}$$

In addition to the obvious relations

$$\mathscr{P}^s_{\pm\mu\nu}(p) \mathscr{P}^{s\nu}_{\pm\lambda}(p) = \mathscr{P}^s_{\pm\mu\nu}(p)$$
$$\mathscr{P}^s_{\pm\mu\nu}(p) \mathscr{P}^{s\nu}_{\mp\lambda}(p) = 0 \tag{3.119}$$
$$(\not{p} \mp m) \mathscr{P}^s_\pm(p) = \mathscr{P}^s_\pm(p)(\not{p} \mp m)$$

and the subsidiary conditions deriving from the wave functions, we have the symmetry relations ($P = \gamma_0$, $T = \gamma_3\gamma_1$)

$$P \mathscr{P}^s_{\pm\mu\nu}(\mathbf{p}) P^{-1} = \prod (-g_{\mu\mu}) \prod (-g_{\nu\nu}) \mathscr{P}^s_{\pm\mu\nu}(-\mathbf{p})$$
$$T \mathscr{P}^{s*}_{\pm\mu\nu}(\mathbf{p}) T^{-1} = \prod (g_{\mu\mu}) \prod (g_{\nu\nu}) \mathscr{P}^s_{\pm\mu\nu}(-\mathbf{p}) \tag{3.120}$$

Again, the left-hand sides of (3.120) are equal. We also note the relations ($\gamma_5 = i\gamma^0\gamma^1\gamma^2\gamma^3$)

$$i\gamma_5 u(p, \lambda) = (-1)^{\frac{1}{2}+\lambda} v(p, -\lambda$$
$$i\gamma_5 v(p, \lambda) = (-1)^{\frac{1}{2}-\lambda} u(p, -\lambda) \tag{3.121}$$
$$i\gamma_5 \mathscr{P}^s_{\pm\mu\nu}(p)(i\gamma_5)^{-1} = \mathscr{P}^s \quad (p$$

The operators $\mathscr{P}^{s}_{\pm\mu\nu}$ are to be constructed from the metric tensor $g_{\mu\nu}$, γ_μ and p_μ. The first equation of (3.120) excludes pseudoquantities like γ_5, and the second makes the coefficients of the independent tensors real. One can then write the projection operators in the form

$$\mathscr{P}^{s}_{\pm\mu\nu}(p) = \Lambda_{\mu\nu}(p)\Lambda_{\pm}(p) = \Lambda_{\pm}(p)\Lambda_{\mu\nu}(p) \tag{3.122}$$

The quantities $\Lambda_{\mu\nu}(p)$ are then determined by the subsidiary conditions.

Further details concerning the projection operators and related matters may be found in refs. 7 and 8. For reference we note the result for spin $\frac{3}{2}$:

$$\mathscr{P}^{\frac{3}{2}}_{\pm\mu\nu}(p) = \frac{m \pm \gamma \cdot p}{2m}\left(g_{\mu\nu} - \tfrac{1}{3}\gamma_\mu\gamma_\nu - \frac{2p_\mu p_\nu}{3m^2} + \frac{p_\mu\gamma_\nu - p_\nu\gamma_\mu}{3m}\right) \tag{3.123}$$

and for spin 2 the operator $M_{\mu\nu}$ of Eq. (3.115) is

$$M^{(2)}_{\mu\nu,\rho\sigma}(p) = \tfrac{1}{2}M_{\mu\rho}M_{\nu\sigma} + \tfrac{1}{2}M_{\mu\sigma}M_{\nu\rho} - \tfrac{1}{3}M_{\mu\nu}M_{\rho\sigma} \tag{3.124}$$

where $M_{\mu\nu}$ is the spin one projector of Eq. (3.114).

PROBLEMS

1. (a) Show that the invariant trace η^{λ}_{λ} of a symmetric second rank spinor ($\eta_{\lambda\mu} = \eta_{\mu\lambda}$) vanishes.
   (b) Show that an antisymmetric second rank spinor $\eta_{\lambda\mu} = -\eta_{\mu\lambda}$ can be written in the form
   $$\eta_{\lambda\mu} = \tfrac{1}{2}\eta^{\lambda}_{\lambda}\epsilon_{\lambda\mu}$$
2. Check that the wave functions (3.108)–(3.109) satisfy the appropriate subsidiary conditions and symmetry relations.

REFERENCES

1. P. A. M. Dirac, *Proc. Roy. Soc.* **A155,** 447 (1936).
2. M. Fierz and W. Pauli, *Helv. Phys. Acta* **12,** 297 (1939).
3. W. Rarita and J. Schwinger, *Phys. Rev.* **60,** 61 (1941).
4. P. Carruthers, *J. Math. Phys.* **9,** 1835 (1968).
5. P. Auvil and J. J. Brehm, *Phys. Rev.* **140,** B135 (1965); *ibid* **145,** 1152 (1966).
6. D. Brudnoy, *Phys. Rev.* **145,** 1229 (1966).
7. C. Fronsdal, *Nuovo Cimento Suppl.* **9,** 416 (1968).
8. R. E. Behrends and C. Fronsdal, *Phys. Rev.* **106,** 345 (1957).

4

THE POINCARÉ GROUP I: REPRESENTATIONS AND STATES

4.1 INTRODUCTION

In Sec. 2.4 we noted that the invariant quantities (j_0, λ) characterizing the unitary representations of the homogeneous Lorentz group do not correspond to the natural physical quantities (mass and spin) characterizing single particle states. The non-unitary representations of Chapter 3 were used to construct equations of motion for field variables (spinors) carrying definite spin and mass. But such representations are not those representing a quantum mechanical symmetry. Very general considerations[1] require symmetry transformations to be represented by unitary (or anti-unitary) transformations. The desired representations of states are obtained by extending the geometrical transformation of relativity to include space-time translations as well as homogeneous Lorentz transformations, as remarked in the introduction to Chapter 2.

This chapter deals almost entirely with the "restricted" Poincaré group, i.e. those ten-parameter transformations continuously connected to the identity. The modifications necessary to describe space and time inversions are treated in Chapter 7 (see also Sec. 3.5). However a full treatment of the associated conceptual problems and phase questions is non-trivial.[2,3]

In Sec. 4.2 we study the geometrical structure of the group and obtain the infinitesimal generators and their commutators. The group invariants (Casimir operators) are shown to be simply related to the mass and spin parameters. Sec. 4.3 deals briefly with the various classes of representations. The latter may be characterized by the structure of various "little groups" in a well-known way. Extensive expositions of this subject have been given by many authors. A partial list of such works is given in refs. 2–8. In Sec. 4.4

the properties of helicity states are studied and their utility in constructing the partial wave expansion for scattering amplitudes is illustrated.

4.2 INFINITESIMAL GENERATORS AND GROUP INVARIANTS

If we represent the transformation $x'^\mu = \Lambda^\mu{}_\nu x^\nu + a^\mu$ by the symbol $g = (\Lambda, a)$ the group multiplication law $g_{21} = g_2 g_1$ is given by

$$(\Lambda_2\Lambda_1, \Lambda_2 a_1 + a_2) = (\Lambda_2, a_2) \cdot (\Lambda_1, a_1) \tag{4.1}$$

The identity element is $(1, 0)$ and the inverse to g is $(\Lambda^{-1}, -\Lambda^{-1}a)$, as is easily verified from (4.1). A pure translation by a is represented by

$$T_a = (1, a) \tag{4.2}$$

The Poincaré group is a composite of translations and homogeneous Lorentz transformations but is not the "direct product" of these operations since they do not commute. This may be verified by conjugating the translation (4.2) by a general group element g

$$gT_a g^{-1} = T_{\Lambda a} \tag{4.3}$$

Thus conjugation of a translation leads to another translation. All translations may be obtained by an appropriate transformation (4.3).

The unitary operators acting on quantum mechanical state vectors imitate (4.1):

$$U(\Lambda_2\Lambda_1, \Lambda_2 a_1 + a_2) = U(\Lambda_2, a_2)U(\Lambda_1, a_1) \tag{4.4}$$

One could modify (4.4) by inserting a phase factor corresponding to the ambiguity of the overall phase of a state vector. We shall always use the natural phase choice exhibited in (4.4). The unitary operators U are related to Hermitian operators P_μ, $M_{\mu\nu}$ by the following expansion

$$U(1 + \omega, a) = 1 + ia^\mu P_\mu - \frac{i}{2}\,\omega^{\mu\nu}M_{\mu\nu} \tag{4.5}$$

for infinitesimal a and ω. The sign conventions have been chosen to agree with special cases considered in earlier chapters (cf. Eqs. (1.5) and (2.21)). In Lagrangian field theories explicit expressions for the operators P_μ and $M_{\mu\nu}$ in terms of field variables are easily found by means of Noether's theorem.[9] In the next section we shall study the behavior of field operators under the transformation (4.5), thereby clarifying the geometrical role of the four-momentum P_μ and the Lorentz angular momentum $M_{\mu\nu}$ as generators of translations and homogeneous Lorentz transformations.

The group composition law (4.4) may be used to obtain the behavior of the operators P_μ and $M_{\mu\nu}$ under Lorentz transformations. As a consequence of (4.4) one easily finds the relation

$$U^{-1}(\Lambda)U(\Lambda', a)U(\Lambda) = U(\Lambda^{-1}\Lambda'\Lambda, \Lambda^{-1}a) \tag{4.6}$$

where $U(\Lambda) \equiv U(\Lambda, 0)$. The following (expected) transformation properties follow directly from (4.6):

$$\begin{aligned} U^{-1}(\Lambda)P_\mu U(\Lambda) &= \Lambda_\mu^\nu P_\nu \\ U^{-1}(\Lambda)M_{\mu\nu}U(\Lambda) &= \Lambda_\mu^\rho \Lambda_\nu^\sigma M_{\rho\sigma} \end{aligned} \tag{4.7}$$

In order to derive the first of these we set $\Lambda' = 1$ in (4.6) and evaluate the derivative $\partial/\partial a_\mu$ at $a = 0$ (cf. Eq. (4.55)). This gives directly

$$\begin{aligned} U^{-1}(\Lambda)\frac{\partial U}{\partial a^\mu}\bigg|_{a=0} U(\Lambda) &= \frac{\partial U(1, \Lambda^{-1}a)}{\partial a^\mu}\bigg|_{a=0} \\ -iU^{-1}(\Lambda)P_\mu U(\Lambda) &= -i(\Lambda^{-1})^\nu{}_\mu P_\nu = -i\Lambda_\mu{}^\nu P_\nu \end{aligned} \tag{4.8}$$

where the last equality follows from Eq. (2.10) (or from the relation $U(1, \Lambda^{-1}a)$ $\exp(i(\Lambda^{-1}a)\cdot P) = \exp(ia \cdot \Lambda P)$).

In order to prove the second of Eqs. (4.7) we set $a = 0$ in (4.6) to obtain

$$U^{-1}(\Lambda)U(\Lambda')U(\Lambda) = U(\Lambda^{-1}\Lambda'\Lambda) \tag{4.9}$$

and consider an infinitesimal transformation $\Lambda' = 1 + \omega'$. The composite transformation $\tilde{\Lambda} = \Lambda^{-1}\Lambda'\Lambda$ is also infinitesimal. We write the latter as

$$\begin{aligned} \tilde{\Lambda} &= 1 + \tilde{\omega} \\ \tilde{\omega} &= \Lambda^{-1}\omega'\Lambda \end{aligned} \tag{4.10}$$

We now write (4.9) in the form

$$U^{-1}(\Lambda)\left(1 - \frac{i}{2}\,\omega'^{\mu\nu}M_{\mu\nu}\right)U(\Lambda) = 1 - \frac{i}{2}\,\tilde{\omega}^{\kappa\lambda}M_{\kappa\lambda} \tag{4.11}$$

and evaluate the derivative $\partial/\partial\omega^{\mu\nu}$ at $\omega = 0$. Using Eqs. (2.10), (4.10), (4.11) and the antisymmetry of $M_{\mu\nu}$ we find the second of Eqs. (4.7) immediately.

Finally it is important to have the set of commutation rules obeyed by the ten operators P_μ, $M_{\mu\nu}$. These are obtained by further specialization of

Eqs. (4.7). From Eq. (4.5) we obtain (with $a = 0$) for (4.7)

$$\begin{aligned} \frac{i}{2}\,\omega^{\mu\nu}[M_{\mu\nu}, P_\rho] &= \omega_\mu{}^\nu P_\nu \\ \frac{i}{2}\,\omega^{\mu\nu}[M_{\mu\nu}, M_{\rho\sigma}] &= \omega_\rho{}^\lambda M_{\lambda\sigma} + \omega_\sigma{}^\lambda M_{\rho\lambda} \end{aligned} \tag{4.12}$$

Computing derivatives with respect to $\omega^{\mu\nu}$ at $\omega = 0$ gives the basic commutation rules

$$\begin{aligned} [M_{\mu\nu}, P_\rho] &= i(g_{\nu\rho}P_\mu - g_{\mu\rho}P_\nu) \\ [M_{\mu\nu}, M_{\rho\sigma}] &= i(g_{\nu\rho}M_{\mu\sigma} - g_{\mu\sigma}M_{\rho\nu} + g_{\mu\rho}M_{\sigma\nu} - g_{\nu\sigma}M_{\mu\rho}) \end{aligned} \tag{4.13a}$$

To complete the set of commutation rules for the generators we note that since two arbitrary translations commute, so do the infinitesimal generators

$$[P_\mu, P_\nu] = 0 \tag{4.13b}$$

These commutation rules are exactly as would be surmised from those abstracted from the prototype differential operators $p_\mu = i\,\partial_\mu$, $m_{\mu\nu} = i(x_\mu\,\partial_\nu - x_\nu\,\partial_\mu)$ (cf. Eq. (2.22)).

The physical content of the commutator $[M_{\mu\nu}, P_\rho]$ is best seen when we break $M_{\mu\nu}$ down into its non-covariant constituents

$$\begin{aligned} \mathbf{J} &= (M_{23}, M_{31}, M_{12}) \\ \mathbf{K} &= (M^{01}, M^{02}, M^{03}) \end{aligned} \tag{4.14}$$

As explained in Chapter 2, $\mathbf{J}$ generate rotations and $\mathbf{K}$ Lorentz transformations. Under a change of reference system, the states transform as $\Psi \to U\Psi$ where for pure rotations and pure Lorentz transformations we have

$$\begin{aligned} U &= 1 + i\theta\mathbf{n}\cdot\mathbf{J} \\ U &= 1 + i\boldsymbol{\alpha}\cdot\mathbf{K} \end{aligned} \tag{4.15}$$

where the infinitesimals $\theta\mathbf{n}$ and $\boldsymbol{\alpha}$ are related to the parameters $\omega_{\mu\nu}$ by

$$\begin{aligned} \theta\mathbf{n} &= (\omega^2{}_3, \omega^3{}_1, \omega^1{}_2) \\ \boldsymbol{\alpha} &= (\omega^{01}, \omega^{02}, \omega^{03}) \end{aligned} \tag{4.16}$$

A positive $\boldsymbol{\alpha}\cdot\hat{r}$ corresponds to a Lorentz transformation in the direction $\hat{r}$ with coordinates transforming infinitesimally as $t' = t - (\boldsymbol{\alpha}\cdot\hat{r})(\mathbf{x}\cdot\hat{r})$, $x' = x - \boldsymbol{\alpha}\cdot\hat{r}t$. The finite transformation of the four vector P^μ in the x

direction is given by (4.7) with $U = e^{i\zeta K^1}$

$$\begin{aligned} U^{-1}(\Lambda)P^0U(\Lambda) &= \cosh\zeta\, P^0 - \sinh\zeta P^1 \\ U^{-1}(\Lambda)P^1U(\Lambda) &= -\sinh\zeta P^0 + \cosh\zeta P^1 \end{aligned} \tag{4.17}$$

with P^2 and P^3 unchanged. As a consequence of (4.17) we see that the state $U(\Lambda)\,|p_0\rangle$ a rest state) corresponds to a state having momentum p^μ

$$\begin{aligned} P^0(U(\Lambda)\,|p_0\rangle) &= U(\Lambda)(\cosh\zeta P^0 - \sinh\zeta P^1)\,|p_0\rangle \\ &= m\cosh\zeta(U(\Lambda)\,|p_0\rangle) \\ P^1(U(\Lambda)\,|p_0\rangle) &= -\, m\sinh\zeta(U(\Lambda)\,|p_0\rangle) \end{aligned} \tag{4.18}$$

($p^\mu = m(\cosh\zeta,\ -\sinh\zeta,\ 0,\ 0)$) in the *original* frame. Changing from the passive interpretation used above to the "active" interpretation, we may generate all the states of momentum **p** by applying Lorentz "boosts" to a standard state, e.g. a rest state, when $m \neq 0$. The transformation $\exp(-i\boldsymbol{\alpha}\cdot\mathbf{K})$ promotes a rest state to one of momentum $p^\mu = m(\cosh\alpha,\ \hat{\alpha}\sinh\alpha)$, in analogy to the way $\exp(-i\theta\mathbf{n}\cdot\mathbf{J})$ rotates a *state* by θ around **n**.

The first line of Eq. (4.13a) may be written as $(i, j, k = 1, 2, 3)$

$$\begin{aligned} [J^i, H] &= 0 \\ [J^i, P^j] &= i\epsilon^{ijk}P^k \\ [K^i, H] &= -iP^i \\ [K^i, P^j] &= -i\,\delta^{ij}H \end{aligned} \tag{4.19}$$

We next construct invariant operators whose eigenvalues may be used to characterize irreducible representations. Clearly these will be Lorentz scalar quantities constructed from the generators P_μ, $M_{\mu\nu}$. It is almost obvious, and immediately verified that the operator

$$P^2 = P_\mu P^\mu \tag{4.20}$$

whose eigenvalues give the mass of a system, is an invariant. On the other hand the invariants of the homogeneous Lorentz group ($M_{\mu\nu}M^{\mu\nu}$, $\epsilon_{\mu\nu\rho\sigma}M^{\mu\nu} \times M^{\rho\sigma}$) are not translation invariant. Rather than deal with $M_{\mu\nu}$ one may instead consider the quantities g_μ and W_μ defined by

$$\begin{aligned} g_\mu &\equiv M_{\mu\lambda}P^\lambda \\ W_\mu &\equiv \tfrac{1}{2}\epsilon_{\mu\nu\lambda\sigma}M^{\nu\lambda}P^\sigma \end{aligned} \tag{4.21}$$

It is possible to express $M_{\mu\nu}$ in terms of the vectors P_μ, g_μ, and W_μ according to the following identity

$$M_{\mu\nu}P^2 = g_\mu P_\nu - g_\nu P_\mu - \epsilon_{\mu\nu\sigma\lambda}W^\sigma P^\lambda \tag{4.22}$$

As a consequence of the relations

$$\begin{aligned} g_\mu P^\mu &= 0 \qquad & P^\mu g_\mu &= 3iP^2 \\ W_\mu P^\mu &= 0 \qquad & P^\mu W_\mu &= 0 \end{aligned} \tag{4.23}$$

the only scalars under the proper inhomogeneous group are $P^\mu P_\mu$, $W^\mu W_\mu$, $g^\mu g_\mu$, and $g^\mu W_\mu$. Of these the last two fail to commute with P_μ and hence the desired Poincaré invariants are P^2 and W^2. For particular classes of representations we shall find other special invariant quantities.

In order to prove Eq. (4.22) we construct the last term using the definition (4.21)

$$\epsilon_{\mu\nu\sigma\lambda}W^\sigma P^\lambda = \tfrac{1}{2}\epsilon_{\mu\nu\sigma\lambda}\epsilon^{\sigma\alpha\beta\gamma}M_{\alpha\beta}P_\gamma P^\lambda \tag{4.24}$$

The contraction of two ϵ symbols on one index is easily related to a generalized Kronecker delta; with our convention that $\epsilon_{0123} = 1$ the necessary relation is

$$\epsilon_{\sigma\mu\nu\lambda}\epsilon^{\sigma\alpha\beta\gamma} = -\delta^{\alpha\beta\gamma}_{\mu\nu\lambda} \tag{4.25}$$

where $\delta^{\alpha\beta\gamma}_{\mu\nu\lambda}$ is ± 1 according to whether an even or odd number of permutations is required to arrange the lower and upper indices in the same order. Using (4.25) to simplify (4.24) leads directly to the identity (4.23).

Next we study the properties of W_μ. If we evaluate W_μ in a particular frame using the correspondence (4.14) we find

$$W^\mu = (W^0, \mathbf{W}) = (\mathbf{J}\cdot\mathbf{P}, \mathbf{J}P^0 + \mathbf{K}\times\mathbf{P}) \tag{4.26}$$

If we apply W^μ to a rest state of spin s, the eigenvalues of W^μ are $m(0, \mathbf{S})$ where $\mathbf{S}$ is the usual spin matrix. Hence the *invariant* $W^2 = -m^2 s(s + 1)$ is simply related to the intrinsic spin of the particle.

4.3 CLASSIFICATION OF REPRESENTATIONS

We have already noted that any momentum state can be obtained from any other by an appropriate Lorentz transformation. Hence the set of all momentum states (supplemented by appropriate spin degrees of freedom) is expected to provide a representation space for the Poincaré operators P_μ, $M_{\mu\nu}$. We now delineate this idea more precisely. The reader should consult refs. 1, 3, and 7 for a discussion of some of the interesting mathematical points involved. The present section follows that of ref. 4 closely.

The momentum eigenstates $|p, \lambda\rangle$ satisfy

$$P_\mu |p, \lambda\rangle = p_\mu |p, \lambda\rangle \tag{4.27}$$

where the index λ labels the spin state. The reader may imagine that λ represents the helicity, i.e. the value of $\mathbf{J} \cdot \hat{p}$. (The helicity description of spin is considered in Sec. 4.4.) We do not include internal symmetry labels since internal symmetry transformations must commute with Poincaré transformations.[11,12]

From the first of Eqs. (4.7) we learn that the state $U(\Lambda)|p, \lambda\rangle$ has momentum $p' = \Lambda p$, so we may write

$$U(\Lambda)|p\lambda\rangle = \sum_{\lambda'} |\Lambda p, \lambda'\rangle C_{\lambda'\lambda} \tag{4.28}$$

In order to specify $C_{\lambda'\lambda}$ precisely we need to relate the momentum states appearing in (4.28) in a definite way. The usual procedure is to refer all momenta p and all states $|p\lambda\rangle$ to a standard fixed momentum $\hat{p}$ and a standard state $|\hat{p}\lambda\rangle$. The three parameter family of transformations L_p satisfies

$$p^\mu = L(p)^\mu{}_\nu \hat{p}^\nu \tag{4.29}$$

where L_p is chosen in a standard way. Corresponding to (4.29) we have the state

$$|p, \lambda\rangle \equiv U[L(p)]\,|\hat{p}\lambda\rangle \tag{4.30}$$

When dealing with massive particles it is convenient to choose $\hat{p} = p_0 = (m, \mathbf{0})$. Suppose the state $|p_0\lambda\rangle$ has component $J_z = \lambda$ along the z axis. The boost $e^{-i\zeta K_z}|p_0\lambda\rangle$ gives a state with momentum $p = m \cosh \zeta$ along the z axis, and $J_z = \lambda$. We now obtain a state having momentum $\mathbf{p} = p(\cos\phi \sin\theta, \sin\phi \sin\theta, \cos\theta)$ by applying the rotation $e^{-i\theta\mathbf{n}\cdot\mathbf{J}}$, where $\mathbf{n}$ is given by (3.98). If we use the Euler angle description (with fixed axes)

$$R(\alpha, \beta, \gamma) = e^{-i\alpha J_z} e^{-i\beta J_y} e^{-i\gamma J_z} \tag{4.31}$$

we easily see that the desired rotation is simply $R(\phi, \theta, -\phi)$ with n given by Eq. (3.98), using the relation

$$\begin{aligned} e^{-i\phi J_z} J_y e^{i\phi J_z} &= \cos\phi J_y - \sin\phi J_x \\ &= \mathbf{n} \cdot \mathbf{J} \end{aligned} \tag{4.32}$$

Hence the state $|\mathbf{p}s\lambda\rangle$ may be written as

$$|\mathbf{p}s\lambda\rangle = e^{-i\theta\mathbf{n}\cdot\mathbf{J}}e^{-i\zeta K_z}|p_0 s\lambda\rangle$$
$$= e^{-i\phi J_z}e^{-i\theta J_y}e^{i\phi J_z}e^{-i\zeta K_z}|p_0 s\lambda\rangle \tag{4.33}$$

since $\mathbf{J}\cdot\mathbf{n}$ commutes with $\mathbf{J}\cdot\hat{p}$ we see that the state (4.33) satisfies

$$\mathbf{J}\cdot\hat{p}\,|\mathbf{p}s\lambda\rangle = \lambda\,|\mathbf{p}s\lambda\rangle \tag{4.34}$$

The helicity λ is especially useful since the helicity of a state does not change under rotations.

Equation (4.33) gives an explicit form for the operators $U[L(p)]$ in the domain $p^2 > 0$. The corresponding 4×4 matrix $L(p)$ is built from the corresponding transformations of Eqs. (2.14) and (2.15):

$$L(p) = R_3(-\phi)R_2(-\theta)R_3(\phi)L_3(-\zeta)$$

$$L(p) = \begin{bmatrix} \cosh\zeta & 0 & 0 & \sinh\zeta \\ -\sin\theta\cos\phi\sinh\zeta & \cos^2\phi\cos\theta+\sin^2\phi & \sin\phi\cos\phi(\cos\theta-1) & -\sin\theta\cos\phi\cosh\zeta \\ -\sin\theta\sin\phi\sinh\zeta & \sin\phi\cos\phi(\cos\theta-1) & \sin^2\phi\cos\theta+\cos^2\phi & -\sin\theta\sin\phi\cosh\zeta \\ \cos\theta\sinh\zeta & \cos\phi\sin\theta & \sin\theta\sin\phi & \cos\theta\cosh\zeta \end{bmatrix} \tag{4.35}$$

We now return to the general analysis of (4.28), and do not restrict our attention to the case $p^2 = m^2 > 0$ just treated. That case (timelike p^μ) corresponds to one of four categories:

$$\begin{aligned} &\text{(a) Timelike} && p^2 > 0 \\ &\text{(b) Spacelike} && p^2 < 0 \\ &\text{(c) Light-like} && p^2 = 0 \\ &\text{(d) Null} && p^\mu = 0 \end{aligned} \tag{4.36}$$

These domains are never connected by a Lorentz transformation.

The notion of the "little group" is very useful. The little group $G(\hat{p})$ of the homogeneous Lorentz group is that subgroup which leaves $\hat{p}$ invariant; for $R \subset G(\hat{p})$ we have

$$R\hat{p} = \hat{p} \tag{4.37}$$

(In the example $\hat{p} = p_0 = (m, \mathbf{0})$, the little group is the three-dimensional rotation group. The other cases are classified below.)

The structure of the little group is independent of the choice of the reference vector $\hat{p}$ as long as one stays within the categories of (4.36). If we choose a vector $\hat{p}_1$ left invariant by R_1 ($R_1\hat{p}_1 = \hat{p}_1$) the little group $G(\hat{p}_1)$ is isomorphic to $G(\hat{p})$. To see this, note that $\hat{p}_1$ is related to $\hat{p}$ by a fixed transformation matrix L_1: $\hat{p}_1 = L_1\hat{p}$. Hence $(L_1^{-1}R_1L_1)\hat{p} = \hat{p}$ and $L_1^{-1}R_1L_1$ is in $G(\hat{p})$.

If we specialize (4.28) so that $\Lambda = R$ and $p = \hat{p}$ we find

$$U(R)\,|\hat{p}\lambda\rangle = \sum_{\lambda'} |\hat{p}\lambda'\rangle C_{\lambda'\lambda} \tag{4.38}$$

i.e. the coefficients $C_{\lambda'\lambda}$ are representation matrices of the little group $G(\hat{p})$: $C_{\lambda'\lambda} = D_{\lambda'\lambda}(R)$ and

$$U(R)\,|\hat{p}\lambda\rangle = \sum_{\lambda'} |\hat{p}\lambda'\rangle D_{\lambda'\lambda}(R) \tag{4.39}$$

It is now possible to give a precise evaluation of the transformation (4.28) in terms of a well-defined little group matrix. We begin by writing

$$\begin{aligned} U(\Lambda)\,|p\lambda\rangle &= U(\Lambda)U[L(p)]\,|\hat{p}\lambda\rangle \\ &= U[L(\Lambda p)]U[L^{-1}(\Lambda p)\Lambda L(p)]\,|\hat{p}\lambda\rangle \end{aligned} \tag{4.40}$$

where the introduction of $U(L(\Lambda p))$ is suggested by (4.28). The key point here is that the transformation $L^{-1}(\Lambda p)\Lambda L(p)$ belongs to the little group $G(\hat{p})$:

$$L^{-1}(\Lambda p)\Lambda L(p)\hat{p} = L^{-1}(\Lambda p)(\Lambda p) = \hat{p}$$

and so Eqs. (4.28) and (4.39) combine to give

$$U(\Lambda)\,|p\lambda\rangle = \sum_{\lambda'} |\Lambda p\lambda'\rangle D_{\lambda'\lambda}[L^{-1}(\Lambda p)\Lambda L(p)] \tag{4.41}$$

where all momentum states are related by Eq. (4.30).

The rotation R is comprised of three separate Lorentz transformations. Further kinematical details are given in refs. 3 and 14. Equation (4.41) shows how the unitarity and irreducibility of the representations of the Poincaré group are directly related to the same properties of the little group.

The operators $U(\Lambda)$ are unitary only if the states are normalized covariantly ($p_0 > 0$)

$$\langle p'\lambda'\,|\,p\lambda\rangle = 2p^0\,\delta(\mathbf{p} - \mathbf{p}') \tag{4.42}$$

With this convention the resolution of the identity is

$$I = \sum_{\lambda} \int \frac{d^3p}{2p_0}\,|p\lambda\rangle\langle p\lambda| \tag{4.43}$$

The latter three equations may be combined to verify unitarity:

$$U^\dagger(\Lambda)U(\Lambda) = \sum_\lambda \int \frac{d^3p}{2p_0} U(\Lambda)\,|p\lambda\rangle\langle p\lambda|\, U^\dagger(\Lambda)$$

$$= \sum_{\lambda\lambda'\lambda''} \int \frac{d^3p}{2p_0} |p'\lambda'\rangle\langle p'\lambda''|\; D_{\lambda'\lambda}(R)D^\dagger(R)_{\lambda\lambda''}$$

$$= \sum_{\lambda'} \int \frac{d^3p'}{2p_0'} |p'\lambda'\rangle\langle p'\lambda'| = I \tag{4.44}$$

To obtain the last line we have used the unitarity of D and the invariance of the volume element $d^3p/2p^0$.

It is frequently convenient to use non-covariant normalization such that $2p^0$ is absent from (4.42). In that case the right hand side of (4.41) should be multiplied by $(E_{\Lambda p}/E_p)^{\frac{1}{2}}$.

We now consider the four classes of representations (4.36) in greater detail. More detailed information about these representations may be found in several of the cited references.

(a) timelike ($p^2 > 0$). The case $\hat{p} = (m, 0)$ has been analyzed following Eq. (4.30). We may generalize slightly by allowing for either sign of p^2, and choose the standard vector as

$$\hat{p}^\mu = \pm(\sqrt{\hat{p}^2}, 0, 0, 0) \tag{4.45}$$

for the (disjoint) representations having $p^0 \gtrless 0$. The little group is the rotation group $SO(3)$ (or $SU(2)$), whose generators are those components of W^μ which survive when acting on $\hat{p}$:

$$\hat{W}^\mu = \pm\sqrt{p^2}(0, \mathbf{J}) \tag{4.46}$$

(cf. Eq. (4.26)). For irreducible representations of the little group $\mathbf{J} \to \mathbf{S}$ as explained following Eq. (4.26).

In accordance with the discussion leading to (4.34) we use a helicity label, with the little group states $|\hat{p}s\lambda\rangle$ satisfying

$$\begin{aligned} P^\mu\,|\hat{p}s\lambda\rangle &= \hat{p}^\mu\,|\hat{p}s\lambda\rangle \\ J^3\,|\hat{p}s\lambda\rangle &= \lambda\,|\hat{p}s\lambda\rangle \\ P^2\,|\hat{p}s\lambda\rangle &= p^2\,|\hat{p}s\lambda\rangle \\ W^2\,|\hat{p}s\lambda\rangle &= -p^2s(s+1)\,|\hat{p}s\lambda\rangle \end{aligned} \tag{4.47}$$

The irreducible representations are labeled by the eigenvalues of P^2, W^2 (with $s = 0, \frac{1}{2}, 1, \ldots$) and the sign of $\hat{p}^0$.

(b) spacelike ($p^2 < 0$). For this case it is convenient to choose $\hat{p}$ as

$$\hat{p}^\mu = (0, 0, 0, \sqrt{-p^2}) \tag{4.48}$$

It is clear that the little group is $SO(2, 1)$. The generators of this group are found by applying W^μ to $|\hat{p}\lambda\rangle$, obtaining

$$\hat{W}^\mu = (J_z, K_y, -K_x, 0)\sqrt{-p^2} = (M_{12}, M_{20}, M_{01}, 0)\sqrt{-p^2} \tag{4.49}$$

The commutation relations are a special case of those for $SO(3, 1)$:

$$\begin{aligned} [M_{12}, M_{20}] &= iM_{01} \\ [M_{20}, M_{01}] &= -iM_{12} \\ [M_{01}, M_{12}] &= iM_{20} \end{aligned} \tag{4.50}$$

This algebra (4.50) also characterizes the groups $SU(1, 1)$ and $SL(2, R)$.

The representations of $SO(2, 1)$ have been studied in great detail.[15] Recently these representations have been used to decompose scattering amplitudes in a manner similar to the familiar expansion in terms of $SO(3)$ eigenfunctions. A review of such techniques is given in ref. 4. Here we only give a qualitative summary. It is still possible to write a formula analogous to (4.47):

$$W^2 |\hat{p}j\mu\rangle = -|p|^2 j(j+1) |\hat{p}j\mu\rangle \tag{4.51}$$

but the parameters j and μ now assume values quite different from the corresponding quantities of the timelike representation.

The unitary representations are as follows:

1. Principal Series:

$$\begin{aligned} &\operatorname{Re} j = -\tfrac{1}{2} \qquad -\infty < \operatorname{Im} j < \infty \\ &\mu = 0, \pm 1, \pm 2, \ldots \\ &\mu = \pm\tfrac{1}{2}, \pm\tfrac{3}{2}, \ldots \end{aligned} \tag{4.52}$$

The restriction on μ follows by requiring that the representations of the covering group $SU(1, 1)$ be single valued.

2. Supplementary Series:

$$\begin{aligned} &-\tfrac{1}{2} < \operatorname{Re} j < 0 \qquad \operatorname{Im} j = 0 \\ &\lambda = 0, \pm 1, \pm 2, \ldots \end{aligned} \tag{4.53}$$

3. Discrete Series:

$$j = -\tfrac{1}{2}, -1, -\tfrac{3}{2}, \ldots$$
$$D^{j+}: \mu = -j, -j+1, -j+2, \ldots \tag{4.54}$$
$$D^{j-}: \mu = j, j-1, j-2, \ldots$$

Cases 1 and 2 should be compared with the corresponding $SO(3, 1)$ representations of Sec. 2.4. We do not list separately the trivial one dimensional representation, which has $\mu = j = 0$.

We now turn to the case (c).

(c) Lightlike ($p^2 = 0$, $p^\mu \neq 0$). The standard vector is chosen to be

$$\hat{p}^\mu = \omega(1, 0, 0, 1) \tag{4.55}$$

where ω is an arbitrary real number. The cases $\omega > 0$ and $\omega < 0$ are distinct. When acting in the subspace of states having $p = \hat{p}$, W^μ reduces to

$$\hat{W}^\mu = \omega(J_z, -\Pi_2, \Pi_1, J_z)$$
$$\Pi_1 = M_{01} - M_{13} \tag{4.56}$$
$$\Pi_2 = M_{02} - M_{23}$$

The little group algebra generated by J_z, Π_1, Π_2:

$$[J_z, \Pi_1] = i\Pi_2$$
$$[J_z, \Pi_2] = -i\Pi_1 \tag{4.57}$$
$$[\Pi_1, \Pi_2] = 0$$

is isomorphic to that of the Euclidian group in two dimensions. (Details may be found in ref. 17.)

The value of the invariant W^2 is written as $-\rho^2$ according to

$$W^2 \, |\hat{p}\rho\lambda\rangle = -\omega^2(\Pi_1^2 + \Pi_2^2) \, |\hat{p}\rho\lambda\rangle$$
$$\equiv -\rho^2 \, |\hat{p}\rho\lambda\rangle \tag{4.58}$$

The operator $\Pi_1^2 + \Pi_2^2$ is a little group invariant and has the value $(\omega/\rho)^2$. The representations are:

1. Principal Series D^ρ:

$$0 < \rho < \infty \qquad \text{Im } \rho = 0$$
$$\lambda = 0, \pm 1, \pm 2, \ldots$$

or (4.59)

$$\lambda = \pm\tfrac{1}{2}, \pm\tfrac{3}{2}, \ldots$$

which is infinite dimensional, or finite (1-dimensional).

2. Discrete Series

$$\begin{aligned} \rho &= 0 \\ \lambda &= 0, \pm\tfrac{1}{2}, \pm 1, \ldots \end{aligned} \tag{4.60}$$

For the discrete series only one value of λ occurs per irreducible representation. Since ρ vanishes, so do Π_1, Π_2 and hence the eigenvalue of J_z is the Casimir operator. We find

$$\begin{aligned} W_\mu \, |\hat{p}0\lambda\rangle &= \hat{p}_\mu J_z \, |\hat{p}0\lambda\rangle \\ &= \lambda P_\mu \, |\hat{p}0\lambda\rangle \end{aligned} \tag{4.61}$$

This relation holds for all states in the manifold and hence we may write

$$W_\mu = \lambda P_\mu \tag{4.62}$$

with λ an invariant.

Finally we consider the null case:

(d) null ($p^\mu = 0$). In this case the full set of homogeneous Lorentz transformations leaves the null vector invariant. The representations of the "little group" are therefore those of the homogeneous Lorentz group $SO(3, 1)$ treated in Sec. 2.4.

4.4 HELICITY STATES AND PARTIAL WAVE EXPANSION OF THE SCATTERING AMPLITUDE

In order to illustrate the utility of the helicity basis described following Eq. (4.30) we discuss in detail the two-particle states and the partial wave analysis of the two body scattering amplitude.[9] As shown in Eq. (4.33) the single particle states $|p, \lambda\rangle$ (omitting the label s) are constructed by subjecting a rest state to a Lorentz transformation and a rotation

$$\begin{aligned} |\mathbf{p}\lambda\rangle &= R(\phi, \theta, -\phi) \, |p\lambda\rangle \\ |p\lambda\rangle &= e^{-i\zeta K_z} \, |p^0\lambda\rangle \end{aligned} \tag{4.63}$$

In these equations the symbol p denotes the magnitude of the three-momentum (aligned along the z axis) whereas p^0 will mean the vector $(m, \mathbf{0})$ for massive particles or the vector (4.55) in the case of massless particles.

Thus far we have not specified the phase relations connecting states of distinct λ. (Such a convention is required to give precise meaning to (4.39).)

For massive particles we make the usual phase convention for rest states

$$(J_x \pm iJ_y)\,|p^0\lambda\rangle = [(s \mp \lambda)(s \pm \lambda + 1)]^{\frac{1}{2}}\,|p_0\lambda \pm 1\rangle \qquad (4.64)$$

This procedure will not work for massless particles. However the latter case is very simple in that (assuming parity conservation) only two values $\lambda = \pm s$ occur. These states may be conveniently related by introducing an operator Y by

$$Y \equiv e^{-i\pi J}\,P \qquad (4.65)$$

where P is the usual parity transformation. It is clear that Y induces a reflection in the x-z plane. This operator leaves the standard vectors $\hat{p}$ of Sec. 4.3 unchanged and is useful for massive states too.

Recalling that λ is the eigenvalue of $\mathbf{J}\cdot\hat{p}$ we see that λ changes sign under space inversion (cf. Eq. (3.112) for an explicit illustration of this). Thus we may write for massless particles

$$Y\,|ps\rangle = \eta\,|p - s\rangle \qquad (4.66)$$

where η is independent of p since Y commutes with Lorentz transformations in the z direction. The constant η is related to the intrinsic parity of the massless particle. For example, the wave functions of photon states with $\lambda = \pm 1$ are conventionally defined to be $\mp(e_x \pm ie_y)/\sqrt{2}$. In that case $\eta = -1$ results from the usual assignment of $\eta_p = -1$ to the electromagnetic four-potential.

We now describe the properties of free two-particle states. Such a state may be obtained by appropriate rotations applied to the direct product $|p_1\lambda_1\rangle\,|p_2\lambda_2\rangle$ (both particles moving along the positive z axis):

$$R(\phi_1, \theta_1, -\phi_1)\,|p_1\lambda\rangle R(\phi_2, \theta_2, -\phi_2)\,|p_2\lambda_2\rangle \qquad (4.67)$$

The states $|p_i\lambda\rangle$ have mass m_i and spin s_i. We are particularly interested in the center of mass frame in which $\mathbf{p}_1 = -\mathbf{p}_2$ (i.e. $\phi_2 = \phi_1 + \pi$, $\theta_2 = \pi - \theta_1$). Rather than using (4.67) directly we follow the slightly different procedure recommended by Jacob and Wick.[9] The desired c.m. state may also be constructed by first giving particle "number 1" momentum p and particle "number 2" momentum $-p$ along the z axis. Then the whole state is rotated to the desired direction. It is conventional to write $\psi_{p\lambda}$ for the state defined by the second of Eqs. (4.63) and $\chi_{p\lambda}$ for the state of particle "number 2"

$$\chi_{p\lambda} \equiv (-1)^{s_2-\lambda_2} e^{-i\pi J_y^{(2)}}\,\psi_{p\lambda_2} \qquad (4.68)$$

The product state (of vanishing total three-momentum) is then

$$\psi_{p\lambda_1\lambda_2} \equiv \psi_{p\lambda_1}(1)\chi_{p\lambda_2}(2) \tag{4.69}$$

and the two particle state with c.m. momentum $\mathbf{p}$ is given by

$$|p\theta\phi\lambda_1\lambda_2\rangle \equiv R(\phi, \theta, -\phi)\psi_{p\lambda_1\lambda_2} = e^{i\lambda\phi}R(\phi, \theta, 0)\psi_{p\lambda_1\lambda_2} \tag{4.70}$$

where $R = \exp(-i\mathbf{n}\theta \cdot \mathbf{J})$, $\mathbf{J} = \mathbf{J}_1 + \mathbf{J}_2$ and $\lambda = \lambda_1 - \lambda_2$.

We now show how to relate the state $|p\theta\phi\lambda_1\lambda_2\rangle$ to components having definite angular momentum. If we subject (4.69) to an arbitrary rotation, the parts having differing J, J_z transform differently.

$$\begin{aligned} R(\alpha, \beta, \gamma)\psi_{p\lambda_1\lambda_2} &= \sum_{JMM'} |pJM\lambda_1\lambda_2\rangle D^J_{MM'}(\alpha, \beta, \gamma)\langle pJM\lambda_1\lambda_2 \mid \psi_{p\lambda_1\lambda_2}\rangle \\ D^J_{MM'} &\equiv \langle JM|\, R(\alpha, \beta, \gamma)\, |JM'\rangle \\ &= e^{-i\alpha M}\, d^J_{MM'}(\beta)e^{-i\gamma M'} \end{aligned} \tag{4.71}$$

Eq. (4.70) shows that only $M' = \lambda_1 - \lambda_2$ occurs, as is obvious. We may now use the orthogonality condition

$$\int dR\, D^{J'*}_{m_1'm_2'}(R)\, D^J_{m_1m_2}(R) = \frac{8\pi^2}{2J+1}\,\delta_{JJ'}\delta_{m_1m_1'}\delta_{m_2m_2'} \tag{4.72}$$

($dR = \sin\beta\, d\beta\, d\alpha\, d\gamma$, $0 \leqslant \alpha, \gamma \leqslant 2\pi$, $0 \leqslant \beta \leqslant \pi$) to derive from (4.71) the result (N_J is a normalization factor to be determined below)

$$|pJM\lambda_1\lambda_2\rangle = \frac{N_J}{2\pi}\int dR\, D^J_{M\lambda}(\alpha\beta\gamma)R(\alpha, \beta, \gamma)\psi_{p\lambda_1\lambda_2} \tag{4.73}$$

The known γ-dependence of D^J_M (Eq. (4.62)) allows (4.73) to be written as follows:

$$\begin{aligned} |pJM\lambda_1\lambda_2\rangle &= N_J\int d\alpha\, d\beta \sin\beta\, D^{J*}_{M\lambda}(\alpha\beta 0)R(\alpha, \beta, 0)\psi_{p\lambda_1\lambda_2} \\ &= N_J\int d\phi\, d\theta \sin\theta\, D^{J*}_{M\lambda}(\phi, \theta, -\phi)R(\phi, \theta, -\phi)\psi_{p\lambda_1\lambda_2} \\ |pJM\lambda_1\lambda_2\rangle &= N_J\int d\phi\, d\theta \sin\theta\, D^{J*}_{M\lambda}(\phi, \theta, -\phi)\,|p\theta\phi\lambda_1\lambda_2\rangle \end{aligned} \tag{4.74}$$

In order to apply these results it is necessary to consider some questions of normalization. The two particle state (4.70) can be regarded as a special case of the usual one normalized to $\delta(\mathbf{p}_1 - \mathbf{p}_1')\,\delta(\mathbf{p}_2 - \mathbf{p}_2')$ (or some multiple of this). This product of delta functions, as well as the two particle S-matrix elements constructed from the two particle states, contain as a factor a four dimensional delta function conserving the total four momentum. Hence it is useful to work in the c.m. frame and regard the states (4.70) to be normalized by the convention

$$\langle p\theta'\phi'\lambda_1'\lambda_2' \mid p\theta\phi\lambda_1\lambda_2\rangle = \delta_2(\theta\phi, \theta'\phi')\delta_{\lambda_1\lambda_1'}\delta_{\lambda_2\lambda_2'} \tag{4.75}$$

where δ_2 is the delta function on the unit sphere

$$\delta_2(\theta\phi, \theta'\phi') = \delta(\cos\theta - \cos\theta')\,\delta(\phi - \phi') \tag{4.76}$$

Now we find that the angular momentum states (4.74) may be normalized as

$$\langle pJ'M'\lambda_1\lambda_2 \mid pJM\lambda_1\lambda_2\rangle = \delta_{JJ'}\delta_{MM'}\delta_{\lambda_1\lambda_1'}\delta_{\lambda_2\lambda_2'} \tag{4.77}$$

provided the normalization factor N_J is chosen to be

$$N_J = \left(\frac{2J+1}{4\pi}\right)^{\frac{1}{2}} \tag{4.78}$$

The following relation is useful:

$$\langle p\theta\phi\lambda_1\lambda_2 \mid pJM\lambda_1'\lambda_2'\rangle = N_J\delta_{\lambda_1\lambda_1'}\delta_{\lambda_2\lambda_2'}D^{J*}_{M\lambda}(\phi, \theta, -\phi) \tag{4.79}$$

The transition matrix for the reaction $ab \to cd$ (for a given energy E) therefore has angular dependence

$$\langle p'\theta\phi\lambda_c\lambda_d|\, T\, |p\theta_0\phi_0\lambda_a\lambda_b\rangle$$

$$= \sum_J \left(\frac{2J+1}{4\pi}\right) D^{J*}_{M\mu}(\phi, \theta, -\phi)\langle\lambda_c\lambda_d|\, T^J\, |\lambda_a\lambda_b\rangle D^J_{M\lambda}(\phi_0, \theta_0, -\phi_0) \tag{4.80}$$

where $\mu = \lambda_c - \lambda_d$, $\lambda = \lambda_a - \lambda_b$. Choosing the initial state along the z axis Eq. (4.80) simplifies to

$$\sum_J \frac{2J+1}{4\pi} \langle\lambda_c\lambda_d|\, T^J\, |\lambda_a\lambda_b\rangle\, d^J_{\lambda\mu}(\theta)e^{i(\lambda-\mu)\phi} \tag{4.81}$$

Details of the calculation of normalization factors may be found in ref. 9. These authors introduce a scattering amplitude f:

$$f_{\lambda_c\lambda_d;\lambda_a\lambda_b} = \frac{1}{p}\sum_J (J + \tfrac{1}{2})\langle\lambda_c\lambda_d|\, T^J\, |\lambda_a\lambda_b\rangle\, d^J_{\lambda\mu}(\theta)e^{i(\lambda-\mu)\phi} \tag{4.82}$$

in terms of which the differential cross section has the usual form $|f_{\lambda_c\lambda_d;\lambda_a\lambda_b}|^2$. In Chapter 10 we shall instead expand invariant amplitudes in terms of (4.81).

Applications of the helicity formalism are common and important for the study of practical and conceptual problems. Further references to such developments are given in Chapter 10.

REFERENCES

1. E. P. Wigner, *Ann. of Math.* **40,** 149 (1939).
2. E. P. Wigner, in "Group Theoretical Concepts and Methods in Elementary Particle Physics," ed. F. Gursey (Gordon and Breach, New York, 1964).
3. F. R. Halpern, "Special Relativity and Quantum Mechanics" (Prentice-Hall, Inc., Englewood Cliffs, N.J.)
4. J. F. Boyce, R. Delbourgo, A. Salam, and J. Strathdee, "Partial Wave Analysis," unpublished report IC/67/9 of the International Centre for Theoretical Physics, Trieste, Italy.
5. T. D. Newton, in "Theory of Groups in Classical and Quantum Physics," ed. T. Kahan (American Elsevier Publ. Co., Inc., New York, 1966).
6. P. T. Matthews, in "Lectures in Theoretical Physics Xa," ed. A. O. Barut (Univ. of Colorado Press, Boulder, Colo., 1968).
7. G. Mackey, "Induced Representations" (W. A. Benjamin, Inc., New York, 1968).
8. I. M. Shirokov, *J. Exptl. Theoret. Phys.* (*U.S.S.R.*) **33,** 861 (1957); *ibid* **33,** 1196 (1957); *ibid* **33,** 1208 (1957). English translations: *Soviet Physics JETP* **6,** 664 (1958); *ibid* **6,** 919 (1958); *ibid* **6,** 929 (1958).
9. M. Jacob and G. C. Wick, *Ann. Phys.* (*N.Y.*) **7,** 404 (1959).
10. N. N. Bogoliubov and D. V. Shirkov, "Introduction to the Theory of Quantized Fields" (John Wiley and Sons, Inc., New York, 1959).
11. W. D. McGlinn, *Phys. Rev. Letters* **12,** 467 (1964).
12. L. O'Raifeartaigh, *Phys. Rev.* **139**B 1052 (1965).
13. M. Hamermesh, "Group Theory" (Addison-Wesley Publ. Co., Reading, Mass., 1962), Chap. 12.
14. G. C. Wick, *Ann. Phys.* (N.Y.) **18,** 65 (1962), Appendix.
15. V. Bargmann, *Ann. Math.* **48,** 568 (1947).
16. L. Sertorio and M. Toller, *Nuovo Cimento* **33,** 413 (1964).
17. J. D. Talman, "Special Functions" (W. A. Benjamin, Inc., New York, 1969), Chap. 11.

5

THE POINCARÉ GROUP II: FIELDS

5.1 INTRODUCTION

The states comprising the representation space for the Poincaré group $\mathscr{P}$ do not provide a simple description of the local properties of quantum mechanical systems, except indirectly through the analytic properties of S-matrix elements. In the present chapter we investigate the properties of local fields transforming irreducibly under the unitary transformations of $\mathscr{P}$. Such fields are of fundamental importance for the construction of theories covariant under $\mathscr{P}$.

As a preliminary step we synthesize results from previous chapters by giving the transformation laws for free fields of spin 0, $\frac{1}{2}$, and 1. This makes contact with the "classical" field transformation rules and shows how the unitary transformations of $\mathscr{P}$ induce non-unitary transformations* in the spin space of the components of the finite spin fields. In Sec. 5.3 local field operators are constructed directly from the transformation law of states, Eq. (4.41). These fields, which we shall refer to as Joos-Weinberg fields,[1,2,3] differ from the Rarita-Schwinger fields of Chapter III in not requiring field equations and subsidiary conditions to project out unwanted spin values. The fields in question transform according to the representations $(j, 0)$ or $(0, j)$ of the homogeneous Lorentz group. The relation between causality and spin-statistics is especially simple in this formalism. In Sec. 5.4 the discrete space-time transformations of space and time inversion and charge conjugation are discussed.

There exist many schemes for describing free high spin particles. Perhaps the most important omission is the Bargmann-Wigner construction of wave functions.[4] In Chapter 6 we shall complete the study of the Rarita-Schwinger fields begun in Chapter 3.

---

* For spin zero the spin transformation is the identity and so is trivially unitary.

5.2 TRANSFORMATION LAWS FOR FREE FIELDS OF SPIN 0, $\frac{1}{2}$, AND 1

Under a homogeneous Lorentz transformation the c-number scalar field $\phi_c(x)$, Dirac spinor field $\psi_c(x)$ and vector field $V_c(x)$ transform as

$$\begin{aligned} \phi_c'(\Lambda x) &= \phi_c(x) \\ \psi_c'(\Lambda x) &= S(\Lambda)\psi_c(x) \\ V_c'^{\mu}(\Lambda x) &= \Lambda^{\mu}_{\nu} V_c^{\nu}(x) \end{aligned} \tag{5.1}$$

where $S(\Lambda)$ is given in Eq. (2.24) for an infinitesimal transformation. (The finite transformation matrix has the form $\exp(-i\omega^{\mu\nu}\sigma_{\mu\nu}/4)$.) The transformation matrices S and Λ in (5.1) give (non-unitary) representations of the homogeneous Lorentz group. More precisely, the Dirac field ψ transforms as the representation $(\frac{1}{2}, 0) \oplus (0, \frac{1}{2})$ (cf. the discussion leading to Eq. (2.70)) as required by parity covariance. In addition the vector field is conventionally chosen to transform exactly as the coordinate x, i.e. as $(\frac{1}{2}, \frac{1}{2})$. (The subsidiary condition (that $\partial^{\mu} V_{\mu}$ vanish) eliminates the spin 0 component.)

The transformation rules for the quantized fields are easily discovered by requiring agreement of (5.1) with the mapping of states $\Psi \to \Psi_{\Lambda} = U(\Lambda)\Psi$ which accompanies $x \to \Lambda x$. The classical fields (5.1) are expectation values

$$\begin{aligned} \langle \Psi_{\Lambda} | \, \varphi(\Lambda x) \, | \Psi_{\Lambda} \rangle &\equiv \varphi_c'(\Lambda x) \\ \langle \Psi_{\Lambda} | \, \Psi(\Lambda x) \, | \Psi_{\Lambda} \rangle &\equiv \psi_c'(\Lambda x) \\ \langle \Psi_{\Lambda} | \, V^{\mu}(\Lambda x) \, | \Psi_{\Lambda} \rangle &\equiv V_c'^{\mu}(\Lambda x) \end{aligned} \tag{5.2}$$

Comparing (5.1) with (5.2) and changing the argument x to $\Lambda^{-1}x$ gives the field transform laws

$$\begin{aligned} U^{-1}(\Lambda)\phi(x)U(\Lambda) &= \phi(\Lambda^{-1}x) \\ U^{-1}(\Lambda)\psi(x)U(\Lambda) &= S(\Lambda)\psi(\Lambda^{-1}x) \\ U^{-1}(\Lambda)V^{\mu}(x)U(\Lambda) &= \Lambda^{\mu}_{\nu}V^{\nu}(\Lambda^{-1}x) \end{aligned} \tag{5.3}$$

Frequently it is useful to use instead of (5.3) the equivalent infinitesimal form. Using Eqs. (2.21), (2.23), and (2.30) to expand the right hand side of

(5.3), we use $U(\Lambda) = 1 - i\omega^{\mu\nu}M_{\mu\nu}/2$ to obtain

$$\begin{aligned}
[\varphi(x), M_{\mu\nu}] &= i(x_\mu\partial_\nu - x_\nu\partial_\mu)\varphi(x) \\
[\psi(x), M_{\mu\nu}] &= [i(x_\mu\partial_\nu - x_\nu\partial_\mu) + \tfrac{1}{2}\sigma_{\mu\nu}]\psi(x) \\
[V^\rho(x), M_{\mu\nu}] &= [i(x_\mu\partial_\nu - x_\nu\partial_\mu)\delta^\rho_\sigma + (m_{\mu\nu})^\rho_\sigma]V^\sigma(x)
\end{aligned} \tag{5.4}$$

where $\frac{1}{2}\sigma_{\mu\nu}$ and $m_{\mu\nu}$ are the spin matrices appropriate for spin $\frac{1}{2}$ and 1.

It is clear that every finite component field will have a similar transformation rule. If we write

$$U^{-1}(\Lambda)\chi_\alpha(x)U(\Lambda) = D_{\alpha\beta}(\Lambda)\chi_\beta(\Lambda^{-1}x) \tag{5.5}$$

(with $D^\dagger \neq D^{-1}$) a standardized form is obtained by writing $D = 1 - \frac{i}{2}\,\omega^{\mu\nu}\,\Sigma_{\mu\nu}$ where $(\Sigma_{\mu\nu})_{\alpha\beta}$ acts on the components of χ. Then the infinitesimal Lorentz transformation law is equivalent to

$$[\chi_\alpha(x), M_{\mu\nu}] = [i(x_\mu\,\partial_\nu - x_\nu\,\partial_\mu)\delta^\beta_\alpha + (\Sigma_{\mu\nu})_{\alpha\beta}]\chi_\beta(x) \tag{5.6}$$

5.3 JOOS-WEINBERG FIELDS

In Chapter 3 the apparatus of spinor calculus was used to construct covariant wave equations for spinors of various ranks. It was found possible to describe particles of spin s by subjecting the spinors to subsidiary conditions. It is possible to construct local fields describing spin s which are subject to no subsidiary conditions at all. The procedure involves a straightforward extension of the transformation law (4.41) for particle states. In the following we consider particles of finite mass. A detailed treatment of the massless case has been presented in ref. 3.

For variety, and to keep close* to the conventions of ref. 2, the treatment differs in two respects from that of Chapter 4. First of all, the standard boost operation is chosen to be a pure Lorentz transformation

$$\begin{aligned}
L^i_j(p) &= \delta_{ij} + \hat{p}_i\hat{p}_j(\cosh\zeta - 1) \\
L^i_0(p) &= L^0_i(p) = \hat{p}_i\sinh\zeta \\
L^0_0(p) &= \cosh\zeta
\end{aligned} \tag{5.7}$$

---

* However, our notation is somewhat different from ref. 2. In particular the representations $(j, 0)$, $(0, j)$ are interchanged.

instead of the sequence (4.33). Further the index λ will be the z-component of spin (σ) in the rest frame. Hence $|p\sigma\rangle = U(L(p))\,|\hat{p}\sigma\rangle$, and the states transform according to

$$U(\Lambda)\,|p\sigma\rangle = \sum_{\sigma'} |\Lambda p\sigma'\rangle D^j_{\sigma'\sigma}(R) \tag{5.8}$$

with $R = L^{-1}(\Lambda p)\Lambda L(p)$ and j the spin in the rest frame.

It is now possible to *define* creation operators $a\dagger(p, \sigma)$ and their adjoints, the annihilation operators $a(p, \sigma)$, satisfying

$$\begin{aligned} a^*(p, \sigma)\,|0\rangle &= |p\sigma\rangle \\ U(\Lambda)a^*(p, \sigma)U^{-1}(\Lambda) &= \sum_{\sigma'} a^*(\Lambda p, \sigma')D^j_{\sigma'\sigma}(R) \\ [a(p, \sigma), a^*(p', \sigma')]_\pm &= 2\omega_p\delta_{\sigma\sigma'}\delta(\mathbf{p} - \mathbf{p}') \\ [a(p, \sigma), a(p', \sigma')]_\pm &= 0 \end{aligned} \tag{5.9}$$

where $A, B_\pm$ denotes $AB \pm BA$ as usual. Using the unitarity of the little group matrix $D^j(R)$ we see that the operators $a(p, \sigma)$ transform as

$$\begin{aligned} U(\Lambda)a(p, \sigma)U^{-1}(\Lambda) &= \sum_{\sigma'} a(\Lambda p, \sigma')D^j_{\sigma'\sigma}(R) \\ &= \sum_{\sigma'} D^j_{\sigma\sigma'}(R^{-1})a(\Lambda p, \sigma'); \\ R^{-1} &= L^{-1}(p)\Lambda^{-1}L(\Lambda p) \end{aligned} \tag{5.10}$$

We shall not consider here the possibility of any "statistics" other than exhibited in Eq. (5.9). If we have "antiparticles" as well as particles, it is necessary to introduce appropriate operators $b(p, \sigma)$, $b^*(p, \sigma)$. The latter are assumed to transform identically to $a(p, \sigma)$ and $a^*(p, \sigma)$.* In Chapter 6 the concept of antiparticle is examined in great detail.

In order to combine particle and antiparticle operators a and b^* in a field having proper transformation properties it is useful to perform a change of basis on one of these sets. The point is that the set $a^*(p, \sigma)$ can be transformed to another set by a matrix transformation C

$$\begin{aligned} A(p, \tau) &\equiv \sum_{\mu} (C^{-1})_{\tau\mu}a^*(p, \mu) \\ a^*(p, \mu) &= \sum_{\tau} C_{\mu\tau}A(p, \tau) \end{aligned} \tag{5.11}$$

† In addition we have $[a(p, \sigma), b(p', \sigma')]_\pm = [a(p, \sigma), b^*(p', \sigma')]_\pm = 0$, etc. See refs. 6–7 for a discussion of this question.

such that the $A(p, \tau)$ transform *identically* to $a(p, \tau)$:

$$U(\Lambda)A(p, \tau)U^{-1}(\Lambda) = \sum_{\tau'} D^j_{\tau\tau'}(R^{-1})A(p, \tau') \tag{5.12}$$

The proof of this rests on the well known matrix C which converts an $SU(2)$ representation matrix D^j to its complex conjugate D^{j*}

$$D^j(R)^* = CD^j(R)C^{-1} \tag{5.13}$$

The $(2j + 1) \times (2j + 1)$ matrix C is unitary and satisfies

$$C^*C = (-1)^{2j} \tag{5.14}$$

An explicit form which we shall sometimes use is

$$C_{\mu\nu} = \xi(-1)^{j+\mu}\delta_{\mu,-\nu} \tag{5.15}$$

where ξ is an arbitrary unimodular phase factor. Equation (5.12) now follows from Eqs. (5.10), (5.11), and (5.13):

If the matrix C appearing in (5.11) is identified with that of Eq. (5.13), the relation (5.12) is easily verified:

$$\begin{aligned} U(\Lambda)A(p, \sigma)U^{-1}(\Lambda) &= \sum (C^{-1})_{\sigma\tau}a^*(\Lambda p, \rho)D^j_{\rho\tau}(R) \\ &= \sum_{\rho} (C^{-1}D^{j*}(R^{-1}))_{\sigma\rho}a^*(\Lambda p, \rho) \\ &= \sum_{\mu\rho} (C^{-1}D^{j*}(R^{-1})C)_{\sigma\mu}C^{-1}_{\mu\rho}a^*(\Lambda p, \rho) \\ &= \sum_{\mu} D^j(R^{-1})_{\sigma\mu}A(\Lambda p, \mu) \end{aligned} \tag{5.16}$$

We are now in a position to construct local fields (defined in space-time) for arbitrary spin. These fields transform as indicated in Eq. (5.5). In the traditional treatment of the subject one "recognizes" the Fourier coefficients to be creation and annihilation operators. Here we follow the opposite approach, in which particle operators are the given elements in terms of which one constructs a local field having the desired transformation properties. This method is very instructive and probably has more direct physical significance than the usual (classically-motivated) treatment.

A primary requirement is that the local fields $\psi(x)$ transform properly under translations

$$\begin{aligned} U(a)\psi(x)U^{-1}(a) &= \psi(x + a) \\ i\partial_\mu\psi(x) &= [\psi(x), P_\mu] \end{aligned} \tag{15.17}$$

The physical significance of the operators $a(p, \sigma)$ requires

$$\begin{aligned}[][a(p, \sigma), P_\mu] &= p_\mu a(p, \sigma)\\ [a^*(p, \sigma), P_\mu] &= -p_\mu a^*(p, \sigma)\end{aligned} \tag{5.18}$$

along with corresponding relations for the antiparticle operators $b(p, \sigma)$, $b^*(p, \sigma)$. It is helpful to give an explicit construction of the momentum operator:

$$P_\mu = \sum_\sigma \int \frac{d^3p}{2\omega_p} p_\mu[a^*(p, \sigma)a(p, \sigma) + b^*(p, \sigma)b(p, \sigma)] \tag{5.19}$$

(Eq. (5.19) is appropriate for the situation in which one deals simultaneously with a particle and its distinct antiparticle. If there is no distinct antiparticle the b^*b term should be omitted.) The commutation rules (5.9) may be used to verify that (5.19) satisfies (5.18). Finally one may use the transformation law for a^*, a (Eq. (5.9)) and the unitarity of the little group matrices to verify that $U(\Lambda)\, P_\mu U^{-1}(\Lambda) = (\Lambda^{-1})_\mu^\nu P_\nu$ in agreement with Eq. (4.7).

It is possible to satisfy the second of Eqs. (5.17) by writing expressions of the form $\int c(p)\, a(p, \sigma) e^{-ip\cdot x}\, d^3p/2\omega_p$, etc. In order to determine the form of the coefficients $c(p)$ we have to study in more detail the behavior of $a(p, \sigma)$ under Lorentz transformations. Since the little group rotation R depends on p the covariant form (5.5) cannot be obtained by setting $c(p) = 1$. Equation (5.5) suggests that we seek some linear combination of $a(p, \sigma)$ with the transformation law

$$U(\Lambda)\tilde{a}(p, \sigma)U^{-1}(\Lambda) = \sum_{\sigma'} D_{\sigma\sigma'}(\Lambda^{-1})\tilde{a}(p, \sigma') \tag{5.20}$$

with D not necessarily unitary. In that case the operator $\chi_1(x)$ defined by

$$\chi_{1\sigma}(x) \equiv \frac{1}{(2\pi)^{\frac{3}{2}}} \int \frac{d^3p}{2\omega_p} e^{-ip\cdot x}\tilde{a}(p, \sigma) \tag{5.21}$$

transforms as (5.5). (In order to show this one notes that $d^2p/2\omega_p$ is invariant and that $p \cdot x = (\Lambda p) \cdot (\Lambda x)$.)

With this motivation it is easy to rewrite the transformation law (5.9) in such a way that (5.20) is obtained. The rotation R characterizing the little group rotation matrix is the product of three homogeneous Lorentz transformations. It is clear then that $D^j_{\sigma\sigma'}(R^{-1})$ may be written as the product of three (non-unitary)$SL(2, C)$ representation matrices of dimension $2j + 1$

$$\begin{aligned} D^j(R^{-1}) &= D^j(L^{-1}(p)\Lambda^{-1}L(\Lambda p))\\ &= D(L^{-1}p)D(\Lambda)D(L(\Lambda p))\end{aligned} \tag{5.22}$$

where the natural choice for $D(\Lambda)$, etc., is one of the representations $(j, 0)$ or $(0, j)$ (cf. Chapter 2). (It is also possible to construct other decompositions, as discussed in ref. 8.)

If (5.22) is used to rearrange the second of Eqs. (5.9), we obtain

$$U(\Lambda)\Big(\sum_\tau D(L(p))_{\sigma\tau}a(p,\tau)\Big)U^{-1}(\Lambda) = \sum_\rho D_{\sigma\rho}(\Lambda^{-1})\Big(\sum_\tau D_{\rho\tau}(L(\Lambda p))a(\Lambda p,\tau)\Big) \tag{5.23}$$

Hence the previously hypothetical quantity $\tilde{a}(p, \sigma)$ exists, and is related to the operators $a(p, \sigma)$ by

$$\tilde{a}(p, \sigma) = \sum_\tau D(L(p))_{\sigma\tau}a(p, \tau) \tag{5.24}$$

Since the operators $A(p, \sigma)$ of Eq. (5.11) transform as $a(p, \sigma)$, we can define quantities $\tilde{a}(p, \sigma)$ in exact analogy to (5.24) which transform according to the rule (5.20).

The entire argument has identical form for the antiparticle operators $b(p, \sigma)$. For example the operator $\hat{B}(p, \sigma)$ allows the construction of the field

$$\tilde{\chi}_{2\sigma}(x) \equiv \frac{1}{(2\pi)^{\frac{3}{2}}}\int \frac{d^3p}{2\omega_p}\, e^{ip\cdot x}\tilde{B}(p, \sigma) \tag{5.25}$$

We can also construct two other fields satisfying (5.17):

$$\begin{aligned} \chi_{2\sigma}(x) &\equiv \frac{1}{(2\pi)^{\frac{3}{2}}}\int \frac{d^3p}{2\omega_p}\, e^{-ip\cdot x}\tilde{b}(p, \sigma) \\ \tilde{\chi}_{1\sigma}(x) &\equiv \frac{1}{(2\pi)^{\frac{3}{2}}}\int \frac{d^3p}{2\omega_p}\, e^{ip\cdot x}\tilde{A}(\rho, \sigma) \end{aligned} \tag{5.26}$$

The fields $\chi_{i\sigma}(x)$, $\tilde{\chi}_{i\sigma}(x)$ $(i = 1, 2)$ have the same behavior under Poincaré transformations. None of these operators taken separately has the property of *locality*, i.e. of commutability at space-like separations. However, it is simple to obtain such fields if one forms simple linear combinations such as

$$\begin{aligned} \psi_\sigma(x) &= \xi\chi_{1\sigma}(x) + \eta\tilde{\chi}_{2\sigma}(x) \\ \varphi_\sigma(x) &= \xi\chi_{1\sigma}(x) + \eta\tilde{\chi}_{1\sigma}(x) \end{aligned} \tag{5.27}$$

and similar constructions with $(1 \leftrightarrow 2)$. The field ψ_σ involves operators of both particles and antiparticles, whereas $\varphi_\sigma(x)$ is made from operators of "particles" alone. We do not introduce fields of the type $\xi\chi_{1\sigma} + \eta\chi_{2\sigma}$ since, as shown in Chapter 6, the two components of this field would not transform in the same way under internal symmetry transformations whose conserved quantum numbers distinguish particle from antiparticle. It is clear that the field φ_σ may be obtained from ψ_σ by making the replacement $b \to a$. This substitution must sometimes be made with care when internal symmetries are included (see Chapter 6).

If the field ψ_σ is written out explicitly one obtains

$$\psi_\sigma(x) = \frac{1}{(2\pi)^{\frac{3}{2}}} \int \frac{d^3p}{2\omega_p} \{\xi D(L(p))_{\sigma\tau} a(p, \tau) e^{-ip\cdot x} + \eta (D(L(p))C^{-1})_{\sigma\tau} b^*(p, \tau) e^{ip\cdot x}\} \quad (5.28)$$

In employing (5.28) we may set the phase ξ occurring in (5.15) equal to unity with no loss of generality. In detail, (5.28) has the form

$$\psi_\sigma(x) = \frac{1}{(2\pi)^{\frac{3}{2}}} \int \frac{d^3p}{2\omega_p} D(L(p))_{\sigma\tau} [\xi a(p, \tau) e^{-ip\cdot x} + \eta (-1)^{j-\tau} b^*(p, -\tau) e^{ip\cdot x}] \quad (5.29)$$

If we use the basic commutators (5.9) the following commutator is obtained:

$$[\psi_\sigma(x), \psi_\tau^\dagger(x')]_\pm = \frac{1}{(2\pi)^3} \int \frac{d^3p}{2\omega_p} [D(L(p)) D^\dagger(L(p))]_{\sigma\tau} (|\xi|^2 e^{-ip\cdot(x-x')} \pm |\eta|^2 e^{ip\cdot(x-x')}) \quad (5.30)$$

In order to understand this expression we consider the simple cases of spin 0 and $\frac{1}{2}$. For spin zero $D = 1$ and the right hand side is a superposition of the invariant delta functions $\Delta(x)$ and $\Delta^{(1)}(x)$. To eliminate the non-causal function $\Delta^{(1)}$ we must choose commutators (minus sign) and set $|\xi|$ equal to $|\eta|$. Moreover $|\xi| = |\eta| = 1$ gives the usual normalization $[\varphi(x), \varphi^\dagger(x')]_- = i\Delta(x - x')$.

For spin $\frac{1}{2}$ we may have a field $\xi_\sigma(x)$ transforming as $D^{(\frac{1}{2},0)}$ and a field $\chi_\sigma(x)$ transforming as $D^{(0,\frac{1}{2})}$. These fields transform under $U^{-1}(\Lambda)(\xi, \chi)U(\Lambda)$ exactly as the corresponding *c*-number spinor fields introduced in Sec. 3.4.† Setting $|\xi|^2 = |\eta|^2 = m$ for both fields and using Eq. (2.64) one finds immediately that the anti-commutator is causal:

$$\begin{aligned} [\xi_\sigma(x), \xi_\tau^\dagger(x')]_+ &= (i\tilde{\sigma}^\mu \partial_\mu) i\,\Delta(x - x') \\ [\chi_\sigma(x), \chi_\tau^\dagger(x')]_+ &= (i\sigma^\mu \partial_\mu) i\,\Delta(x - x') \end{aligned} \quad (5.31)$$

† A detailed discussion of the field theory of two-component spinor fields has been given in Ref. 9.

The phases ξ and η of the fields ξ_σ and χ_σ are not completely free if we include both fields (constructed from the same particle operators) in the same theory. From our study of the Dirac equation (Sec. 3.4) we know that parity covariance always requires that ξ_σ and χ_σ occur together. Hence we require that the anti-commutator $[\xi_\sigma, \chi_\tau^\dagger]_+$ be causal in order that the full Dirac field $\psi = \text{col}\,(\xi(x), \chi(x))$ have this essential property. Hence we write $(D = D(L(p))$

$$\xi_\sigma(x) \equiv \frac{1}{(2\pi)^{\frac{3}{2}}} \int \frac{d^3p}{2\omega_p} [\xi D^{(\frac{1}{2},0)}_{\sigma\nu} a(p, \nu) e^{-ip\cdot x} + \eta (D^{(\frac{1}{2},0)} C^{-1})_{\sigma\nu} b^*(p, \nu) e^{ip\cdot x}]$$

$$\chi_\tau(x) \equiv \frac{1}{(2\pi)^{\frac{3}{2}}} \int \frac{d^3p}{2\omega_p} [\xi' D^{(0,\frac{1}{2})}_{\tau\rho} a(p, \rho) e^{-ip\cdot x} + \eta' (D^{(0,\frac{1}{2})} C^{-1})_{\tau\rho} b^*(p, \rho) e^{ip\cdot x}] \tag{5.32}$$

If we recall that $D^{(0,\frac{1}{2})} D^{(\frac{1}{2},0)} = 1$ for the boost $L(p)$ we find

$$[\xi_\sigma(x), \chi_\tau^\dagger(x')]_+ = \frac{1}{(2\pi)^3} \int \frac{d^3p}{2\omega_p} [\xi\xi'^* e^{-ip\cdot(x-x')} + \eta\eta'^* e^{ip\cdot(x-x')}] \tag{5.33}$$

which requires $\eta\eta'^* = -\xi\xi'$ to obtain $\Delta(x - x')$ (cf. Eq. (5.53) below). Hence a sensible convention would be to set $\xi = \eta = \eta' = -\eta' = \sqrt{m}$. It is now straight-forward to show that the commutators (5.29), (5.31) are equivalent to

$$[\psi_\alpha(x), \bar\psi_\beta(x')]_+ = i(i\gamma^\mu_{\alpha\beta}\partial_\mu + \delta_{\alpha\beta} m)\, \Delta(x - x') \tag{5.34}$$

where $\bar\psi(x) = \text{col}\,(\chi^\dagger(x), \xi^\dagger(x))$ and γ^μ is given by (3.51). Subsequently we shall generalize this result to see that the antiparticle operator of a $D^{(0,j)}$ field must carry a phase $(-1)^{2j}$ with respect to the antiparticle operator in the $D^{(j,0)}$ field. We thus learn that causality plays a decisive role in leading to the well known result that the intrinsic parity of a state having one antiparticle and one particle is odd for Fermions.

The examples just considered suggest several general rules which we shall verify below. The first conclusion is that locality, relativity and quantization by (5.9) require the spin-statistics connection, i.e. quantization by commutators for integral spins and by anti-commutators for half-integral spin. The second point, which emerges simultaneously with the first, is that particle and antiparticle coordinates are joined in a definitely prescribed way (apart from a constant phase) in a local field. (The consequences of this intimate connection of particle and antiparticle are explored in detail in Chapters 6 and 7.) These results are with good reason frequently regarded as the finest achievements of local relativistic quantum field theory.

The problem of "spin and statistics" may also be approached in a more general (and mathematically refined) way which presumably remains valid for interacting systems (at least for fields having low spin values). Textbooks on axiomatic field theory[6,10] should be consulted for this development. Other approaches to the problem emphasize the principles of positive definite probabilities, energies, etc.[11,12] Recently it has been noticed[13] that the situation is quite different for infinite-component fields. A critique of experimental tests of the spin-statistics connection may be found in ref. 14. The principles involved in this area of research are of profound consequence and deserve intensive study.

We now return to Eq. (5.30) and consider the case of general spin. Clearly we need to express $DD^\dagger$ as a function of p in order to understand the structure of the commutator. Comparison with Eq. (2.64) suggests that $DD^\dagger$ might be expressible as a "spin-tensor" matrix contracted with p_μ. Since D is one of the representations $D^{(j,0)}$, $D^{(0,j)}$ this generalized spin-tensor will be a matrix of dimension $2j + 1$.

We recall the four-vector matrices σ^μ, $\tilde{\sigma}^\mu$ which satisfy

$$\begin{aligned} D^\dagger \sigma^\mu D &= \Lambda^\mu_\nu \sigma^\nu \\ D\tilde{\sigma}^\mu D^\dagger &= \Lambda^\mu_\nu \tilde{\sigma}^\nu \end{aligned} \tag{5.35}$$

where $D = D^{(\frac{1}{2},0)}$. If we contract with p_μ, we obtain

$$\begin{aligned} D^\dagger \sigma^\mu P_\mu D &= \Lambda^\mu_\nu p_\mu \sigma^\nu = (\Lambda^{-1} p)_\nu \sigma^\nu \\ D\tilde{\sigma}^\mu p_\mu D^\dagger &= (\Lambda p)_\nu \tilde{\sigma}^\nu \end{aligned} \tag{5.36}$$

If we let p be the rest frame four-vector and Λ a pure boost $L(p)$, $\sigma^\mu p_\mu$ and $\tilde{\sigma}^\mu p_\mu$ reduce to mI, giving $DD^\dagger = p_\nu \tilde{\sigma}^\nu / m$ in agreement with Eq. (2.64). (If we change $\Lambda \to \Lambda^{-1}$ in the first of Eqs. (5.36) and use $D^{(\frac{1}{2},0)}(\Lambda^{-1}) = D^{(0,\frac{1}{2})}(\Lambda)$ we see that the first of Eqs. (2.64) is obtained.)

In order to generalize Eq. (5.35) we recall the construction of (3.18), which may be generalized slightly to $\eta_{\dot{a}} \sigma^{\mu \dot{a} b} \xi_b$. The basis set $\eta_{\dot{a}} \xi_b$ transforms as $D^{(\frac{1}{2},0)*} \otimes D^{(\frac{1}{2},0)}$ and the spin tensor $\sigma^{\mu \dot{a} b}$ converts this basis to a standard tensor form. In direct analogy we introduce a spinor basis $v_{\dot{a}_1 \cdots \dot{a}_{2j}} u_{b_1 \cdots b_{2j}}$ for the representation $D^{(j,0)*} \otimes D^{(j,0)}$. Here the u's and v's are traceless spinors having total symmetry in their indices, e.g. $u_{b_1 \ldots b_2}$ ($\sim D^{(j,0)}$) satisfies

$$\begin{aligned} u_{b_1' \cdots b_{2j}'} &= D_{b_1'}{}^{b_1} \cdots D_{b_{2j}'}{}^{b_{2j}} u_{b_1 \cdots b_2} \\ u_{b_1 b_2 \cdots b_{2j}} &= u_{b_2 b_1 \cdots b_{2j}} = u_{b_3 b_2 b_1 \cdots b_{2j}} = \cdots \\ \epsilon^{b_1 b_2} u_{b_1 b_2 \cdots b_{2j}} &= 0 \end{aligned} \tag{5.37}$$

with similar expressions for $v_{a_1 \ldots a_{2j}}$.

We can construct a tensor basis for this representation as follows:

$$T^{\mu_1 \cdots \mu_{2j}} = \left(\prod_{i=1}^{2j} \sigma^{\mu_i \dot{a}_i b_i} \right) v_{\dot{a}_1 \cdots \dot{a}_{2j}} u_{b_1 \cdots b_2} \tag{5.38}$$

T is clearly symmetric in its four-vector indices because of the symmetry of the spinors. Tracelessness follows from the identity (3.36) and the last of (5.37). Since the representation in question is equivalent to (j, j) we see that the latter may be represented by the set of traceless symmetric tensors of rank $2j$. By construction we know that the (irreducible) tensor $T^{\mu_1 \cdots \mu_{2j}}$ transforms as

$$T'^{\mu_1 \cdots \mu_{2j}} = \Lambda^{\mu_1}_{\nu_1} \Lambda^{\mu_2}_{\nu_2} \cdots \Lambda^{\mu_{2j}}_{\nu_{2j}} T^{\nu_1 \cdots \nu_{2j}} \tag{5.39}$$

A more useful basis for our purpose is the (j, m) basis (Eq. (2.42)) in terms of which the $D^{(j_1, j_2)}$ matrices were defined in Chapter 2. Denoting the basis for $D^{(j,0)}$ by u_τ and that of $D^{(j,0)*}$ by v^* ($\sigma, \tau = -j, \ldots, +j$) we write (5.38) in the form

$$T^{\mu_1 \cdots \mu_{2j}} = \sum_{\sigma\tau} v^*_\sigma t^{\mu_1 \cdots \mu_{2j}}_{\sigma\tau} u_\tau \equiv v^\dagger t^{\mu_1 \cdots \mu_{2j}} u \tag{5.40}$$

Under an $SL(2, C)$ transformation $u \to D^{(j,0)} u$, $v^\dagger \to v^\dagger D^{(j,0)\dagger}$ and Eq. (5.39) shows that the $(2j + 1) \times (2j + 1)$ matrices $t^{\mu_1 - \mu_{2j}}$ transform as

$$D^{(j,0)}(\Lambda)^\dagger t^{\mu_1 \cdots \mu_{2j}} D^{(j,0)}(\Lambda) = \Lambda^{\mu_1}_{\nu_1} \cdots \Lambda^{\mu_{2j}}_{\nu_{2j}} t^{\nu_1 \cdots \nu_{2j}} \tag{5.41}$$

We have now succeeded in generalizing the first of Eqs. (5.35). To extend the second of these relations we recall the definition of $\tilde{\sigma}$ (Eq. (3.24)), and define a tensor $\tilde{T}$ by

$$\tilde{T}^{\mu_1 \cdots \mu_{2j}} \equiv \left(\prod_{i=1}^{2j} \sigma^{\mu_i}_{a_i \dot{b}_i} \right) v'^{a_1 \cdots a_{2j}} u'^{\dot{b}_1 \cdots \dot{b}_2} \tag{5.42}$$

Here $u'^{b_1 \cdots b_{2j}}$ transforms as $D^{(0,j)}$ and $v'^{a_1 \cdots a_{2j}}$ as $D^{(0,j)*}$ (cf. the definitions (3.17)); u' and v' are traceless and symmetric in spinor indices and hence $\tilde{T}$ has the same property as T in its four-vector indices. Again we go to a (j, m)

basis and the transformation $u' \to D^{(0,j)}u'$, $v'^\dagger \to v'^\dagger D^{(0,j)\dagger}$ leads to

$$D^{(0,j)}(\Lambda)^\dagger \tilde{t}^{\mu_1\cdots\mu_{2j}} D^{(0,j)}(\Lambda) = \Lambda^{\mu_1}_{\nu_1} \cdots \Lambda^{\mu_{2j}}_{\nu_{2j}} \tilde{t}^{\nu_1\cdots\nu_{2j}} \tag{5.43}$$

Hence $\tilde{t}$ transforms under $D^{(0,j)}$ exactly as t transforms under $D^{(j,0)}$. If we change Λ to Λ^{-1} in (5.43) and use $D^{(0,j)}\,(\Lambda)^\dagger = D^{(j,0)}(\Lambda^{-1})$ we find

$$D^{(j,0)}(\Lambda)\tilde{t}^{\mu_1\cdots\mu_{2j}} D^{(j,0)}(\Lambda)^\dagger = (\Lambda^{-1})^{\mu_1}_{\nu_1} \cdots (\Lambda^{-1})^{\mu_{2j}}_{\nu_{2j}} \tilde{t}^{\nu_1\cdots\nu_{2j}} \tag{5.44}$$

The relations (5.41), besides having intrinsic interest, allow the evaluation of the quantities $DD^\dagger$ occurring in the field commutator (5.30). Proceeding as in Eq. (5.36), we define matrices $\Pi(p)$ and $\tilde{\Pi}(p)$ by

$$\begin{aligned} \Pi(p) &\equiv t^{\mu_1\cdots\mu_{2j}} p_{\mu_1} \cdots p_{\mu_2} \\ \tilde{\Pi}(p) &\equiv \tilde{t}^{\mu_1\cdots\mu_{2j}} p_{\mu_1} \cdots p_{\mu_2} \end{aligned} \tag{5.45}$$

These matrices obey the following relations:

$$\begin{aligned} D^{(j,0)}(\Lambda)^\dagger \Pi(p) D^{(j,0)}(\Lambda) &= \Pi(\Lambda^{-1}p) \\ D^{(0,j)}(\Lambda)^\dagger \tilde{\Pi}(p) D^{(0,j)}(\Lambda) &= \tilde{\Pi}(\Lambda^{-1}p) \end{aligned} \tag{5.46}$$

or equivalently

$$\begin{aligned} D^{(0,j)}(\Lambda) \Pi(p) D^{(0,j)}(\Lambda)^\dagger &= \Pi(\Lambda p) \\ D^{(j,0)}(\Lambda) \tilde{\Pi}(p) D^{(j,0)}(\Lambda)^\dagger &= \tilde{\Pi}(\Lambda p) \end{aligned} \tag{5.47}$$

If we set $p = p_0 = (m, \mathbf{0})$ we find

$$\begin{aligned} \Pi(p_0) &= t^{00\cdots0} m^{2j} \\ \tilde{\Pi}(p_0) &= \tilde{t}^{00\cdots0} m^{2j} \end{aligned} \tag{5.48}$$

Choosing Λ to be a rotation ($\Lambda p_0 = p_0$) we see that $t^{0\cdots\cdot}$, $\tilde{t}^{0\cdots\cdot}$ are invariant under rotations. Hence $t^{0\cdots\cdot}$ and $\tilde{t}^{0\cdots\cdot}$ are multiples of the identity matrix. Choosing the multiple to be unity gives the relations

$$\begin{aligned} D^{(0,j)}(\Lambda) D^{(0,j)}(\Lambda)^\dagger &= \Pi(\Lambda p_0)/m^{2j} \\ D^{(j,0)}(\Lambda) D^{(j,0)}(\Lambda)^\dagger &= \tilde{\Pi}(\Lambda p_0)/m^{2j} \end{aligned} \tag{5.49}$$

Finally if we take Λ to be the pure boost $L(p)$ occurring in (5.30) ($L(p)p_0 = p$) we find the desired generalization of (2.64):

$$\begin{aligned} D^{(0,j)}(L(p)) D^{(0,j)}(L(p))^\dagger &= \frac{\Pi(p)}{m^{2j}} = t^{\mu_1\cdots\mu_{2j}} p_{\mu_1} \cdots p_{\mu_{2j}}/m^{2j} \\ D^{(j,0)}(L(p)) D^{(j,0)}(L(p))^\dagger &= \tilde{\Pi}(p)/m^{2j} = \tilde{t}^{\mu_1\cdots\mu_{2j}} p_{\mu_1} \cdots p_{\mu_{2j}}/m^{2j} \end{aligned} \tag{5.50}$$

For a pure boost $D^\dagger = D$ and the left hand sides of (5.50) become D^2, so that

$$\begin{aligned} \exp(-2\zeta\hat{p}\cdot\mathbf{J}) &= t^{\mu_1\cdots\mu_{2j}}p_{\mu_1}\cdots p_{\mu_{2j}}/m^{2j} \\ \exp(2\zeta\hat{p}\cdot\mathbf{J}) &= \tilde{t}^{\mu_1\cdots\mu_{2j}}p_{\mu_1}\cdots p_{\mu_{2j}}/m^{2j} \end{aligned} \tag{5.51}$$

Here $\mathbf{J}$ are the standard angular momentum matrices. The matrices t and $\tilde{t}$ may be computed from the formulas (5.51), as shown in ref. 2, where explicit forms are given.

If we replace p_μ by $i\partial_\mu$ when acting on $e^{ip\cdot x}$ we can rewrite Eq. (5.30) in the form

$$\begin{aligned} [\psi_\sigma(x), \psi_\tau^\dagger(x')]_\pm &= \frac{i(-i)^{2j}}{m^{2j}}\tilde{t}^{\mu_1\cdots\mu_{2j}}\partial_{\mu_1}\cdots\partial_{\mu_2} \\ &\left[\frac{-i}{(2\pi)^3}\int\frac{d^3p}{2\omega_p}(|\xi|^2 e^{-ip\cdot(x-x')} \pm (-1)^{2j}|\eta|^2 e^{ip\cdot(x-x')})\right] \end{aligned} \tag{5.52}$$

for $D^{(j,0)}$ fields. For $D^{(0,j)}$ fields $\tilde{t}$ is replaced by t. The integral (5.52) is a sum of the (causal) function $\Delta(x)$

$$\Delta(x) = \frac{-i}{(2\pi)^3}\int\frac{d^3p}{2\omega_p}(e^{-ip\cdot x} - e^{ip\cdot x}) \tag{5.53}$$

and the (non-causal) function $\Delta^{(1)}$. The only way to obtain (5.53) from (5.52) is by satisfying the conditions

$$\begin{aligned} |\xi|^2 &= |\eta|^2 \\ \pm(-1)^{2j} &= -1 \end{aligned} \tag{5.54}$$

The latter restriction implies that quantization by commutators is required for Bosons ($j = 0, 1, 2, \ldots$); anticommutators are implied for Fermions ($j = \frac{1}{2}, \frac{3}{2}, \frac{5}{2}, \ldots$).

The condition $|\xi|^2 = |\eta|^2$ requires the particle and antiparticle components to have equal amplitudes. As remarked earlier, this implies many interesting "crossing" relations characteristic of field theory. The normalization of the constant $|\xi|^2$ may be chosen to make the commutator (5.52) coincide with the appropriate canonical commutation relation. For $j = 0$ we found $|\xi| = 1$, while for $j = \frac{1}{2}$, $|\xi|^2 = m$ is called for. In general we choose $|\xi|^2 = m^{2j}$ and find

$$[\psi_\sigma(x), \psi_\tau^\dagger((x')]_\pm = i(-i)^{2j}\tilde{t}^{\mu_1\cdots\mu_{2j}}\partial_{\mu_1}\cdots\partial_{\mu_{2j}}\Delta\, x - x') \tag{5.55}$$

Thus far we have only considered fields for which the antiparticle is distinct from the particle. For such fields the vanishing of the commutators $[\psi_\sigma(x), \psi_\tau(x')]_\pm$ and $[\psi_\sigma^\dagger(x), \psi^\dagger(x')]_\pm$ is automatic. We can also construct fields for which the antiparticles are the same as the particles, by changing $b(p, \sigma)$ to $a(p, \sigma)$ in (5.28) as suggested by (5.27). This "self-conjugate" field has a commutator $[\varphi_\sigma(x), \varphi_\tau^\dagger(x')]_\pm$ of the same structure as (5.52) and hence leads to the same conclusions (5.54) and (5.55). In the present case we also should recheck the commutator $[\varphi_\sigma(x), \varphi_\tau(x')]$ in light of the results of Chapter 6 concerning internal symmetries. In problem (5.3) it is found that this commutator is indeed causal (neglecting internal symmetry) provided the restrictions (5.54) are obeyed. The same result is derived in Sec. 6.9 using the Rarita-Schwinger fields.

It is also of interest to check the locality of commutators of fields constructed from the same operators but transforming as $D^{(j,0)}$ and $D^{(0,j)}$. This is important because in parity-conserving theories the field $D^{(0,j)}$ must occur if $D^{(j,0)}$ does. The Dirac field is the most common example of this phenomenon.

If we repeat the analysis leading to Eq. (5.55) for the $2j + 1$ fields $\chi_\sigma(x)$, which transform as $D^{(0,j)}$, we find the result

$$[\chi_\sigma(x), \chi_\tau^\dagger(x')]_\pm = i(-1)^{2j} t^{\mu_1 \cdots \mu_{2j}} \partial_{\mu_1} \cdots \partial_{\mu_{2j}} \Delta(x - x') \tag{5.56}$$

The above results are insensitive to the relative phase of the creation and annihilation parts of the fields ψ and χ. However, when we construct fields ψ and χ from the same a's and b's, a calculation similar to (5.33) shows that an extra relative phase factor $(-1)^{2j}$ must occur in χ in order that $[\chi(x), \psi^\dagger(x')]_\pm$ be causal. We shall use the following conventions for the fields ψ and χ:

$$\begin{aligned}
\psi_\sigma(x) &= \frac{m^{2j}}{(2\pi)^{\frac{3}{2}}} \int \frac{d^3p}{2\omega_p} \{D^{(j,0)}_{\sigma\nu}(L(p)) a(p, \nu) e^{-ip\cdot x} \\
&\quad + (D^{(j,0)}(L(p))C^{-1})_{\sigma\nu} b^*(p, \nu) e^{ip\cdot x}\} \\
\chi_\sigma(x) &= \frac{m^{2j}}{(2\pi)^{\frac{3}{2}}} \int \frac{d^3p}{2\omega_p} \{D^{(0,j)}_{\sigma\nu}(L(p)) a(p, \nu) e^{-ip\cdot x} \\
&\quad + (-1)^{2j}(D^{(0,j)}(L(p))C^{-1})_{\sigma\nu} b^*(p, \nu) e^{ip\cdot x}\} \\
U^{-1}(\Lambda)\psi_\sigma(x)U(\Lambda) &= \sum_\tau D^{(j,0)}_{\sigma\tau}(\Lambda)\psi_\tau(\Lambda^{-1}x) \\
U^{-1}(\Lambda)\chi_\sigma(x)U(\Lambda) &= \sum_\tau D^{(0,j)}_{\sigma\tau}(\Lambda)\chi_\tau(\Lambda^{-1}x)
\end{aligned} \tag{5.57}$$

The commutators for the fields χ, ψ are now

$$[\chi_\sigma(x), \psi_\tau^\dagger(x')]_\pm = [\psi_\sigma(x), \chi_\tau^\dagger(x')]_\pm = im^{2j}\delta_{\sigma\tau}\Delta(x - x') \tag{5.58}$$

To conclude this section we derive the covariant propagators for the fields

ψ and χ. As usual we begin with the vacuum expectation value of the time-ordered product, for example,

$$\langle T\psi_\sigma(x)\psi_\tau^\dagger(x')\rangle_0 = \theta(x - x')\langle\psi_\sigma(x)\psi_\tau^\dagger(x')\rangle_0 + (-1)^{2j}\theta(x' - x)\langle\psi_\tau^\dagger(x')\psi_\sigma(x)\rangle_0 \tag{5.59}$$

where $\theta(x)$ is unity for $x_0 > 0$ and zero for $x_0 < 0$. From Eq. (4.57) and the definitions (5.50) we find

$$\begin{aligned}\langle\psi_\sigma(x)\psi_\tau^\dagger(x')\rangle_0 &= \frac{1}{(2\pi)^3}\int\frac{d^3p}{2\omega_p}\,\tilde{\Pi}_{\sigma\tau}(p)e^{-ip\cdot(x-x')}\\ \langle\psi_\tau^\dagger(x')\psi_\sigma(x)\rangle_0 &= \frac{1}{(2\pi)^3}\int\frac{d^3p}{2\omega_p}\,\tilde{\Pi}_{\sigma\tau}(p)e^{ip\cdot(x-x')}\end{aligned} \tag{5.60}$$

If we recall that $\tilde{\Pi}$ depends on p as indicated in Eq. (5.50) and introduce the functions $\Delta_\pm(x)$ by

$$i\,\Delta_\pm(x) = \frac{\pm 1}{(2\pi)^3}\int\frac{d^3p}{2\omega_p}\,e^{\mp ip\cdot x} \tag{5.61}$$

we can represent (5.59) as

$$\begin{aligned}&\langle T\psi_\sigma(x)\psi_\tau^\dagger(x')\rangle_0\\ &\quad = i[\theta(x - x')\tilde{\Pi}_{\sigma\tau}(i\partial)\,\Delta_+(x - x') - \theta(x' - x)\tilde{\Pi}_{\sigma\tau}\,\Delta_-(x - x')]\\ &\quad = \tilde{\Pi}_{\sigma\tau}(i\partial)\,\Delta_c(x) + \text{non-covariant terms}\end{aligned} \tag{5.62}$$

where the causal spin zero propagation function is

$$\begin{aligned}\Delta_c(x) &= i[\theta(x)\,\Delta_+(x) - \theta(-x)\,\Delta_-(x)]\\ &= \int\frac{d^4p}{(2\pi)^4}\,e^{-ip\cdot x}\,\frac{i}{p^2 - m^2 + i\varepsilon}\end{aligned} \tag{5.63}$$

From (5.62) it is clear that the definition (5.59) was not sufficiently accurate at the (ambiguous) point $x = 0$ to assure covariance. Discarding the non-covariant terms and calling the result a T^* product we find

$$\langle T^*\psi_\sigma(x)\psi_\tau^\dagger(x')\rangle_0 = \tilde{\Pi}_{\sigma\tau}(i\partial)\,\Delta_c(x) \tag{5.64}$$

Going through the same computation for the other fields gives

$$\begin{aligned}\langle T^*\chi_\sigma(x)\chi_\tau^\dagger(x')\rangle_0 &= \Pi_{\sigma\tau}(i\partial)\,\Delta_c(x)\\ \langle T\chi_\sigma(x)\psi_\tau^\dagger(x')\rangle_0 = \langle T\psi_\sigma(x)\chi_\tau^\dagger(x')\rangle_0 &= \delta_{\sigma\tau}m^{2j}\,\Delta_c(x - x')\end{aligned} \tag{5.65}$$

Defining momentum space propagators in analogy to (5.63):

$$\langle T^*\psi_\sigma(x)\psi_\tau^\dagger(x')\rangle_0 \equiv \int \frac{d^4p}{(2\pi)^4}\, e^{-ip\cdot(x-x')}\,\Delta_{\sigma\tau}^{\psi\psi}(p) \tag{5.66}$$

etc., gives

$$\begin{aligned} \Delta^{\psi\psi}(p) &= \frac{i\tilde{\Pi}(p)}{p^2 - m^2 + i\varepsilon} \\ \Delta^{\chi\chi}(p) &= \frac{i\Pi(p)}{p^2 - m^2 + i\varepsilon} \\ \Delta^{\psi\chi}(p) = \Delta^{\chi\psi}(p) &= \frac{im^{2j}}{p^2 - m^2 + i\varepsilon} \end{aligned} \tag{5.67}$$

Further discussion of these questions may be found in ref. 2.

5.4 PARITY, TIME REVERSAL, AND CHARGE CONJUGATION

The discrete transformations associated with parity, time reversal and charge conjugation are of basic importance and are treated in all basic textbooks on field theory. An extensive treatment of these topics, with special attention to internal symmetries, is given in Chapter 6. In this section we give a brief discussion, emphasizing the interesting relations among the $D^{(j,0)}$ and $D^{(0,j)}$ fields revealed by a study of the discrete transformations. We begin by considering the transformation of particle states under the discrete transformations P, T, and C. It is then noticed that the phases occurring in the particle operator transformations are restricted by the requirement that the transformed field have causal commutators with the original field.

The parity transformation is defined by†

$$\begin{aligned} Pa(\mathbf{p}, \sigma)P^{-1} &= \eta_P a(-\mathbf{p}, \sigma) \\ Pb(\mathbf{p}, \sigma)P^{-1} &= \eta_P'^* b(-\mathbf{p}, \sigma) \end{aligned} \tag{5.68}$$

where $|\eta_p| = |\eta_p'| = 1$. We recall that the index σ referred to fixed-axis spin quantization in the development of Sec. 5.3. (Since space inversions commute with rotations, P does not change σ.) The rule (5.68) should be contrasted with the transformation law (6.125) for helicity particle operators. If we use

† Here and elsewhere we assume that the vacuum is invariant under P, C, and T; $P\,|0\rangle = |0\rangle$, etc.

(5.68) to compute $P\psi_\sigma(x)P^{-1}$ from (5.57) we find

$$P\psi_\sigma(x)P^{-1} = \frac{m^{2j}}{(2\pi)^{\frac{3}{2}}} \int \frac{d^3p}{2\omega_p} \{\eta_P D^{(0,j)}_{\sigma\tau}(L(p))a(p, \sigma)e^{-ip\cdot x_p} + \eta'_P(D^{(0,j)}(L(p))C^{-1})_{\sigma\tau} b^*(p, \sigma)e^{ip\cdot x_p}\} \quad (5.69)$$

where $x^\mu_P = (x^0, -\mathbf{x})$. In order to derive (5.69) we have changed $\mathbf{p} \to -\mathbf{p}$ and noted that for the pure boost matrices $D^{(j,0)}(L(-\mathbf{p})) = D^{(0,j)}(L(\mathbf{p}))$. It will be noticed that (5.69) is very similar to $\chi_\sigma(x_P)$ (cf. Eq. (5.57)), in fact if we write

$$\eta'_P = (-1)^{2j}\eta_P \quad (5.70)$$

then (5.69) may be written in the form

$$P\psi_\sigma(\mathbf{x}, t)P^{-1} = \eta_P\chi_\sigma(-\mathbf{x}, t) \quad (5.71$$

A similar calculation leads to

$$P\chi_\sigma(\mathbf{x}, t)P^{-1} = \eta_P\psi_\sigma(-\mathbf{x}, t) \quad (5.72)$$

It is also apparent that any relation differing from (5.70) would lead to fields $P\psi P^{-1}$ not having a causal commutator with $\psi^\dagger$. At present no one knows how to build causal relativistic theories in the presence of such a transformation.

It is not always true that (5.68) is a physical transformation (for example a left handed neutrino has no right handed counterpart). In that case it is perfectly consistent to have a field of type $D^{(j,0)}$ without necessitating its parity-induced counterpart $D^{(0,j)}$.

In order to describe time-inversion we introduce an antiunitary operator T whose effect on the angular momentum $\mathbf{J}$ and rotation operator $R = \exp(i\theta\mathbf{n}\cdot\mathbf{J})$ is given by

$$\begin{aligned} T\mathbf{J} &= -\mathbf{J}T \\ TR &= RT \end{aligned} \quad (5.73)$$

From these equations we learn that $T\,|p\sigma\rangle$ has $J_z = -\sigma$ and that this set of states transforms as the complex conjugate representation of R:

$$\begin{aligned} RT\,|p\sigma\rangle &= TR\,|p\sigma\rangle \\ &= \sum_{\sigma'} D^*_{\sigma'\sigma}(R)T\,|p\sigma'\rangle \end{aligned} \quad (5.74)$$

Thus we are led (recall Eqs. (5.13)–(5.15)) to the particle operator transformation law

$$Ta(\mathbf{p}, \sigma)T^{-1} = \eta_T \sum_{\sigma'} C_{\sigma\sigma'} a(-\mathbf{p}, \sigma')$$
$$Tb(\mathbf{p}, \sigma)T^{-1} = \eta_T'^{*} \sum_{\sigma'} C_{\sigma\sigma'} b(-\mathbf{p}, \sigma') \tag{5.75}$$

Equations (5.75) now allow one to compute $T\psi_\sigma(x)T^{-1}$:

$$T\psi_\sigma(x)T^{-1} = \frac{m^{2j}}{(2\pi)^{\frac{3}{2}}} \int \frac{d^3p}{2\omega_p} \{\eta_T (D^{(j,0)*}(L(-\mathbf{p}))C)_{\sigma\tau} a(p, \tau) e^{-ip\cdot x_T} + \eta_T' D^{(j,0)*}_{\sigma\tau}(L(-\mathbf{p}))b^*(\mathbf{p}, \tau)e^{ip\cdot x_T}\} \tag{5.76}$$

where $x_T^\mu = (-x_0, \mathbf{x})$. Since $D^{(j,0)}(L(\mathbf{p})) = \exp(\zeta\hat{p}\cdot\mathbf{J}) = D^{(0,j)}(L(-\mathbf{p}))$ we can use $C\mathbf{J}C^{-1} = -\mathbf{J}^*$ (this follows from (5.13)) to establish the identity

$$D^{(j,0)}(L(-\mathbf{p}))^* = CD^{(j,0)}(L(\mathbf{p}))C^{-1} \tag{5.77}$$

This allows Eq. (5.76) to be written as

$$T\psi_\sigma(\mathbf{x}, t)T^{-1} = \eta_T \sum C_{\sigma\tau}\psi_\tau(\mathbf{x}, -t) \tag{5.78}$$

provided the phase η_T' is given by

$$\eta_T' = \eta_T \tag{5.79}$$

Again, other phase choices would not lead to an acceptable time inversion operator.

The field χ_σ transforms as

$$T\chi_\sigma(\mathbf{x}, t)T^{-1} = \eta_T \sum_\tau C_{\sigma\tau}\chi_\tau(\mathbf{x}, -t) \tag{5.80}$$

Charge conjugation simple replaces particle by antiparticle without changing the spin or momentum (provided such antiparticles exist!):

$$Ca(p, \sigma)C^{-1} = \eta_C b(p, \sigma)$$
$$Cb(p, \sigma)C^{-1} = \eta_C'^{*} a(p, \sigma) \tag{5.81}$$

A simple calculation analogous to those leading to (5.71) and (5.78) establishes the following relations:

$$C\psi_\sigma(x)C^{-1} = \eta_C \sum_\tau C^{-1}_{\sigma\tau}\chi_\tau^\dagger(x)$$
$$C\chi_\sigma(x)C^{-1} = (-1)^{2j} \sum_\tau (C^{-1})_{\sigma\tau}\psi_\tau^\dagger(x) \tag{5.82}$$

$$\eta_C = \eta_C' \tag{5.83}$$

It will be noted that the operations CP and T do not change fields of the type ψ into those of type χ.

If one is dealing with a self-conjugate field ($a(p, \sigma) = b(p, \sigma)$) then further restrictions on the phases η_C, η_P, η_T follow immediately. Setting $a = b$ in Eqs. (5.68), (5.75), and (5.81) shows that

$$\begin{aligned} \eta_C^* &= \eta_C' \\ \eta_P^* &= \eta_P' \\ \eta_T^* &= \eta_T' \end{aligned} \tag{5.84}$$

Combining these relations with the causal restrictions (5.70), (5.79), and (5.83) gives

$$\begin{aligned} \eta_P^2 &= (-1)^{2j} \\ \eta_C^2 &= 1 \\ \eta_T^2 &= 1 \end{aligned} \tag{5.85}$$

In Chapter 6 these considerations are extended to fields "carrying" internal symmetries.

In theories with parity conservation it is useful to combine the fields ψ, χ into a single $2(2j + 1)$-component field Ψ

$$\Psi = \begin{pmatrix} \psi \\ \chi \end{pmatrix} \tag{5.86}$$

in exact analogy to the construction of the four-component Dirac field in Chapter 3. Under Lorentz transformations Ψ transforms as

$$\begin{aligned} U^{-1}(\Lambda)\Psi(x)U(\Lambda) &= \mathscr{D}^j(\Lambda)\Psi(\Lambda^{-1}x) \\ \mathscr{D}^j(\Lambda) &= \begin{pmatrix} D^{(j,0)}(\Lambda) & 0 \\ 0 & D^{(0,j)}(\Lambda) \end{pmatrix} \end{aligned} \tag{5.87}$$

i.e. as a direct sum $(j, 0) \oplus (0, j)$. Under space inversion the representation is not reducible, however, since

$$\begin{aligned} P\Psi(\mathbf{x}, t)P^{-1} &= \eta_P \gamma_0 \Psi(-\mathbf{x}, t) \\ \gamma_0 &= \begin{pmatrix} 0 & 1 \\ 1 & 0 \end{pmatrix} \end{aligned} \tag{5.88}$$

in analogy to (3.55). We can also construct the adjoint $\bar{\Psi} = \Psi^\dagger \gamma_0$ and build scalars $\bar{\Psi}\Psi$, etc.

Other interesting generalizations of the spin $\frac{1}{2}$ case are easily derived. For example, Eqs. (3.49), written in the form

$$\begin{aligned} \Pi(i\partial)\xi &= m\chi \\ \tilde{\Pi}(i\partial)\chi &= m\xi \end{aligned} \tag{5.89}$$

suggest the generalization

$$\begin{aligned} \Pi(i\partial)\psi &= m^{2j}\chi \\ \tilde{\Pi}(i\partial)\chi &= m^{2j}\psi \end{aligned} \tag{5.90}$$

Equations (5.90) are easily verified. Application of $\Pi(i\partial)$ to $\psi(x)$ has the effect of placing $\Pi(p)$ under the integral sign in Eq. (5.57). Using Eq. (5.50) and $D^{(j,0)}(L(p))D^{(0,j)}(L(p)) = 1$ we get the first of Eqs. (5.90). The second relation is derived in the same way.

Defining generalized gamma-matrices $\gamma^{\mu_1\cdots\mu_{2j}}$ by

$$\gamma^{\mu_1\cdots\mu_{2j}} \equiv \begin{pmatrix} 0 & \tilde{t}^{\mu_1\cdots\mu_{2j}} \\ t^{\mu_1\cdots\mu_{2j}} & 0 \end{pmatrix} \tag{5.91}$$

and using (5.50) we may write the pair of Eqs. (5.90) in the form

$$\gamma^{\mu_1\cdots\mu_{2j}}(i\partial_{\mu_1})\cdots(i\partial_{\mu_{2j}})\Psi = m^{2j}\Psi \tag{5.92}$$

Finally we give the covariant propagator $\langle T^*\Psi_\sigma(x)\Psi_\tau^\dagger(x')\rangle_0$ for the field Ψ This is a $[2(2j+1)]^2$ matrix whose form in momentum space is

$$\begin{gathered} \frac{iM(p)}{p^2 - m^2 + i\epsilon} \\ M(p) = \begin{pmatrix} m^{2j} & \tilde{\Pi}(p) \\ \Pi(p) & m^{2j} \end{pmatrix} \end{gathered} \tag{5.93}$$

REFERENCES

1. H. Joos, *Fortschritte d. Physik* **10,** 65 (1962).
2. S. Weinberg, *Phys. Rev.* **133,** B1318 (1964).
3. S. Weinberg, *Phys. Rev.* **134,** B882 (1964).
4. V. Bargmann and E. P. Wigner, *Proc. Nat. Acad. Sci.* **34,** 211 (1946).
5. T. Araki, *J. Math. Phys.* **2,** 267 (1961).
6. R. Jost, "The General Theory of Quantized Fields" (American Mathematical Society, Providence, R.I., 1965).

7. E. P. Wigner, "Group Theory" (Academic Press, New York, 1959), p. 288.
8. G. Feldman and P. T. Matthews, *Ann. Phys.* (*N.Y.*) **40,** 19 (1966).
9. L. M. Brown, *Phys. Rev.* **111,** 957 (1958).
10. R. F. Streater and A. S. Wightman, "PCT, Spin and Statistics and All That" (W. A. Benjamin, New York, 1964).
11. W. Pauli, *Rev. Mod. Phys.* **13,** 203 (1941).
12. W. Pauli, *Prog. Theor. Phys.* **5,** 526 (1950).
13. G. Feldman and P. T. Matthews, *Phys. Rev.* **151,** 1176 (1966).
14. A. Messiah and O. W. Greenberg, *Phys. Rev.* **136,** B248 (1964).

6

INTERNAL SYMMETRIES, ANTIPARTICLE CONJUGATION, AND SPACE-TIME INVERSIONS

6.1 INTRODUCTION

The remarkable connection between the mathematical operation of Hermitian conjugation of fields and the particles and antiparticles described by those fields has long been appreciated. The simplest expression of this relation occurs for a single "real" (Hermitian) or "complex" (non-Hermitian) scalar field. Only one type of quantum is associated with the real field, while the complex field, having twice as many degrees of freedom, describes particles which carry opposite charges associated with changes of phase of the field. The existence of interactions apparently invariant under transformations of various continuous internal symmetry groups suggests that various sets of fields be considered together as a basis for irreducible representations of the group in question. The antiparticle concept is considerably enriched by consideration of the relations between antiparticle conjugation and internal symmetry transformations.[1–5] The natural generalizations of the primitive distinctions between real and complex fields are (1) real $\rightarrow$ self-conjugate multiplet (2) complex $\rightarrow$ pair-conjugate multiplet. The self-conjugate field is characterized by a "generalized reality condition," which relates the adjoint of a field component linearly to another field component. The term multiplet refers to a set of particle states (or the associated field, fundamental or not) associated with an irreducible representation of the internal symmetry group. When the antiparticles of the members of a representation constitute a distinct multiplet we call the multiplet in question pair-conjugate (P-C.) When the antiparticles of the particles belong to the same multiplet we call the multiplet in question self-conjugate (S-C).

It should be stressed that the distinction between S-C and P-C multiplets depends upon the symmetry group in question. For example when the symmetry is $SU(2)_I \times U(1)_Y$ (isospin-hypercharge) the pion triplet is self-conjugate but the $\bar{K}$ doublet is pair-conjugate with respect to the K doublet. (The K and $\bar{K}$ doublets are distinguished by their hypercharge.) However, if the symmetry group is enlarged to $SU(3)$, the group operations transform K's into $\bar{K}$'s, $\bar{K}$'s into K's, etc. so that the set of 8 mesons (π, K, $\bar{K}$, η) comprising the regular representation of $SU(3)$ can be regarded as self-conjugate. Also note that in the presence of electromagnetism isospin symmetry is broken $(m(\pi^0) \neq (m(\pi^{\pm}))$ and the pion triplet cannot strictly be regarded as self conjugate. We describe the $SU(2)$ symmetry associated with isospin conservation in great detail.

The properties of states and fields transforming as various representations of the approximate symmetry group $SU(3)$ have been described in ref. 6.

In order to exhibit clearly the structural relations of fields which "carry" internal symmetries we first treat spin zero fields. Then the arbitrary spin case is analyzed in Sec. (6.9) using the Rarita-Schwinger description of arbitrary spin particles. This analysis follows closely the development given in ref. 5. Special attention is given to the notion of the self-conjugate field. It is noted that such fields are local only for integral isospins.[4,5] Antiparticle conjugation is studied in Sec. (6.3) while the parity and time-reversal transformations are given in Sec. (6.4) and Sec. (6.5). These concepts can be subjected to considerable refinement. The reader is referred to refs. 1–3 for a description of some interesting recent work along these lines. In Sec. (6.6) we describe the properties of various composite transformations. The results are essentially the same as when internal symmetry is omitted completely.[7] Constraints among phase factors in the case of S-C fields are noted. In particular it is noted that self-conjugate isospinor fields change sign under transformation by the operator $(TCP)^2$, in contradiction to the TCP theorem[8–11] These unusual, unacceptable properties of certain types of self-conjugate fields are discussed in Sec. (6.8), where we study the restrictions on the allowed representations of internal symmetry groups carried by fields. The result[4,5,8–15] may be stated as follows: *only potentially real representations of self-conjugate internal symmetry multiplets are allowed by the postulates of local relativistic field theory.* We tacitly understand that the representations considered are irreducible. A representation is potentially real if there exists a non-singular transformation to a basis for which the representation matrices are real.[16] For $SU(2)$ this means that isospinor representations $(2I = \text{odd})$

are not allowed for self-conjugate particles (i.e. the K mesons *have* to carry a quantum number distinct from isospin). In $SU(3)$ only $p = q$ representations $D(p, q)$ are permitted for S-C multiplets.

The theorem states that we can always represent self-conjugate multiplets by a collection of Hermitian fields. This result seems obvious because it was previously always tacitly *assumed*, and therefore the question never arose in Lagrangian-oriented field theory, in which one cannot construct self-conjugate isospinor fields as a local superposition of real fields. The question is not trivial since we can give an explicit construction of a free self-conjugate isospinor field.

6.2 SPIN ZERO ISOSPIN MULTIPLETS: PARTICLE AND FIELD OPERATORS

To introduce our ideas we follow a phenomenological, particle-oriented approach. Only zero-spin bosons are considered, in order to expose the structure of the theory in its simplest form. The arbitrary spin case is treated in Sec. (6.9). We consider two types of isospin multiplets: (1) *pair-conjugate multiplets*, in which the antiparticles of the members of an isospin multiplet constitute a distinct isospin multiplet; and (2) *self-conjugate multiplets*, in which the antiparticles belong to the same isospin multiplet. We shall abbreviate these types by *PCM* and *SCM*, respectively. The K mesons (K^+, K^0), $(\bar{K}^0, K^-)$ are the most familiar case of a *PCM*, while the pions (π^+, π^0, π^-) are a good example of an *SCM*.

A. Particle aspects

To describe a *PCM* we introduce $2(2I + 1)$ boson operators $a_\alpha^*(k)$ and $a_\alpha(k)$ to create and annihilate the "particles" and $2(2I + 1)$ independent boson operators $b_\alpha^*(k)$ and $b_\alpha(k)$ to create and annihilate the "antiparticles." Here α denotes the I_3 eigenvalue (a_α^*, b_α^* increase I_3 by α) and hence runs from $-I$ to $+I$ in integral steps. We shall omit the four-vector k, whenever convenient.

$$\begin{aligned} [a_\alpha(k), a_\beta^*(k')] &= [b_\alpha(k), b_\beta^*(k')] = \delta_{\alpha\beta}\delta_{kk'} \\ [a_\alpha(k), a_\beta(k')] &= [b_\alpha(k), b_\beta(k')] = 0 \\ [a_\alpha(k), b_\beta(k')] &= [a_\alpha(k), b_\beta^*(k')] = 0 \end{aligned} \tag{6.1}$$

We employ box normalization with unit volume. It is understood (unless otherwise noted) that k is a physical four-vector with $k_0 \geqslant m$, m being the meson mass. From our point of view it is quite arbitrary* which multiplet is called "particles" and which "antiparticles." A full understanding of this distinction requires a detailed study of the charge conjugation (preferably antiparticle conjugation) operation. This theory is given in Sec. 6.3.

Thus for each k we have particle states $|I\alpha\rangle$ and antiparticle states $|\bar{I}\alpha\rangle$ defined by

$$\begin{aligned} |I\alpha\rangle &= a_\alpha^* \,|0\rangle \\ |\bar{I}\alpha\rangle &= b_\alpha^* \,|0\rangle \end{aligned} \tag{6.2}$$

where $|0\rangle$ denotes the vacuum state. In order to discuss conveniently the group properties of the theory we require the various $|I\alpha\rangle$ to be related to each other by isospin operators chosen to obey the standard (Condon-Shortley) phase relation.[17] (For a given α, the phase of a_α or b_α can be arbitrarily adjusted.) With the standard definition $I_\pm = I_1 \pm iI_2$, we have

$$\begin{aligned} I_3\,|I\alpha\rangle &= \alpha\,|I\alpha\rangle \\ I_3\,|\bar{I}\alpha\rangle &= \alpha\,|\bar{I}\alpha\rangle \\ I_\pm\,|I\alpha\rangle &= \Gamma_\pm(\alpha)\,|I, \alpha \pm 1\rangle \\ I_\pm\,|\bar{I}\alpha\rangle &= \Gamma_\pm(\alpha)\,|\bar{I}, \alpha \pm 1\rangle \\ \Gamma_\pm(\alpha) &= [(I \mp \alpha)(I \pm \alpha + 1)]^{\frac{1}{2}} \\ \Gamma_\pm(\alpha \mp 1) &= \Gamma_\mp(\alpha) \end{aligned} \tag{6.3}$$

An explicit construction of these operators is easily given:

$$\begin{aligned} I_\pm &= \sum_{k,\alpha} \Gamma_\mp(\alpha)[a_\alpha^*(k)a_{\alpha\mp1}(k) + b_\alpha^*(k)b_{\alpha\mp1}(k)] \\ I_3 &= \sum_{k,\alpha} \alpha[a_\alpha^*(k)a_\alpha(k) + b_\alpha^*(k)b_\alpha(k)] \end{aligned} \tag{6.4}$$

Using the last of relations (6.3) one easily verifies that $(I_\pm)^* = I_\mp$. A simpler form of the isospin operators is given as follows. Define $(2I + 1)$ component vectors

$$\begin{aligned} \mathbf{a}(k) &= \text{col}\,[a_\alpha(k)] \\ \mathbf{b}(k) &= \text{col}\,[b_\alpha(k)] \end{aligned} \tag{6.5}$$

Then, if t_i is the usual $(2I + 1) \times (2I + 1)$ isospin matrix, we have

$$\mathbf{I} = \sum_k [\mathbf{a}^\dagger(k)t_i\mathbf{a}(k) + \mathbf{b}^\dagger(k)t_i\mathbf{b}(k)] \tag{6.6}$$

---

* Once having made a choice, one must adhere to it in order to avoid phase errors.

We cite the obvious commutation relations among the I_i, a, and b:

$$\begin{aligned} [I_i, I_j] &= i\epsilon_{ijk} I_k \\ [a_\alpha(k), I_3] = \alpha a_\alpha(k), &\qquad [b_\alpha(k), I_3] = \alpha b_\alpha(k) \\ [a_\alpha(k), I_\pm] &= \Gamma_\mp(\alpha) a_{\alpha\mp 1}(k) \\ [b_\alpha(k), I_\pm] &= \Gamma_\mp(\alpha) b_{\alpha\mp 1}(k) \end{aligned} \tag{6.7}$$

These equations show that the a_α^* or the states $|I\alpha\rangle$ constitute a standard basis for the irreducible representation $D^{(I)}$ of the isospin group $SU(2)$. Introducing the unitary operator $O(\boldsymbol{\lambda})$ we find

$$\begin{aligned} O(\boldsymbol{\lambda}) &\equiv \exp(i\boldsymbol{\lambda} \cdot \mathbf{I}) \\ O(\boldsymbol{\lambda})\,|I\alpha\rangle &= \sum_\beta |I\beta\rangle D_{\beta\alpha}(\boldsymbol{\lambda}) \end{aligned} \tag{6.8}$$

where the D's are representation matrices $\exp(i\boldsymbol{\lambda} \cdot \mathbf{t})$ in the notation of Edmonds[18]. The antiparticle states $|\bar{I}\alpha\rangle$ transform exactly as $|I\alpha\rangle$ under $0(\boldsymbol{\lambda})$.

The operators a_α^*, a_α must transform as

$$\begin{aligned} O(\boldsymbol{\lambda}) a_\alpha^* O(\boldsymbol{\lambda})^{-1} &= \sum_\beta a_\beta^* D_{\beta\alpha}^{(I)}(\boldsymbol{\lambda}) \\ O^{-1}(\boldsymbol{\lambda}) a_\alpha O(\boldsymbol{\lambda}) &= \sum_\beta D_{\alpha\beta}^{(I)}(\boldsymbol{\lambda}) a_\beta \end{aligned} \tag{6.9}$$

The antiparticle operators b_α^*, b_α transform exactly as a_α^*, a_α.

For an *SCM* the antiparticle multiplet is not distinct, and so there are only $2(2I + 1)$ operators a_α^*, a_α. The description of the *SCM* is obtained from that of the *PCM* by systematically setting b equal to zero in Eqs. (6.1)–(6.9).

The four-momentum operators are *defined* to be

$$\begin{aligned} P_\mu &= \sum_{k\alpha} k_\mu (a_\alpha^*(k) a_\alpha(k) + b_\alpha^*(k) b_\alpha(k)), \qquad (PC) \\ P_\mu &= \sum_{k\alpha} k_\mu a_\alpha^*(k) a_\alpha(k) \qquad (SC) \end{aligned} \tag{6.10}$$

It follows from (6.7) that P_μ commutes with the isospin operator. This was tacitly assumed in the labeling of the state vectors by k, I, and α. The states (or creation operators) are assumed to have the standard behavior under transformations of the Poincaré group.

B. Construction of field operators

In order to investigate the local behavior of a quantum-mechanical system in space and time, we introduce field operators. We require that the field operators transform according to irreducible representations of the various invariance groups under consideration ($SU(2)$, Poincaré group, etc.). From such fields we can systematically construct all observables. Although the field itself is generally taken to be the primary concept, we prefer to regard it as the (essentially) unique construction possessing the desired group transformation properties simultaneously in the space-time coordinates χ and the space of the internal symmetries (isospin, in this case). Having constructed such fields, we can investigate the consistency of various assumptions common to relativistic field theories, such as locality, causality, etc.

The positive-energy wave functions are†

$$f_k(x) = (2\omega)^{-\frac{1}{2}} e^{-ik\cdot x} \tag{6.11}$$

where $k \cdot x = \omega x^0 - \mathbf{k} \cdot \mathbf{x}$ defines our metric. The f_k are orthogonal within the usual Klein-Gordon inner product

$$(f_{k'}, f_k) \equiv i \int d^3x f^*_{k'}(x) \overleftrightarrow{\partial}_0 f_k(x) = \delta_{kk'} \tag{6.12}$$

the integration running over a box of unit volume.

There are four independent operators transforming appropriately under Lorentz transformations and according to isospin rotations:

$$\begin{aligned} \chi_{1\alpha} &= \sum_k a_\alpha(k) f_k(x), \quad & \chi_{2\alpha} &= \sum_k b_\alpha(k) f_k(x) \\ \chi^*_{1\alpha} &= \sum_k a^*_\alpha(k) f^*_k(x), \quad & \chi^*_{2\alpha} &= \sum_k b^*_\alpha(k) f^*_k(x) \end{aligned} \tag{6.13}$$

A priori one might expect quantities like $\sum_k a_\alpha(k) f^*_k(x)$, but these do not behave properly under translations as do $\chi_{i\alpha}(x)$, $\chi^*_{i\alpha}(x)$:

$$[\chi_{i\alpha}(x), P_\mu] = i\, \partial_\mu \chi_{i\alpha}(x) \tag{6.14}$$

$\chi_{i\alpha}(x)$ transforms as $D^{(I)*}$, $\chi^*_{i\alpha}$ as $D^{(I)}$, under isospin rotations. This property follows trivially from (6.9).

For reasons of causality, theories are not generally described separately in terms of the $\chi_{i\alpha}$ but in terms of linear combinations of them. Clearly a superposition $\alpha\chi_1 + \beta\chi_2$ will be noncausal. Thus judicious superposition of pairs of the type (χ_1, χ_1^*), (χ_2, χ_2^*), (χ_1, χ_2^*), (χ_2, χ_1^*) with coefficients of equal magnitude is called for.

Since $\chi_{i\alpha}$ and $\chi^*_{i\alpha}$ transform only equivalently and not identically under

---

† In the present chapter the normalization has been chosen differently from that of Chapter V.

$SU(2)$, we must transform the basis vectors in a well-known way.[16] $\chi_{i\alpha}$ transforms as $D^{(I)*}$, so $\chi'_{i\alpha}$, defined by

$$\chi'_{i\alpha} = \sum_{\beta} C_{\alpha\beta}\chi_{i\beta} = \eta^*_\alpha \chi_{i-\alpha} \tag{6.15}$$

$$\begin{aligned} C_{\alpha\beta} &= \eta^*_\alpha \delta_{\alpha,-\beta} \\ \eta_\alpha &= \xi(-1)^{I-\alpha}, \qquad |\xi| = 1 \end{aligned} \tag{6.16}$$

transforms as $D.^{(I)}$ The phase factor ξ is arbitrary, but independent of α.

The properties of η_α are summarized by

$$\begin{aligned} \eta^*_\alpha \eta_\alpha &= 1 \\ \eta^*_\alpha \eta_{-\alpha} &= (-1)^{2I} \end{aligned} \tag{6.17}$$

We now construct four fields transforming as $D^{(I)}$

$$\begin{aligned} \varphi^*_{1\alpha} &= \chi^*_{1\alpha} + \chi'_{1\alpha} = \chi^*_{1\alpha} + \eta^*_\alpha \chi_{1-\alpha} \\ \varphi^*_{2\alpha} &= \chi^*_{2\alpha} + \chi'_{2\alpha} = \chi^*_{2\alpha} + \eta^*_\alpha \chi_{2-\alpha} \\ \psi^*_{1\alpha} &= \chi^*_{1\alpha} + \chi'_{2\alpha} = \chi^*_{1\alpha} + \eta^*_\alpha \chi_{2-\alpha} \\ \psi^*_{2\alpha} &= \chi^*_{2\alpha} + \chi'_{1\alpha} = \chi^*_{2\alpha} + \eta^*_\alpha \chi_{1-\alpha} \end{aligned} \tag{6.18}$$

φ_1 and φ_2 are completely independent because of Eqs. (6.1), so we need only consider one of them and drop the identifying index. We thus obtain the field operator for *SC* multiplets:

$$\begin{aligned} \varphi_\alpha(x) &= \sum_k (a_\alpha(k) f_k(x) + \eta_\alpha a^*_{-\alpha}(k) f^*_k(x)) \\ O^{-1}(\boldsymbol{\lambda})\varphi_\alpha(x) O(\boldsymbol{\lambda}) &= \sum_\beta D_{\alpha\beta}(\boldsymbol{\lambda})\varphi_\beta(x) \end{aligned} \tag{6.19}$$

Either set, $\{\psi_{1\alpha}, \psi^*_{1\alpha}\}$ or $\{\psi_{2\alpha}, \psi^*_{2\alpha}\}$ may be used to describe *PC* particles. Arbitrarily choosing (as in Eq. (6.2)) the operators a, a^* to correspond to the particles, we take $\psi_\alpha \equiv \psi_{1\alpha}$ to destroy "particles":

$$\psi_\alpha(x) = \sum_k (a_\alpha(k) f_k(x) + \eta_\alpha b^*_{-\alpha}(k) f^*_k(x)) \tag{6.20}$$

$\psi_\alpha(x)$ transforms identically to $\varphi_\alpha(x)$ in Eq. (6.19).

The set $\{\psi_{2\alpha}, \psi^*_{2\alpha}\}$ is not independent of the set $\{\psi_{1\alpha}, \psi^*_{1\alpha}\}$. This is shown by the relation

$$\psi_{2,-\alpha}(x) = \eta_{-\alpha}(\chi^*_{1\alpha}(x) + (-1)^{2I}\eta^*_2 \chi_{2-\alpha}(x)) \tag{6.21}$$

For integral I, the right-hand side of (6.21) is just $\eta_{-\alpha}\psi^{\alpha *}$. For half-integral I, we can obtain equivalence by changing the sign of the b operators in the ψ_2 fields. Note that the independence of the two pieces of the field allows us to change the phase of the operators in this way. Hence the theories described by $\{\psi_{1\alpha}, \psi^*_{1\alpha}\}$ or $\{\psi_{2\alpha}, \psi^*_{2\alpha}\}$ are identical.

Thus *PC* bosons are described by the $2(2I + 1)$ independent operators $\{\psi_\alpha, \psi^*_\alpha\}$. It therefore seems strange that the *SC* bosons, which entail half as many degrees of freedom, involve the same number of field variables $\{\varphi_\alpha, \varphi^*_\alpha\}$. The latter set is not independent, however, since $\varphi_{-\alpha}$ is *not* independent of φ^*_α:

$$\varphi_{-\alpha}(x) = \eta_{-\alpha}(\chi^*_{1\alpha}(x) + (-1)^{2I}\eta^*_\alpha \chi_{1-\alpha}(x)) \tag{6.22}$$

For integral I we have the expected proportionality

$$\varphi_{-\alpha}(x) = \eta_{-\alpha}\varphi^*_\alpha(x), \qquad I = 0, 1, 2, \ldots \tag{6.23}$$

For half-integral I, however, the relation is not local because of the minus sign.

The isotopic spin operators are expressed by means of the conventional isospin matrices t_i as

$$I_i = i \int d^3x \sum_{\alpha\beta} \psi^*_\alpha(x) \overleftrightarrow{\partial_0} (t_i)_{\alpha\beta} \psi_\beta(x) \qquad (PC) \tag{6.24}$$

$$I_i = \tfrac{1}{2} i \int d^3x \sum_{\alpha\beta} \varphi^*_\alpha(x) \overleftrightarrow{\partial_0} (t_i)_{\alpha\beta} \varphi_\beta(x) \qquad (SC) \tag{6.25}$$

One can easily check the commutation relations of these operators from those of the fields given in the next section.

C. Connection between locality and isospin for self-conjugate fields

The requirement that the field operators transform irreducibly under $SU(2)$ inevitably brings all components of the field into the theory. In particular, various observables, transforming as some tensor, need all components of the fields for their construction. In the case of *SC* fields the requirement of $SU(2)$ invariance thus introduces redundant fields into the theory as shown by Eq. (6.22). For integral isospin these dependent fields can be eliminated using the local relation (6.23). However, as we now show, this process is non-local for half-integral isospins. Thus isospin invariants like $\sum_\alpha \varphi^*_\alpha(x)\varphi_\alpha(x)$, which appear local, are non-local when expressed in terms of independent fields (e.g. with $\alpha > 0$).

We recall the definitions of the well-known functions

$$\begin{aligned} i\,\Delta_+(x-y) &= \sum_k f_k(x) f_k^*(y) \\ i\,\Delta_-(x-y) &= -\sum_k f_k^*(x) f_k(y) \\ \Delta(x) &= \Delta_+(x) + \Delta_-(x) \\ \Delta^{(1)}(x) &= i(\Delta_+(x) - \Delta_-(x)) \end{aligned} \tag{6.26}$$

$\Delta(x)$ vanishes for spacelike x, in contrast to $\Delta^{(1)}(x)$, which only decays exponentially outside the light cone. Recall that Δ is odd, $\Delta^{(1)}$ even, and that $\dot{\Delta}(t=0) = -\delta(\mathbf{x})$, $\Delta(t=0) = 0$, $\dot{\Delta}^{(1)}(t=0) = 0$. From (6.26) and the orthogonality relations (6.12), we find that

$$\begin{aligned} i\int d^3x' f_k^*(x')\overleftrightarrow{\partial_0'}\,\Delta(x'-x) &= -if_k^*(x) \\ i\int d^3x' f_k^*(x')\overleftrightarrow{\partial_0'}\,\Delta^{(1)}(x'-x) &= f_k^*(x) \end{aligned} \tag{6.27}$$

We thus find (cf. (6.13)) the Klein-Gordon inner products

$$\begin{aligned} (\chi_{1\alpha}(x'), \Delta(x'-x)) &= -i\chi_{1\alpha}^*(x) \\ (\chi_{1\alpha}(x'), \Delta^{(1)}(x'-x)) &= \chi_{1\alpha}^*(x) \end{aligned} \tag{6.28}$$

Thus, for integral I, (6.22) becomes

$$\begin{aligned} \varphi_{-\alpha}(x) &= i\eta_{-\alpha}([\chi_{1\alpha}(x') + \eta_\alpha^*\chi_{1-\alpha}^*(x')], \Delta(x'-x)) \\ \varphi_{-\alpha}(x) &= -\eta_{-\alpha}\int d^3x'\,\varphi_\alpha^*(x')\overleftrightarrow{\partial_0'}\,\Delta(x'-x) \qquad I = 0, 1, 2, \ldots \end{aligned} \tag{6.29}$$

Evaluating the integral at $t' = t$ recovers Eq. (6.23).

For half-integral I the extra minus sign is compensated for by using $\Delta^{(1)}$:

$$\begin{aligned} \varphi_{-\alpha}(x) &= \eta_{-\alpha}([\chi_{1\alpha}(x') + \eta_\alpha^*\chi_{1-\alpha}^*(x')], \Delta^{(1)}(x'-x)) \\ \varphi_{-\alpha}(x) &= i\eta_{-\alpha}\int d^3x'\,\varphi_\alpha^*(x')\overleftrightarrow{\partial_0'}\,\Delta^{(1)}(x'-x), \quad I = \tfrac{1}{2}, \tfrac{3}{2}, \tfrac{5}{2}, \ldots \end{aligned} \tag{6.30}$$

Thus $\varphi_{-\alpha}(x)$ is a non-local superposition of $\varphi_\alpha(x)$. Moreover, contributions come from spacelike regions $(x - x')^2 < 0$. Setting $t = t'$ in (6.30) gives

$$\varphi_{-\alpha}(x) = -i\eta_{-\alpha}\int d^3x'\,\dot{\varphi}_\alpha^*(\mathbf{x}', t)\,\Delta^{(1)}(\mathbf{x}' - \mathbf{x}, 0) \tag{6.31}$$

(It is clear that the time derivative supplies the requisite minus sign.) In this equation the entire integral is in a spacelike region relative to x!

For *PC* fields the particle-operator commutators (6.1) and definitions (6.26) give

$$\begin{aligned} [\psi_\alpha(x), \psi_\beta(x')] &= 0 \\ [\psi_\alpha(x), \psi_\beta^*(x')] &= i\delta_{\alpha\beta}\,\Delta(x - x') \end{aligned} \tag{6.32}$$

The momentum density conjugate to $\psi_\alpha(x)$ is $\dot{\psi}_\alpha^*(x)$ and the theory is entirely standard.

For *SC* fields the fact that the positive and negative frequency components of the field involve dependent operators brings about an entanglement for the case of half-integral isospin. We find the results

$$\begin{aligned} [\varphi_\alpha(x), \varphi_\beta^*(x')] &= i\delta_{\alpha\beta}\,\Delta(x - x') \\ [\varphi_\alpha(x), \varphi_\beta(x')] &= \delta_{\alpha,-\beta}\eta_\beta \sum_k [f_k(x)f_k^*(x') - (-1)^{2I} f_k^*(x) f_k(x')] \end{aligned} \tag{6.33}$$

Thus the second commutator is non-causal when $2I$ is an even integer:

$$\begin{aligned} [\varphi_\alpha(x), \varphi_\beta(x')] &= i\delta_{\alpha-\beta}\eta_\beta\,\Delta(x - x'); \qquad I = 0, 1, 2, \ldots \\ [\varphi_\alpha(x), \varphi_\beta(x')] &= \delta_{\alpha-\beta}\eta_\beta\,\Delta^{(1)}(x - x'); \qquad I = \tfrac{1}{2}, \tfrac{3}{2}, \tfrac{5}{2}, \ldots \end{aligned} \tag{6.34}$$

The non-commutativity of $\varphi_{-\alpha}$ with φ_α can be regarded as a consequence of the dependence of $\varphi_{-\alpha}$ on φ_α^*.

Thus locality restricts the isospin of self-conjugate fields to integral values. Such representations are equivalent to real ones generated by the irreducible Cartesian tensors, and a real basis can always be chosen. Taking $\xi = (-1)^I$ leads to the *SC* condition

$$\varphi_\alpha^*(x) = (-1)^\alpha \varphi_{-\alpha}(x), \qquad I \text{ integral} \tag{6.35}$$

which is the standard phase choice for spherical tensors.

The theoretical reasons for rejecting field theories of self-conjugate isofermions are very similar to those used to arrive at the spin-statistics connection. The similarity is quite striking in the analysis of ref. 19, which showed that quantization of bosons with anticommutators (or Fermions with commutators) led to non-causal commutators.

An instructive comparison can be made between the charged scalar theory and the $I = \frac{1}{2}$ *SCIF* theory. Both theories involve the same number of degrees of freedom. The free charged scalar theory is described by fields ϕ

and ϕ^* and a Hamiltonian H

$$\phi(x) = \sum_k (a_k f_k(x) + b_k^* f_k^*(x))$$
$$H = \sum_k \omega_k (a_k^* a_k + b_k^* b_k) \tag{6.36}$$

The same Boson operators a and b can be used to construct the isospinor $\phi_\pm = \phi_{\pm\frac{1}{2}}$ by identifying $a_{\frac{1}{2}} = a$, $a_{-\frac{1}{2}} = b$:

$$\phi_+ = \phi = \sum_k (a_k f_k + b_k^* f_k^*)$$
$$\phi_- = \sum_k (b_k f_k - a_k^* f_k^*) \neq \phi_+^* \tag{6.37}$$

(Here the phase η_α was chosen as $(-1)^{\frac{1}{2}-\alpha}$.) The energy is the same as given in Eq. (2.10). ϕ_- is not independent of ϕ_+^* but is not locally related to the latter. Note that the operators a, b of the charged scalar theory suffice to construct an isospin algebra

$$I_+ = \sum_k a_k^* b_k, \qquad I_- = \sum_k b_k^* a_k, \qquad I_3 = \tfrac{1}{2} \sum_k (a_k^* a_k - b_k^* b_k) \tag{6.38}$$

The energy is invariant under $\mathbf{I}$ in the charged scalar theory, though $\mathscr{H}(x)$ is not, and the fields ϕ, ϕ^* do not give a local representation of the isospin algebra. Thus the charged scalar theory is not isospin invariant. Only I_3 (which generates gauge transformations) leaves the charged scalar theory invariant.

6.3 ANTIPARTICLE CONJUGATION FOR ISOSPIN MULTIPLETS

Experience shows that many laws of nature are invariant under antiparticle conjugation. This seems to be the case for the strong and electromagnetic interactions. Even when this operation is not a symmetry, a proper formulation of the transformation is essential to study the asymmetry itself. Our definition is based on the type of transformation which leads to a symmetry, namely, that to the behavior of a particle state with four-momentum k, ($k^2 = m^2$), additive quantum numbers ν (such as I_3, charge, hypercharge, . . .) corresponds an identical evolution of an antiparticle state with four-momentum k and opposite additive quantum numbers $-\nu$. Here we only consider isospin variables, so that the set of variables describing a particle state is (k, I, α).

As in Sec. 6.2, we begin with the particle approach, defining a unitary operator C which maps particle operators into antiparticle operators, preserving the commutator algebra. Our discussion of the PC case is standard, except that a careful treatment is given of the phase factors which occur. By requiring that C map causal fields into causal fields (all having definite isospin-transformation properties), we can restrict the phase factors entering the theory. Virtues of different phase choices are discussed. G conjugation is also discussed. A general proof is given that the G parity of a self-conjugate, integral-isospin Boson is $\eta_c(-1)^I$, where η_c is the charge parity of the $I_3 = 0$ member of the multiplet.

The antiparticle transformation C maps $|I\alpha\rangle$ onto $|\bar{I} - \alpha\rangle$ and vice versa. To be useful, we also require that the algebra of the field operators ψ_α be preserved under the transformation.

The most general possibility is therefore

$$\begin{aligned} C\,|I\alpha\rangle &= \lambda_\alpha\,|\bar{I} - \alpha\rangle \\ C\,|\bar{I}\alpha\rangle &= \mu_\alpha\,|I - \alpha\rangle \end{aligned} \tag{6.39}$$

where λ_α and μ_α depend on α. The a's and b's have to transform as

$$\begin{aligned} Ca_\alpha^* C^{-1} &= \lambda_\alpha b_{-\alpha}^* \\ Cb_\alpha^* C^{-1} &= \mu_\alpha a_{-\alpha}^* \end{aligned} \tag{6.40}$$

We require that C be unitary, which preserves the operator structure and requires that

$$\begin{aligned} \lambda_\alpha^* \lambda_\alpha = \mu_\alpha^* \mu_\alpha &= 1 \\ [a_\alpha, a_\beta^*] \xrightarrow[C]{} \lambda_\alpha^* \lambda_\beta [b_{-\alpha}, b_{-\beta}^*] &= \delta_{\alpha\beta} \end{aligned} \tag{6.41}$$

Two applications of C yields

$$\begin{aligned} C^2 a_\alpha^* C^{-2} &= \lambda_\alpha \mu_{-\alpha} a_\alpha^* \\ C^2 b_\alpha^* C^{-2} &= \mu_\alpha \lambda_{-\alpha} b_\alpha^* \end{aligned} \tag{6.42}$$

From this we cannot yet conclude that $C^2 = \xi I$, since $\lambda_\alpha \mu_{-\alpha}$ is not 1, *a priori*. Consider the effect of C on ψ_α:

$$C\psi_\alpha(x)C^{-1} = \sum_k \lambda_\alpha^* b_{-\alpha}(k) f_k(x) + \eta_\alpha \mu_{-\alpha} a_\alpha^*(k) f_k^*(x) \tag{6.43}$$

We wish to formulate our theory in terms of fields with definite transformation properties under $SU(2)$. There are only four which will be local: ψ_α, ψ^*_α, $\psi_{-\alpha}$, $\psi^*_{-\alpha}$. Inspection shows that only ψ^*_α has the right structure to qualify:

$$\psi^*_\alpha(x) = \sum_k a^*_\alpha(k) f^*_k(x) + \eta^*_\alpha b_{-\alpha}(k) f_k(x) \tag{6.44}$$

Requiring $C\psi_\alpha C^{-1} = \zeta_\alpha \psi^*_\alpha$ gives the constraints

$$\begin{aligned} \zeta_\alpha &= \eta_\alpha \mu_{-\alpha} \\ \zeta_\alpha \eta^*_\alpha &= \lambda^*_\alpha \end{aligned} \tag{6.45}$$

Elimination of ζ_α gives

$$\mu_{-\alpha} = \lambda^*_\alpha, \qquad \zeta_\alpha = \eta_\alpha \lambda^*_\alpha \tag{6.46}$$

Therefore (6.42) simplifies to the desired form ($\lambda_\alpha \mu_{-\alpha} = 1$):

$$\begin{aligned} C^2 a^*_\alpha C^{-2} &= a^*_\alpha \\ C^2 b^*_\alpha C^{-2} &= b^*_\alpha \\ C^2 &= \xi I \end{aligned} \tag{6.47}$$

the last step following from Schur's lemma. ξ can now be shown to be unity. From Eqs. (6.46) and (6.39)

$$\begin{aligned} C\,|I\alpha\rangle &= \lambda_\alpha\,|\bar{I} - \alpha\rangle \\ C\,|\bar{I}\alpha\rangle &= \lambda^*_{-\alpha}\,|\bar{I} - \alpha\rangle \\ C^2\,|I\alpha\rangle &= \lambda_\alpha C\,|\bar{I} < \alpha\rangle = \lambda_\alpha \lambda^*_\alpha\,|I\alpha\rangle \\ &= |I\alpha\rangle \end{aligned} \tag{6.48}$$

To summarize, the preservation of the operator algebra and state vector norms require $C^{-1} = C^\dagger$. Further requiring that C map ψ_α linearly to ψ^*_α implied $C^2 = 1$, so $C = C^{-1}$:

$$\begin{aligned} C = C^{-1} &= C^\dagger \\ C a^*_\alpha C^{-1} &= \lambda_\alpha b^*_{-\alpha} \\ C b^*_\alpha C^{-1} &= \lambda^*_{-\alpha} a^*_{-\alpha} \end{aligned} \tag{6.49}$$

ψ_α transforms as

$$\begin{aligned} \psi^c_\alpha \equiv C\psi_\alpha C^{-1} &= \zeta_\alpha \psi^*_\alpha \\ [\psi^C_\alpha(x), \psi^{*C}_\beta(x')] &= \zeta_\alpha \zeta^*_\beta [\psi^*_\alpha(x), \psi_\beta(x')] \\ &= i\delta_{\alpha\beta}\,\Delta(x - x') \end{aligned} \tag{6.50}$$

Equation (6.50) verifies the preservation of the equal-time commutation relations under C.

A convenient choice of λ_α is simply $\eta_C\eta_\alpha$, where η_C is arbitrary. This makes ψ_α transform nicely, but the particle operators inherit a phase. In this case the transformation rules are

$$\begin{aligned} C\psi_\alpha C^{-1} &= \eta_C\psi_\alpha^* \\ Ca_\alpha^* C^{-1} &= \eta_C^*\eta_\alpha b_{-\alpha}^* \\ Cb_\alpha^* C^{-1} &= \eta_C\eta_{-\alpha}^* a_{-\alpha}^* \end{aligned} \tag{6.51}$$

The isospin operator transforms as

$$\begin{aligned} CI_kC^{-1} &= (\psi^*, t_k\psi^*) = -(\psi, \tilde{t}_k\psi) \\ CI_1C^{-1} &= -I_1 \\ CI_2C^{-1} &= +I_2 \\ CI_3C^{-1} &= -I_3 \end{aligned} \tag{6.52}$$

This result is also easily obtained from Eqs. (6.7). $\tilde{t}_k$ denotes the transpose of t_k. C is *not* the same as rotation by π about the 2 axis, although its effect on I is the same. This remark suggests[4] the introduction of the useful composite operation

$$G \equiv C \exp(i\pi I_2) \tag{6.53}$$

which is essentially charge conjugation but which has a simple effect on isospin states. Clearly, $R_2 = \exp(i\pi T_2)$ preceding C will give $\mathbf{I} \to \mathbf{I}$ under $G = CR_2$:

$$\begin{aligned} G\mathbf{I}G^{-1} &= \mathbf{I} \\ G &\equiv CR_2 \end{aligned} \tag{6.54}$$

Thus G maps isospin multiplets into antiparticle isospin multiplets.

Apart from a phase change, G conjugation simply interchanges the a and b operators. The SC field φ_α and PC field ψ_α transform as

$$\begin{aligned} G\varphi_\alpha G^{-1} &= \eta_C(-1)^{I+\alpha}\varphi_{-\alpha}^* \\ G\psi_\alpha G^{-1} &= \eta_C(-1)^{I+\alpha}\psi_{-\alpha}^* \end{aligned} \tag{6.55}$$

If we employ the standard choice (6.35) for SC fields we obtain $G\varphi_\alpha G^{-1} = \eta_C(-1)^I\varphi_\alpha$ so that the SC field ϕ has definite G parity $\eta_C(-1)^I$. (For such fields $\eta_C = \pm 1$.) The physical implications of associated selection rules are well known.

For self-conjugate multiplets the operators a and b coincide so that the phases on the right hand side of Eq. (6.51) are equal. Hence we find the constraint $\eta_C^2 = (-1)^{2I}\eta_\alpha^2$.

In the preceding analysis the phase factor η_α (or ξ) was kept general, apart from the compulsory variation $(-1)^\alpha$. For most applications it is helpful to make a universal choice which is (1) simple and easy to remember, (2) allows the neutral member ($I_3 = 0$) of a self-conjugate multiplet to be represented by a real field, and (3) agrees with $SU(3)$ phase conventions. If we employ the Gell-Mann, Nishijima formula $Q_\alpha = \alpha + \frac{1}{2}Y$, Y the hypercharge, and write $\xi = (-1)^{I-\frac{1}{2}Y}$ we obtain

$$\eta_\alpha = (-1)^{Q_\alpha} \tag{6.56}$$

so that the phase change associated with antiparticle conjugation of a state simply depends on its charge.

6.4 PARITY TRANSFORMATIONS

In analogy to Eq. (6.39) we define the parity transformation by

$$\begin{aligned} P\,|I\alpha\mathbf{k}\rangle &= \eta_P^*\,|I\alpha - \mathbf{k}\rangle \\ P\,|\bar{I}\alpha\mathbf{k}\rangle &= \eta'_P\,|\bar{I}\alpha - \mathbf{k}\rangle \end{aligned} \tag{6.57}$$

Here space-time variables change, but internal variables are unchanged. The creation operators must transform as

$$\begin{aligned} Pa_\alpha^*(\mathbf{k})P^{-1} &= \eta_P^* a_\alpha^*(-\mathbf{k}) \\ Pb_\alpha^*(\mathbf{k})P^{-1} &= \eta'_P b_\alpha^*(-\mathbf{k}) \end{aligned} \tag{6.58}$$

Again we ask that the field transform sensibly:

$$\begin{aligned} P\psi_\alpha(\mathbf{x}, t)P^{-1} &= \sum_k \eta_P a_\alpha(-\mathbf{k}) f_k(x) + \eta_\alpha \eta'_P b_\alpha^*(-\mathbf{k}) f_k^*(x) \\ &= \eta_P \sum_k \left[a_\alpha(\mathbf{k}) f_k(-\mathbf{x}, t) + (\eta_\alpha \eta'_P/\eta_P) b_\alpha^*(\mathbf{k}) f_k^*(-\mathbf{x}, t)\right] \end{aligned} \tag{6.59}$$

If the phases η_P, η'_P are correlated we can produce $\psi_\alpha(-\mathbf{x}, t)$ on the right-hand side. To prevent the two components of $\psi_\alpha(-\mathbf{x}, t)$ from getting out of phase requires $\eta_P = \eta'_P$ independently of η_α, as might be expected. This gives

$$\begin{aligned} Pa_\alpha(\mathbf{k})P^{-1} &= \eta_P a_\alpha(-\mathbf{k}) \\ Pb_\alpha(\mathbf{k})P^{-1} &= \eta_P^* b(-\mathbf{k}) \\ P\psi_\alpha(\mathbf{x}, t)P^{-1} &= \eta_P \psi_\alpha(-\mathbf{x}, t) \end{aligned} \tag{6.60}$$

The various commutation rules are preserved by P even before the restriction $\eta'_P = \eta_P$.

An interesting distinction arises now: P^2 is not necessarily the identity, although usually we can take it to be. We see that

$$\begin{aligned} P^2 a_\alpha(k)P^{-2} &= \eta_P^2 a_\alpha(k) \\ P^2 b_\alpha(k)P^{-2} &= \eta_P^{*2} b_\alpha(k) \end{aligned}$$

Whenever we can choose $\eta_P = \pm 1$ then $P^2 = 1$. The choice of η_P depends on the nature of the interactions of the ψ particle. For *SC* multiplets $a_\alpha = b_\alpha$ so that Eq. (6.60) shows that $\eta_P = \pm 1$ for such particles.

6.5 TIME-REVERSAL TRANSFORMATION

The common name of the transformation under consideration is misleading. The notion that an operation reverses the *sense* of time (i.e. the direction of time) leads to an unobservable type of situation. In order for a symmetry to exist we must be able (at least theoretically) to compare the transformed state and its behavior, with the original state. This is generally possible for P and C (except when these operations are not symmetry operations: e.g. for neutrinos $P\,|\nu_L\rangle$ leads to a $|\nu_R\rangle$ which doesn't exist.) The best we can do is to compare a "*motion*-reversed state" moving *forwards* in time with the original state. The idea of time running backwards seems at best a formal mathematical device, full of dangerous heuristic overtones.

From ordinary quantum mechanics we recall that the time reversal operation T is antiunitary. Momentum states transform as

$$\begin{aligned} T\,|I\alpha\mathbf{k}\rangle &= \eta_n^*\,|I\alpha - \mathbf{k}\rangle \\ T\,|\bar{I}\alpha\mathbf{k}\rangle &= \eta_T'^*\,|\bar{I}\alpha - \mathbf{k}\rangle \end{aligned} \tag{6.61}$$

i.e. for operators

$$\begin{aligned} Ta_\alpha(\mathbf{k})T^{-1} &= \eta_T a_\alpha(-\mathbf{k}) \\ Tb_\alpha(\mathbf{k})T^{-1} &= \eta'_T b_\alpha(-\mathbf{k}) \end{aligned} \tag{6.62}$$

There are two features to notice: (1) For spin 0 this differs from P only by anti-linearity; (2) T reverses $\mathbf{k}$ but *not* the I_3 component. When the phase $\eta_T' = \eta_\alpha^* \eta_T^* / \eta_\alpha$ the field transforms as

$$T\psi_\alpha(\mathbf{x}, t)T^{-1} = \eta_T \psi_\alpha(\mathbf{x}, -t) \tag{6.63}$$

6.6 COMPOSITE DISCRETE TRANSFORMATIONS

It is interesting to consider[5,7] discrete transformations which correspond to various combinations of the C, P, and T operations in the previous sections. When the constituents of such a composite transformation are not symmetry transformations it is useful to consider the entire transformation to be the useful construct.[1–3] Taking the prototype spin zero transformation to be

$$\begin{aligned} C\psi_\alpha(x)C^{-1} &= \eta_C \psi_\alpha^*(x) \\ T\psi_\alpha(\mathbf{x}, t)T^{-1} &= \eta_T \psi_\alpha(\mathbf{x}, -t) \\ P\psi_\alpha(\mathbf{x}, t)P^{-1} &= \eta_P \psi_\alpha(-\mathbf{x}, t) \end{aligned} \tag{6.64}$$

one easily constructs the following table:

$$\begin{aligned} C^2\psi_\alpha(x)C^{-2} &= \psi_\alpha(x) \\ T^2\psi_\alpha(x)T^{-2} &= \psi_\alpha(x) \\ P^2\psi_\alpha(x)P^{-2} &= \eta_P^2 \psi_\alpha(x) \\ (CP)\psi_\alpha(\mathbf{x}, t)(CP)^{-1} &= \eta_C \eta_P \psi_\alpha^*(-\mathbf{x}, t) \\ (PC)\psi_\alpha(\mathbf{x}, t)(PC)^{-1} &= \eta_C \eta_P^* \psi_\alpha^*(-\mathbf{x}, t) \\ (PT)\psi_\alpha(x)(PT)^{-1} &= \eta_P \eta_T \psi_\alpha(-x) \\ (TP)\psi_\alpha(x)(TP)^{-1} &= \eta_P^* \eta_T \psi_\alpha(-x) \\ (CT)\psi_\alpha(\mathbf{x}, t)(CT)^{-1} &= \eta_C \eta_T \psi_\alpha^*(\mathbf{x}, -t) \\ (TC)\psi_\alpha(\mathbf{x}, t)(TC)^{-1} &= \eta_C^* \eta_T^* \psi_\alpha^*(\mathbf{x}, -t) \\ (CP)^2\psi_\alpha(x)(CP)^{-2} &= \psi_\alpha(x) \\ (PC)^2\psi_\alpha(x)(PC)^{-2} &= \psi_\alpha(x) \\ (PT)^2\psi_\alpha(x)(PT)^{-2} &= \psi_\alpha(x) \\ (TP)^2\psi_\alpha(x)(TP)^{-2} &= \psi_\alpha(x) \\ (CT)^2\psi_\alpha(x)(CT)^{-2} &= (\eta_C^* \eta_T^*)^2 \psi_\alpha(x) \\ (TC)^2\psi_\alpha(x)(TC)^{-2} &= (\eta_C \eta_T)^2 \psi_\alpha(x) \end{aligned} \tag{6.65}$$

Clearly any operation in which C, P, and T occur *paired* leads to the same field operator. However the phase differs according to the order and the number of times the various operators occur. Many of the transformations are of simpler form than the individual components (6.64). Assuming that the fields ψ_α comprise a complete set we note that several transformations are proportional to the identity, independent of the choice of phases: $C^2 = T^2 = I$, $(CP)^2 = (PC)^2 = I$, and $(TP)^2 = (PT)^2 = I$. In addition we have phase dependent results; for real η_P we have: $P^2 = I$, $CP = PC$, and $TP = PT$. Further, if $\eta_C\eta_T$ is real we have in addition: $(CT)^2 = (TC)^2 = I$, and $CT = TC$.

The mathematical origin of these patterns is due to the complex conjugation operation associated (though in diverse ways) with the definition of the C and T operations.

In contrast to the foregoing discrete transformations the combined transformation TCP leaves invariant any local relativistic field theory. The proof of this symmetry hinges crucially on the property of locality. Although TCP-invariance is widely accepted, little attention seems to have been given the structural aspects of the transformation. Here we draw attention to the simple and interesting dependence of the TCP operation on the order of its constituent operations. From Eqs. (6.65) we can work out the phases ω_i defined by

$$\Theta_i\psi_\alpha(x)\Theta_i^{-1} = \omega_i\psi_\alpha^*(-x) \tag{6.66}$$

where the six operators Θ_i and their associated phases are given by

$$\begin{aligned}
\Theta_1 &= CPT & \omega_1 &= \eta_P\eta_C\eta_T \\
\Theta_2 &= TCP & \omega_2 &= \eta_T^*\eta_C^*\eta_P^* \\
\Theta_3 &= PTC & \omega_3 &= \eta_P^*\eta_C^*\eta_T^* \\
\Theta_4 &= TPC & \omega_4 &= \eta_T^*\eta_P\eta_C^* \\
\Theta_5 &= PCT & \omega_5 &= \eta_P^*\eta_C\eta_T \\
\Theta_6 &= CTP & \omega_6 &= \eta_P^*\eta_T\eta_C
\end{aligned} \tag{6.67}$$

Hence $\Theta_i\Theta_j$ applied to $\psi_\alpha(x)$ merely represents a gauge transformation of the latter, with phase factor $(\omega_i\omega_j)^*$. Thus we have an Abelian gauge group. The following relations among the ω_i should be noted:

$$\omega_1 = \omega_2^* = \omega_3^*, \qquad \omega_4 = \omega_5^* = \omega_6^*; \qquad \begin{aligned} \omega_1 &= \eta_P\eta_C\eta_T \\ \omega_4 &= \eta_P\eta_C^*\eta_T^* \end{aligned} \tag{6.68}$$

Table 6.1 The multiplication table of the phase factors associated with various TCP transformations Θ_i is given

	1	2	3	4	5	6
1	$(\bar{\omega}_1)^2$	1	1	$(\bar{\eta}_P)^2$	$(\eta_C\eta_T)^2$	$(\eta_C\eta_T)^2$
2	1	ω_1^2	ω_1^2	$(\bar{\eta}_C\bar{\eta}_T)^2$	$(\bar{\eta}_P)^2$	$(\bar{\eta}_P)^2$
3	1	ω_1^2	ω_1^2	$(\bar{\eta}_C\bar{\eta}_T)^2$	$(\bar{\eta}_P)^2$	$(\bar{\eta}_P)^2$
4	$(\bar{\eta}_P)^2$	$(\bar{\eta}_C\bar{\eta}_T)^2$	$(\bar{\eta}_C\bar{\eta}_T)^2$	ω_4^2	1	1
5	$(\eta_C\eta_T)^2$	$(\bar{\eta}_P)^2$	$(\bar{\eta}_P)^2$	1	$\bar{\omega}_4^2$	$\bar{\omega}_4^2$
6	$(\eta_C\eta_T)^2$	$(\bar{\eta}_P)^2$	$(\bar{\eta}_P)^2$	1	$\bar{\omega}_4^2$	$\bar{\omega}_4^2$

The multiplication table of the phases is given in Table 6.1. The corresponding operator multiplications are given by

$$\begin{aligned}
\Theta_1^2 &= CPT \cdot CPT \\
\Theta_1\Theta_2 &= CPT \cdot TCP = CP(T^2)CP = (CP)^2 = I \\
\Theta_2\Theta_1 &= TCP \cdot CPT = T(CP)^2T \;=\; T^2 \;= I \\
\Theta_1\Theta_3 &= CPT \cdot PTC = \; C(PT)^2C \;=\; C^2 \;= I \\
\Theta_3\Theta_1 &= PTC \cdot CPT = (PT)^2 \qquad = I \\
\Theta_2^2 &= TCP \cdot TCP \\
\Theta_3^2 &= PTC \cdot PTC
\end{aligned} \tag{6.69}$$

If we specialize to self-conjugate multiplets the foregoing relations simplify considerably. Although self-conjugate isospinor fields were shown in Sec. (6.3) to give non-local theories we shall include them here in order to exhibit their peculiar behavior under the *TCP* transformation.

We first summarize the constraints put on the phase factors η_C, η_P, η_T by the self-conjugate condition $a_\alpha = b_\alpha$ (cf. Eqs. (6.51), (6.60), and (6.62).)

$$\begin{aligned}
\eta_P^2 &= 1 \\
\eta_C^2 &= (-1)^{2I}\eta_\alpha^2 \\
\eta_T^2 &= (\eta_2^*)^2
\end{aligned} \tag{6.70}$$

We note further that $(\eta_C\eta_P\eta_T)^2 = (-1)^{2I}$ and further that

$$\Theta_i^2\varphi_\alpha(x)\Theta_i^{-2} = (-1)^{2I}\varphi_\alpha(x) \tag{6.71}$$

for $i = 1$ to 6. For $2I =$ odd integer this result violates a general result of local relativistic quantum field theory[10] that one always *can* choose $\Theta^2 = I$ for Boson fields. Hence the *TCP* transformation also exposes the anomalous character of self-conjugate isospinor fields. We return to this question in Sec. (6.8).

6.7 PHASE ARBITRARINESS AND GAUGE TRANSFORMATIONS

In the present section we analyze in more detail the arbitrariness of the phases occurring in the T, C, and P transformations. First consider parity:

$$P\psi_\alpha(\mathbf{x}, t)P^{-1} = \eta_P\psi_\alpha(-\mathbf{x}, t)$$
$$\eta_P \equiv e^{i\phi_P} \tag{6.72}$$

We see that P can be written as the product of a gauge transformation and an "intrinsic" parity transformation P_0 defined by

$$P_0\psi_\alpha(\mathbf{x}, t)P_0^{-1} = \psi_\alpha(-\mathbf{x}, t)$$

To produce the phase changes we consider the operators U_θ and V_θ defined by

$$U_\theta = \exp(i\theta N), \qquad N = \sum_{k\alpha} a_{k\alpha}^* a_{k\alpha}$$
$$V_\theta = \exp(i\theta \bar{N}), \qquad \bar{N} = \sum_{k\alpha} b_{k\alpha}^* b_{k\alpha} \tag{6.73}$$

The effect of U and V on the particle operators is

$$U_\theta a_\alpha^* U_\theta^{-1} = e^{i\theta}a_\alpha^*$$
$$V_\theta b_\alpha^* V_\theta^{-1} = e^{i\theta}b_\alpha^* \tag{6.74}$$

The operator $W_\theta \equiv U_{-\theta}V_\theta$ transforms $\psi_\alpha(x)$ as

$$W_\theta\psi_\alpha(x)W_\theta^{-1} = e^{i\theta}\psi_\alpha(x) \tag{6.75}$$

Hence the full transformation may be written as

$$P = W_{\phi_P}P_0 = P_0W_{\phi_P} \tag{6.76}$$

where W_ϕ is defined by

$$W_\phi = \exp(i\phi(N - \bar{N})) \tag{6.77}$$

We can extract the phases η_C and η_T in a similar way.

$$C\psi_\alpha(x)C^{-1} = \eta_C\psi_\alpha^*(x) = e^{i\phi_C}\psi_\alpha^*(x)$$
$$C\psi_\alpha^*(x)C^{-1} = \eta_C^*\psi_\alpha(x) = e^{-i\phi_C}\psi_\alpha(x) \tag{6.78}$$

Defining C_0 to correspond to $\eta_C = 1$ and requiring W_{φ_C} to have the effect

$$W_{\varphi_C}\psi_\alpha(x)W_{\varphi_C}^{-1} = e^{i\phi_C}\psi_\alpha(x) \tag{6.79}$$

we can write $C = C_0 W_{\phi_C}$.

To treat time reversal we write the phase factor η_C as $e^{i\phi_T}$ and define the anti-linear operator T_0 by

$$T_0\psi_\alpha(\mathbf{x}, t)T_0^{-1} = \psi_\alpha(\mathbf{x}, -t)$$

Then we may write

$$T = W_{\phi_T}T_0 = T_0 W_{-\varphi_T} \tag{6.80}$$

There are infinitely many phase choices ϕ_P, ϕ_C, ϕ_T which correspond to the same *TCP* transformation. Thus if we consider the latter alone, the detailed structure of the theory is not at stake compared with behavior under one of the constituent transformations. Corresponding to the transformations Θ_i the phases ϕ_i of $\eta_i = \exp(i\phi_i)$ are

$$\begin{aligned}
\phi_1 &= \phi_P + \phi_C + \phi_T \\
\phi_2 &= -(\phi_P + \phi_C + \phi_T) \\
\phi_3 &= -(\phi_P + \phi_C + \phi_T) \\
\phi_4 &= \phi_P - \phi_C - \phi_T \\
\phi_5 &= -(\phi_P - \phi_C - \phi_T) \\
\phi_6 &= -(\phi_P - \phi_C - \phi_T)
\end{aligned} \tag{6.81}$$

Thus only two phase angles specify the ϕ_i, i.e. $\xi = \phi_P + \phi_C + \phi_T$ and $\eta = \phi_P - \phi_C - \phi_T$. In detail we write $\phi_1 = \xi = -\phi_2 = -\phi_2$; $\phi_4 = \eta = -\phi_5 = -\phi_6$. If we write the number difference operator $N - \bar{N}$ as $\tilde{Y}$, we find

$$\begin{aligned}
\Theta_0\tilde{Y} \equiv C_0P_0T_0\tilde{Y} &= C_0\tilde{Y}P_0T \\
&= -\tilde{Y}\Theta_0
\end{aligned} \tag{6.82}$$

$$\therefore \quad \{\Theta_0, \tilde{Y}\} = 0 \Rightarrow \{\Theta_i, \tilde{Y}\} = 0$$

Hence the antilinearity of Θ_0 gives

$$\Theta_i = \Theta_0 e^{i\phi_i\tilde{Y}} = e^{i\tilde{Y}\phi_i}\Theta_0 \tag{6.83}$$

The following useful relations should be noted.

$$\begin{aligned}
\Theta_0^2 &= C_0^2P_0^2T_0^2 = I \\
\Theta_i^2 &= \exp(2i\phi_i\tilde{Y})
\end{aligned} \tag{6.84}$$

It is now possible to reduce the group property of $\Theta_i\Theta_j$ to that of the Abelian group of phase addition. Defining Z_i by $Z_i = \exp(i\varphi_i \tilde{Y})$ we find

$$\begin{aligned}\Theta_i\Theta_j &= \Theta_0 Z_i \Theta_0 Z_j = \Theta_0^2 Z_i Z_j \\ &= \exp[i\tilde{Y}(\phi_i + \phi_j)]\end{aligned} \tag{6.85}$$

Thus the product of (antilinear) Θ_i, Θ_j gives a gauge transformation by the "hypercharge" $\tilde{Y}$, where the phase is $\phi_i + \phi_j$.

$$\Theta_i\Theta_j \to \phi_i + \phi_j \quad (\text{mod } 2\pi) \tag{6.86}$$

To each operator (antilinear) we associate a number (ϕ_i), and to operator multiplication corresponds (linear) addition $\phi_i + \phi_j$.

6.8 ALLOWED REPRESENTATIONS OF INTERNAL SYMMETRY GROUPS FOR SELF-CONJUGATE MULTIPLETS

In Secs. (6.3) and (6.7) it was noted that contradictions with local field theory occur for self-conjugate isospinor fields. In this section we analyze this question in more detail, and consider more general internal symmetry groups.

The operations C, P, T, and Θ have the following effect on the fields

$$\begin{aligned} C\varphi_\alpha(x)C^{-1} &= \eta_C \varphi_\alpha^*(x) \\ P\varphi_\alpha(x)P^{-1} &= \eta_P \varphi_\alpha(-\mathbf{x}, t) \\ T\varphi_\alpha(x)T^{-1} &= \eta_T \varphi_\alpha(\mathbf{x}, -t) \\ \Theta\varphi_\alpha(x)\Theta^{-1} &= \eta \varphi_\alpha^*(-x) \end{aligned} \tag{6.87}$$

It will be noticed that in the limit of perfect isospin symmetry there is no preferred axis in isospace and that the operators defined in Eq. (6.87) are scarcely unique. For example the operator $P' = PO(\boldsymbol{\lambda})$ where $O(\boldsymbol{\lambda})$ is an arbitrary isospin rotation, is "just as good" an inversion operator as P. This is the point of view taken in refs. 5 and 6, for example. However our choice has many advantages; e.g. P commutes with I, Θ anticommutes with I, etc. And when the symmetry breaking due to electromagnetism is taken into account the fields are "already" oriented in a useful direction (determined by the isovector part of the electromagnetic current).

As already noted the phases are not completely free but are constrained by the self-conjugacy conditions. In momentum space the self-conjugacy condition is simple for any I:

$$\begin{aligned}\varphi_\alpha(p) &= \eta_\alpha \varphi^*_{-\alpha}(-p),\\ \varphi_\alpha(p) &\equiv \int d^4x e^{ip\cdot x}\varphi_\alpha(x)\end{aligned} \tag{6.88}$$

In terms of $\varphi_\alpha(p)$ Eqs. (6.87) become

$$\begin{aligned}C\varphi_\alpha(p)C^{-1} &= \eta_C \varphi^*_\alpha(-p)\\ P\varphi_\alpha(p)P^{-1} &= \eta_P \varphi_\alpha(p')\\ T\varphi_\alpha(p)T^{-1} &= \eta_T \varphi_\alpha(p')\\ \Theta\varphi_\alpha(p)\Theta^{-1} &= \eta \varphi^*_\alpha(-p)\end{aligned} \tag{6.89}$$

where the components of p' are p^0, $-\mathbf{p}$. (In this section $-p$ is $(-p_0, -\mathbf{p})$.) If we now apply Eqs. (6.89) (and the corresponding adjoint equations) to the *SC* condition, Eq. (6.88), we obtain the relations (6.70)–(6.71). This procedure is somewhat simpler than that given before based on the particle operator transformations.

We note that the parity is ± 1. Consider now the case $2I =$ even. When a real basis is chosen η_C and η_T become real and also ± 1. This is extremely useful in theories with charge conjugation invariance, because η_C can then be identified with the charge parity of the neutral component of the multiplet ($\alpha = 0$). Otherwise theories are more complicated than need be, with no increase in generality.

Within the context of local field theory the *TCP* operator gives the most general definition of antiparticle conjugation. For if $\psi^*_\nu(p)\,|0\rangle$ creates a particle state with mass m, momentum p and additive internal quantum number ν then $\Theta\psi^*_\nu(p)\Theta^{-1} \equiv \psi^*_\nu(p)^\Theta$ creates a state with the *same* p and m but opposite ν. If the antiparticle state is in the same multiplet as the original state we expect $\psi^{*\Theta}_\nu(p) \sim \psi_\nu(-p) \sim \psi^*_{-\nu}(p)$ or $\psi_\nu(p) \sim \psi^*_{-\nu}(-p)$. We return to this point below.

For a general internal symmetry group, consider a field transforming according to the irreducible representation D. A pair conjugate field ψ_α (α denoting the appropriate quantum numbers) is constructed from particle operators a^*_α transforming like D and antiparticle operators b^*_α which transform like D^* in the combination $a + b^*$, with appropriate wave functions and phase factors. For self-conjugate multiplets there is no b

independent of a, and therefore there must be a transformation matrix converting D^* to D, if the field is to transform irreducibly. Formally we have

$$\varphi_\alpha = \chi_\alpha + \chi_\alpha^c; \qquad \chi_\alpha = \sum_k a_\alpha(k) f_k(x), \qquad \chi_\alpha^c = \sum_\beta C_{\alpha\beta}\chi_\beta^* \tag{6.90}$$

φ_α^* transforms as D only if

$$\tilde{C} D^* \tilde{C}^{-1} = D \tag{6.91}$$

where $\tilde{C}$ is the transpose of C.

Next consider the implications of causality. The fields satisfy the first of Eq. (6.33) only if C is unitary

$$[\varphi_\alpha(x), \varphi_\beta^*(x')] = i\,\delta_{\alpha\beta}\,\Delta(x - x') \Rightarrow C^\dagger C = 1 \tag{6.92}$$

Next consider the commutator

$$[\varphi_\alpha(x), \varphi_\beta(x')] = \sum_k [f_k(x) f_k^*(x') C_{\beta\alpha} - f_k^*(x) f_k(x') C_{\alpha\beta}] \tag{6.93}$$

This commutator is causal only if C is symmetric, in which case Eq. (6.93) is proportional to $\Delta(x - x')$.

Next we have the key theorem[16]: If a unitary, symmetric matrix C transforms D^* to D, then D is equivalent to a real representation. In the case of $SU(2)$, $C_{\alpha\beta} = \eta_\alpha\,\delta_{\alpha-\beta}$ is symmetric for $2I =$ even and antisymmetric for $2I =$ odd. For $SU(3)$, the equivalence of D^* to D is only possible for representations $D(p, q)$ with $p = q$, e.g. **1**, **8**, **27**, etc. (These representations have symmetrical weight diagrams.) The charge conjugation matrix $C_{\alpha\beta}$ can be chosen as $(-1)^{Q_\alpha}\,\delta_{\alpha-\beta}$ where $Q_\alpha = -Q_{-\alpha}$ is integral so that $C_{\alpha\beta}$ is symmetric. Hence none of the allowed $SU(3)$ representations is eliminated by the symmetry condition.

Similar results follow if one assumes the existence of a "standard" *TCP* operation. This approach involves the following assumptions:

(1) The fields $\varphi_\alpha(p)$ transform irreducibility under the unitary transformations of the internal symmetry group:

$$O^{-1}\varphi_\alpha(p) O = \sum_\beta D_{\alpha\beta}\varphi_\beta(p) \tag{6.94}$$

(2) There is an antilinear *TCP* operation Θ:

$$\Theta\varphi_\alpha(p)\Theta^{-1} = \eta\varphi_\alpha^*(-p) \tag{6.95}$$

such that $\Theta^2 = I$;

(3) There is a generalized reality condition

$$\varphi_\alpha(p) = C_{\alpha\beta}(p)\varphi_\beta^*(-p) \tag{6.96}$$

for self-conjugate fields.

Relation Eq. (6.96) is equivalent to the translation-invariant configuration space connection

$$\varphi_\alpha(x) = \int d^4x' C_{\alpha\beta}(x - x')\varphi^*_\beta(x') \tag{6.97}$$

and is a generalization of Eqs. (6.29) and (6.30). Applying Θ to Eq. (6.96) gives

$$\varphi^*_\alpha(-p)C^*_{\alpha\beta}(p)\varphi_\beta(p)(\eta^*)^2 \tag{6.98}$$

Comparing this with the adjoint of Eq. (6.96) yields

$$C_{\alpha\beta}(p) = C_{\alpha\beta}(-p)(\eta^*)^2 \tag{6.99}$$

Substitution of Eq. (6.98) in Eq. (6.96) gives

$$C(p)C^*(p) = (\eta^*)^2 I \tag{6.100}$$

Now the isospin transformation rule Eq. (6.94) and the *SC* condition Eq. (6.96) may be used to show that C transforms D^* to D:

$$\begin{aligned} O^{-1}\varphi_\alpha(p)O &= \sum_\beta D_{\alpha\beta}\varphi_\beta(p) = \sum_{\beta\gamma} D_{\alpha\beta}C_{\beta\gamma}(p)\varphi^*_\gamma(-p) \\ &= \sum_\beta C_{\alpha\beta}(p)O^{-1}\varphi^*_\beta(-p)O = \sum_{\beta\gamma} C_{\alpha\beta}(p)D^*_{\beta\gamma}\varphi^*_\gamma(-p) \end{aligned}$$

$$C(p)D^*C^{-1}(p) = D \tag{6.101}$$

For the normal Θ operation $\Theta^2 = I$ and $(\eta^*)^2 = 1$. Thus, from Eq. (6.100) and the unitary of C, we see that a symmetric matrix C transforms D^* to D, so that only real representations are allowed by assumptions 1–3. It is clear what happens in the *SCIF* case: assumption 2 is wrong because Θ^2 changes the sign of $\varphi_\alpha(p)$. When $(\eta^*)^2 = -1$, instead of Eq. (6.99) and Eq. (6.100) we have $\tilde{C} = -C$, which is precisely Wigner's criterion for "pseudo-real" representations. For $SU(2)$ this property characterizes the isospinor representations ($2I =$ odd) as is appropriate for this case.

The second proof, based on the *TCP* theorem, is somewhat more general than the first calculation, which employed free fields. It is also mathematically very simple, once one has proved assumption 2. The free-field approach allows a more detailed insight into the role of the physical role played by locality and causality.

6.9 INTERNAL SYMMETRIES AND T, C, P TRANSFORMATIONS FOR RARITA-SCHWINGER FIELDS

In the present section we consider the technical changes in the foregoing analysis which are necessitated by spin. Particle states with isospin I, spin S, momentum p, and helicity λ are created by the operators $a^*_\alpha(p, \lambda)$. For antiparticle states the same task is accomplished by the operators $b^*_\alpha(p, \lambda)$. We only consider particles with finite rest mass, and employ the standard spin-statistics assumption. We require that the a^*_α and b^*_α create isospin multiplets whose members are connected by the Condon-Shortley phase convention, in order that the standard apparatus of the theory of the rotation group can be used intact.

The Rarita-Schwinger field[20] may be expanded in the form

$$\psi^\alpha_\mu(x) = \sum_{p\lambda} [a_\alpha(p, \lambda)\chi_\mu(p, \lambda)f_p(x) + \eta_\alpha b^*_{-\alpha}(p, \lambda)\chi^c_\mu(p, \lambda)f^*_p(x) \tag{6.102}$$

Here ψ^α_μ is a Rarita-Schwinger field—a four-component spinor symmetric in its four-vector indices $\boldsymbol{\mu} = \mu_1 \cdots \mu_k$, where $k = S$ for Bosons and $S - \frac{1}{2}$ for Fermions. ψ^α_μ obeys appropriate wave equations and subsidiary conditions. The spin S wave functions were discussed in detail in Sec. (3.5). χ^c_μ denotes the conjugate wave function of the antiparticle, and $f_p(x)$ is $N_p e^{-ip\cdot x}$, where N_p is $(2E)^{-\frac{1}{2}}$ or $(M/E)^{\frac{1}{2}}$ for Bosons and Fermions, respectively.

The Dirac field can be expended in terms of wave functions and operators referring to a fixed axis or in terms of helicity wave functions and operators. Letting i designate the components of the Dirac field $\psi^\alpha_i(x)$,

$$\psi^\alpha_i(x) = \sum_{p\lambda} [a_\alpha(p, \lambda)u_i(p, \lambda)f_p(x) + \eta_\alpha b^*_{-\alpha}(p, \lambda)v_i(p, \lambda)f^*_p(x)] \tag{6.103}$$

where λ could be a helicity label or a component(s) of S_z. We have the relations

$$\begin{aligned} \sum_s a(p, s)u(p, s) &= \sum_\lambda a(p, \lambda)u(p, \lambda) \\ \sum_s b^*(p, s)v(p, s) &= \sum_\lambda b^*(p, \lambda)v(p, \lambda) \end{aligned} \tag{6.104}$$

where the wave functions were given in Sec. (3.6). From the orthogonality relations we find

$$\begin{aligned} a(p, \lambda) &= \sum_s M_{\lambda s} a(p, s) \\ b^*(p, \lambda) &= \sum_s R^*_{\lambda s} b^*(p, s) \end{aligned} \tag{6.105}$$

where a little calculation gives

$$\begin{aligned} M_{\lambda s} &= \bar{u}(p, \lambda)u(p, s) = \chi^\dagger_\lambda(\hat{p})\chi_s(\hat{z}) \\ &= \chi^\dagger_\lambda(\hat{z})e^{i\hat{n}\cdot\boldsymbol{\sigma}\theta/2}\chi_s(\hat{z}) \\ R^*_{\lambda s} &= (-1)^{\lambda-s}\bar{v}(p, \lambda)v(p, s) = (-1)^{\lambda-s}\chi^\dagger_{-\lambda}(\hat{p})\chi_{-s}(\hat{z}) \\ &= (\chi^\dagger_\lambda(\hat{p})\chi_s(\hat{z}))^* = M^*_{\lambda s} \end{aligned} \tag{6.106}$$

We have used the second of Eqs. (3.100) to establish that $R = M$, i.e. that the particle and antiparticle helicity operators are related in exactly the same way to the fixed axis operators. The matrix $M_{\lambda s}$ is given by

$$M = e^{i\hat{n}\cdot\sigma\theta/2} = \begin{pmatrix} \cos\frac{\theta}{2} & e^{-i\phi}\sin\frac{\theta}{2} \\ -e^{i\phi}\sin\frac{\theta}{2} & \cos\frac{\theta}{2} \end{pmatrix} \tag{6.107}$$

Thus the transformations (6.104) are, not surprisingly, unitary and unimodular.

Using the notation $+$, $-$ for $(\lambda, s) = +\frac{1}{2}$ or $-\frac{1}{2}$ and introducing subscripts h and s to distinguish helicity from z-axis operators, we have

$$\begin{aligned} a_h(p, +) &= \cos\frac{\theta}{2}\, a_s(p, +) + e^{-i\phi}\sin\frac{\theta}{2}\, a_s(p, -) \\ a_h(p, -) &= -e^{i\phi}\sin\frac{\theta}{2}\, a_s(p, +) + \cos\frac{\theta}{2}\, a_s(p, -) \end{aligned} \tag{6.108}$$

Corresponding expressions hold for the antiparticle operators. For $-p$, we obtain the appropriate relations by the substitutions $\theta \to \pi - \theta$, $\phi \to \pi + \phi$. These are useful in discussing the parity transformation. Employing completeness relations for the spinors u and v given in Sec. 3.6 gives for the anticommutators:

$$\begin{aligned} \{\psi_j^\alpha(x), \bar{\psi}_k^\beta(x')\} &= i(i\gamma\cdot\partial + m)_{jk}\delta_{\alpha\beta}\,\Delta(x - x') \qquad (PC, SC) \\ \{\psi_j^\alpha(x), \psi_k^\beta(x')\} &= 0 \qquad (PC) \\ &= \eta_\beta\delta_{\alpha-\beta}\sum_{p\lambda}\frac{1}{2E}[u_j(p,\lambda)v_k(p,\lambda)e^{-ip\cdot(x-x')} \\ &\qquad + (-1)^{2I}v_j(p,\lambda)u_k(p,\lambda)e^{ip\cdot(x-x')}] \\ &= \eta_\beta\delta_{\alpha-\beta}C_{1kl}(i\gamma\cdot\partial + m)_{jl}\begin{cases} i\Delta(x-x') & 2I = \text{even} \\ \Delta^{(1)}(x-x') & 2I = \text{odd} \end{cases} \end{aligned} \tag{6.109}$$

In Eq. (6.109) the matrix C_1 is the usual charge-conjugation matrix. The self-conjugate isospinor-spinor fields suffer from the same troubles as the zero-spin case. (The usual Majorana field[21] corresponds to the special case $I = 0$.)

For the PC vector-meson field we have

$$V_\mu^\alpha(x) = \sum_{p\lambda}[a_\alpha(p,\lambda)e_\mu(p,\lambda)f_p(x) + \eta_\alpha b_{-\alpha}^*(p,\lambda)e_\mu^*(p,\lambda)f_p^*(x)] \tag{6.110}$$

where $e_\mu(p, \lambda)$ is the vector wave function for helicity λ, transverse to p, thus guaranteeing $\partial^\mu V^\alpha_\mu = 0$. The commutators are

$$[V^\alpha_\mu(x), V^{\beta *}_\nu(x')] = i\delta_{\alpha\beta} M_{\mu\nu}(-i\partial)\,\Delta(x - x') \qquad (PC, SC)$$

$$[V^\alpha_\mu(x), V^\beta_\nu(x')] = 0 \qquad (PC)$$

$$\begin{aligned}[V^\alpha_\mu(x), V^\beta_\nu(x')] &= \eta_\beta \delta_{\alpha-\beta} \sum_{p\lambda} [e_\mu(p, \lambda) e^*_\nu(p, \lambda) e^{-ip\cdot(x-x')} \\ &\qquad - (-1)^{2I} e^*_\mu(p, \lambda) e_\nu(p, \lambda) e^{ip\cdot(x-x')}]/2E \\ &= -\eta_\beta \delta_{\alpha-\beta} M_{\mu\nu}(-i\partial) \begin{cases} i\Delta(x - x') & 2I = \text{even} \\ \Delta^{(1)}(x - x') & 2I = \text{odd} \end{cases}\end{aligned} \qquad (6.111)$$

The operator $M_{\mu\nu}$ is $(g_{\mu\nu} + \partial_\mu \partial_\nu / m^2)$, obtained by replacing p_μ in Eq. (3.114) by $-i\partial_\mu$.

The extension to arbitrary spin is now quite elementary using the wave functions and projection operators of Sec. 3.5. For spin S mesons the only change from the vector field is to replace μ by a set of μ_i $(i = 1, 2, \ldots, S)$ and the polarization vector e_μ by the spin-S wave function $e_{\boldsymbol{\mu}}(p, \lambda)$. Equations (6.10)–(6.11) are changed by the replacements $(\mu, \nu) \to (\boldsymbol{\mu}, \boldsymbol{\nu})$, $e_\mu \to e^s_{\boldsymbol{\mu}}$, $-M_{\mu\nu} \to (-1)^S M_{\boldsymbol{\mu\nu}}$.

The case of a spin-S Fermion is quite similar to that of spin $\frac{1}{2}$, but we give a few details. The field $\psi^\alpha_{\boldsymbol{\mu}}(x)$ has the expansion

$$\psi^\alpha_{\boldsymbol{\mu}}(x) = \sum_{p\lambda} [a_\alpha(p, \lambda) u^S_{\boldsymbol{\mu}}(p, \lambda) f_p(x) + \eta_\alpha b^*_{-\alpha}(p, \lambda) v^S_{\boldsymbol{\mu}}(p, \lambda) f^*_p(x)] \qquad (6.112)$$

where $v^S_{\boldsymbol{\mu}}(p, \lambda) = C_1 \bar{u}^{ST}_{\boldsymbol{\mu}}(p, \lambda)$.

For either PC or SC fields we find

$$\begin{aligned}\{\psi^\alpha_{\boldsymbol{\mu}j}(x), \bar{\psi}^\beta_{\boldsymbol{\nu}k}(x')\} &= \sum_{p\lambda} \frac{M}{E} [u^S_{\boldsymbol{\mu}j}(p, \lambda) \bar{u}^S_{\boldsymbol{\nu}k}(p, \lambda) e^{-ip\cdot(x-x')} \\ &\qquad + v^S_{\boldsymbol{\mu}j}(p, \lambda) \bar{v}^S_{\boldsymbol{\nu}k}(p, \lambda) e^{ip\cdot(x-x')}] \\ &= (-1)^{S-\frac{1}{2}} \sum_{p\lambda} [(\mathscr{P}^S_{+\boldsymbol{\mu\nu}})_{jk} e^{-ip\cdot(x-x')} - (\mathscr{P}^S_{-\boldsymbol{\mu\nu}})_{jk} e^{ip\cdot(x-x')}] \\ &= i(-1)^{S-\frac{1}{2}} [\Lambda_{\boldsymbol{\mu\nu}}(-i\partial)(i\gamma\cdot\partial + m)]_{jk}\,\Delta(x - x')\end{aligned} \qquad (6.113)$$

on making use of Eqs. (3.118) and (3.122).

For PC fields we find

$$\{\psi^\alpha_{\boldsymbol{\mu}j}(x), \psi^\beta_{\boldsymbol{\nu}k}(x')\} = 0 \qquad (6.114)$$

while for SC fields a calculation similar to that of Eq. (6.109) gives

$$\{\psi^{\alpha}_{\mu j}(x), \psi^{\beta}_{\nu k}(x')\} = \eta_{\beta}\delta_{\alpha-\beta}(-1)^{S-\frac{1}{2}}$$
$$\times\, C_{1kl}[\Lambda_{\mu\nu}(-i\partial)(i\gamma\cdot\partial + m)]_{jl}\begin{cases} i\Delta(x-x') & 2I = \text{even} \\ \Delta^{(1)}(x-x') & 2I = \text{odd}\end{cases} \tag{6.115}$$

In all these cases we expect that for SC fields $\psi^{-\alpha}$ is dependent on $\psi^{\alpha *}$. For $2I =$ even, this dependence is a simple proportionality, in which case the quantities $[\psi, \psi]_{\pm}$ are not independent of $[\psi, \psi^*]_{\pm}$. For $2I =$ odd, the dependence is more intricate. The relation between $\psi^{-\alpha}$ and $\psi^{\alpha *}$ is essentially the same as derived in Sec. (6.2) for zero spin. We do not repeat details of the calculation. Let $\psi^{\alpha c}_{\mu}(x)$ be $\psi^{\alpha *}_{\mu}(x)$ for mesons and $C_1\bar{\psi}^{\alpha T}_{\mu}(x)$ for Fermions. An elementary calculation gives the SC conditions:

$$\psi^{-\alpha}_{\mu}(x) = \eta_{-\alpha}\psi^{\alpha c}_{\mu}(x) \qquad 2I = \text{even}$$
$$\psi^{-\alpha}_{\mu}(x) = i\eta_{-\alpha}\int d^3x'\,\psi^{\alpha c}_{\mu}(x')\overset{\leftrightarrow}{\partial_0'}\Delta^{(1)}(x'-x) \qquad 2I = \text{odd} \tag{6.116}$$

Although the second relation is non-local in configuration space, the general expression has a space-time translation-invariant structure. Thus, in momentum space, one can write a linear connection between $\psi^{-\alpha}_{\mu}(p)$ and $\psi^{\alpha c}_{\mu}(p)$.

We have shown that self-conjugate Fermions of any spin give rise to non-causal commutators. It should be amply clear that the associated field theory is non-local.

Our discussion of C, P, and T transformations is a straightforward generalization of the results for zero-spin particles. We omit details of arguments essentially the same as given there. The idea is first to define the discrete operations by particle mappings and then to require that the associated field have a sensible transformation law. Results are given first for PC multiplets; equating particle and antiparticle operators yields consistency conditions among the phases which must be satisfied in the SC case.

Requiring that antiparticle conjugation C maps $\psi^{\alpha}_{\mu}(x)$ into $\psi^{\alpha}_{\mu}{}^{*}(x)$ with an α-independent phase, we obtain for the particle transformations

$$Ca_{\alpha}(p, \lambda)C^{-1} = \eta_C\eta^*_{\alpha}b_{-\alpha}(p, \lambda)$$
$$Cb_{\alpha}(p, \lambda)C^{-1} = \eta^*_C\eta_{-\alpha}a_{-\alpha}(p, \lambda) \tag{6.117}$$

For SC multiplets the phases on the right-hand side are equal, giving the constraint

$$\eta_C^2 = \eta_\alpha \eta_{-\alpha} = (-1)^{2I} \eta_\alpha^2 \tag{6.118}$$

The field transformation law follows easily.

$$C\psi^\alpha(x)C^{-1} = \eta_C C_1 \bar{\psi}_\mu^{\alpha T}(x) \tag{6.119}$$

The matrix C_1 is $i\gamma_0\gamma_2$ for Fermions and unity for mesons. The bar-transpose operation is the usual one for Fermions; for Bosons it simply means taking the Hermitian adjoint.

The parity transformation is somewhat more intricate because of the more complicated behavior of the wave functions under parity. For spin 0 we had

$$\begin{aligned} Pa_\alpha(\mathbf{p})P^{-1} &= \eta_P a_\alpha(-\mathbf{p}) \\ Pb_\alpha(\mathbf{p})P^{-1} &= \eta_P^* b_\alpha(-\mathbf{p}) \\ P\psi_\alpha(\mathbf{x}, t)P^{-1} &= \eta_P \psi_\alpha(-\mathbf{x}, t) \end{aligned} \tag{6.120}$$

In the usual discussions of the Dirac field, one uses wave functions and operators whose spin component is referred to a fixed axis. We review this before going over to the helicity formalism. We note the well-known sign difference of the particle and antiparticle transformations:

$$\begin{aligned} Pa_\alpha(\mathbf{p}, s)P^{-1} &= \eta_P a_\alpha(-\mathbf{p}, s) \\ Pb_\alpha(\mathbf{p}, s)P^{-1} &= -\eta_P^* b_\alpha(-\mathbf{p}, s) \\ P\psi_\alpha(\mathbf{x}, t)P^{-1} &= \eta_P P\psi_\alpha(-\mathbf{x}, t) \end{aligned} \tag{6.121}$$

where $P = \gamma_0$.

If we use the helicity basis, the transformation laws for the operators $a_\alpha(p, \lambda)$ and $b_\alpha(p, \lambda)$ follow from (6.121) and (6.108):

$$\begin{aligned} Pa_\alpha(\mathbf{p}, \lambda)P^{-1} &= \eta_P(-1)^{\frac{1}{2}-\lambda} e^{-2i\lambda\phi} a_\alpha(-\mathbf{p}, -\lambda) \\ Pb_\alpha(\mathbf{p}, \lambda)P^{-1} &= -\eta_P^*(-1)^{\frac{1}{2}-\lambda} e^{-2i\lambda\phi} b_\alpha(-\mathbf{p}, -\lambda) \end{aligned} \tag{6.122}$$

We note that the helicity changes sign, although s does not, as expected. This detailed structure could have been surmised from the identities (3.103), which guarantee that the spinor-field parity-transformation law is the same as shown in (6.121).

For spin 1, inspection of (3.107) leads us to the definition

$$\begin{aligned} Pa_\alpha(\mathbf{p}, \lambda)P^{-1} &= \eta_P(-1)^{1-\lambda} e^{-2i\lambda\phi} a_\alpha(-\mathbf{p}, -\lambda) \\ Pb_\alpha(\mathbf{p}, \lambda)P^{-1} &= \eta_P^*(-1)^{1-\lambda} e^{-2i\lambda\phi} b_\alpha(-\mathbf{p}, -\lambda) \end{aligned} \tag{6.123}$$

This gives the usual field-transformation law

$$PV^{\alpha}_{\mu}(\mathbf{x}, t)P^{-1} = \eta_P(-g_{\mu\mu})V^{\alpha}_{\mu}(-\mathbf{x}, t) \tag{6.124}$$

The symmetry relations for higher spin wave functions summarized in Eq. (3.11) lead to the definition for arbitrary spin (including the previous cases):

$$\begin{aligned} Pa_{\alpha}(\mathbf{p}, \lambda)P^{-1} &= \eta_P(-1)^{S-\lambda}e^{-2i\lambda\phi}a_{\alpha}(-\mathbf{p}, -\lambda) \\ Pb_{\alpha}(\mathbf{p}, \lambda)P^{-1} &= (-1)^{2S}\eta_P^*(-1)^{S-\lambda}e^{-2i\lambda\phi}b_{\alpha}(-\mathbf{p}, -\lambda) \end{aligned} \tag{6.125}$$

where the factor $(-1)^{2S}$ is especially to be noted. With these conventions the spin-S field transforms as

$$P\psi^{\alpha}_{\mu}(\mathbf{x}, t)P^{-1} = \eta_P\Pi(-g_{\mu\mu})P\psi^{\alpha}_{\mu}(-\mathbf{x}, t) \tag{6.126}$$

The over-all sign has been chosen to give η_P its conventional value. P is again γ_0 for Fermions and unity for mesons.

In the special case of SC multiplets the consistency of Eqs. (6.125) requires that

$$\eta_P^2 = (-1)^{2S} \tag{6.127}$$

For SC mesons the parity factor is thus ± 1, while for Majorana particles it is $\pm i$.

The discussion of time reversal proceeds in a similar way.

For spin $\frac{1}{2}$ we first work in the usual basis, and then give results for the helicity basis. The result (3.97) indicates the necessity for a spin-dependent phase factor:

$$\begin{aligned} Ta_{\alpha}(\mathbf{p}, s)T^{-1} &= \eta_T(-1)^{\frac{1}{2}+s}a_{\alpha}(-\mathbf{p}, -s) \\ Tb_{\alpha}(\mathbf{p}, s)T^{-1} &= (\eta_T^*\eta_{\alpha}^*/\eta_{\alpha})(-1)^{\frac{1}{2}+s}b_{\alpha}(-\mathbf{p}, -s) \end{aligned} \tag{6.128}$$

The Dirac field transforms as

$$T\psi^{\alpha}(\mathbf{x}, t)T^{-1} = \eta_T T\psi^{\alpha}(\mathbf{x}, -t) \tag{6.129}$$

where the matrix $T = \gamma_3\gamma_1$ is real.

The transformation of the helicity operators follows directly from Eqs. (6.128) and (6.108):

$$\begin{aligned} Ta_{\alpha}(\mathbf{p}, \lambda)T^{-1} &= \eta_T e^{2i\lambda\phi}a_{\alpha}(-\mathbf{p}, \lambda) \\ Tb_{\alpha}(\mathbf{p}, \lambda)T^{-1} &= (\eta_T^*\eta_{\alpha}^*/\eta_{\alpha})e^{2i\lambda\phi}b_{\alpha}(-\mathbf{p}, \lambda) \end{aligned} \tag{6.130}$$

For spin 1, Eqs. (3.107) suggest that the transformations of a and b are formally identical to (6.130). The field transformation law is

$$TV^{\alpha}_{\mu}(\mathbf{x}, t)T^{-1} = \eta_T g_{\mu\mu}V^{\alpha}_{\mu}(\mathbf{x}, -t) \tag{6.131}$$

The general case is now clear from the preceding examples and the second of Eqs. (3.112); Eq. (6.130) summarizes the form of the T transformation for any spin. The field undergoes the transformation

$$T\psi_{\mu}^{\alpha}(\mathbf{x}, t)T^{-1} = \eta_T \Pi(g_{\mu\mu}) T\psi_{\mu}^{\alpha}(\mathbf{x}, -t) \tag{6.132}$$

For the special case of SC fields the consistency of (6.130) implies

$$\eta_T^2 = (\eta_\alpha^*)^2 \tag{6.133}$$

Combining Eqs. (6.118), (6.127), and (6.133) leads to the SC consistency conditions

$$\begin{aligned} \eta_P^2 &= (-1)^{2S} \\ (\eta_C \eta_T)^2 &= (-1)^{2I} \end{aligned} \tag{6.134}$$

The fact that η_P is imaginary for isospinor Majorana particles has no physical significance for single valued operators. One might argue that "since half-integral spin particles and antiparticles have the opposite parity, they can't be in the same multiplet" but this is not the case. To see this point we consider the actual meaning of the above statement for spin $\frac{1}{2}$, beginning with PC particles and later specializing to the SC case. If the particle operator $a_\alpha(\mathbf{p})$ transforms to $\eta_P a_\alpha(-\mathbf{p})$ under P then the antiparticle operator $b_\alpha(\mathbf{p})$ transforms to $-\eta_P^* b_\alpha(-\mathbf{p})$ in order that $\psi_\alpha(\mathbf{x}) \rightarrow \eta_P \gamma_0 \psi_\alpha(-\mathbf{x})$. Thus the two-particle state with one particle and one antiparticle has definite (negative) parity. If we specialize to SC Fermions (i.e. $b_\alpha = a_\alpha$) this is still true. But there is no conflict between the a and b transformations when $b = a$ because $-\eta_P^* = \eta_P$ since η_P is purely imaginary.

Finally we study the composite transformations with emphasis on their dependence on operator order. The phases are kept absolutely general (except for SC multiplets, for which consistency restrictions arise) to exhibit an aspect of the theory usually obscured by a remark of the type "clearly we can choose phases so that . . . ."

In order to handle all spins at the same time, we rewrite the C, P, and T transformations in a more convenient matrix form. Let i ($i = 1, 2, 3, 4$) be the spinor index, when appropriate. Then the transformations in question are

$$\begin{aligned} C\psi_{\mu i}^{\alpha}(x)C^{-1} &= \eta_C \psi_{\mu j}^{\alpha *}(x) C_{ji} \\ P\psi_{\mu i}^{\alpha}(\mathbf{x}, t)P^{-1} &= \eta_P \Pi(-g_{\mu\mu}) \psi_{\mu j}^{\alpha}(-\mathbf{x}, t) P_{ji} \\ T\psi_{\mu i}^{\alpha}(\mathbf{x}, t)T^{-1} &= \eta_T \Pi(g_{\mu\mu}) \psi_{\mu j}^{\alpha}(\mathbf{x}, -t) T'_{ji} \end{aligned} \tag{6.135}$$

Here summation over repeated indices is understood. For Fermions the real matrices $C = C_1\gamma_0$, $C_1 = -i\gamma_2$, $P = \gamma_0$, $T = \gamma_3\gamma_1$, while for Bosons $C = P = T = I$ and no spinor index occurs. T' is the transpose of T: $T' = -T$.

The matrices C, P, and T have been put to the right of the fields in (6.135) so that the matrix products occur in the same order as the operator products which induce the transformation. Since C, P, and T are real, the conjugate fields also transform in this way except that each phase η_i is replaced by η_i^*. Thus far the various phases η_C, η_P, η_T, η_α are completely general, subject to the unimodular condition. Hence four arbitrary phases occur in the behavior of the PC field; for SC fields η_P and one of η_C or η_T are restricted by Eqs. (6.134).

We begin by discussing bilinear products. We find easily the relations for the squares:

$$\begin{aligned} C^2\psi C^{-2} &= \psi \\ T^2\psi T^{-2} &= (-1)^{2S}\psi \\ P^2\psi P^{-2} &= \eta_P^2\psi \end{aligned} \tag{6.136}$$

The various field labels (α, μ, x) are the same on both sides of this equation. The phases η_C and η_T drop out since C and T involve conjugation in one form or another. In the PC case η_P^2 is an arbitrary number of the form $\exp(2i\varphi_P)$, $\eta_P = \exp(i\varphi_P)$ so that P^2 is a gauge transformation generated by the difference in number of particles and antiparticles. For SC multiplets this operator vanishes, so no such gauge transformation is possible. However, in this case η_P^2 is fixed to be $(-1)^{2S}$. Thus, introducing the Fermi counting operator F_s, which is $+1$ for states with an odd number of Fermions and 0 for states with an even number of Fermions or any number of Bosons, we have

$$\begin{aligned} C^2 &= I \\ T^2 &= (-1)^F \\ P^2 &= \begin{cases} (-1)^F & SC \text{ multiplets} \\ \exp(2i\varphi_P(N - \bar{N})) & PC \text{ multiplets} \end{cases} \end{aligned} \tag{6.137}$$

The products CP and PC differ in general

$$\begin{aligned} CP\psi^\alpha_i(\mathbf{x}, t)(CP)^{-1} &= \eta_P\eta_C\Pi(-g_{\mu\mu})\psi^{\alpha*}_{\mu j}(-\mathbf{x}, t)(CP)_{ji} \\ PC\psi^\alpha_{\mu i}(\mathbf{x}, t)(PC)^{-1} &= \eta_C\eta_P^*\Pi(-g_{\mu\mu})\psi^{\alpha*}_{\mu j}(-\mathbf{x}, t)(PC)_{ji} \end{aligned} \tag{6.138}$$

The product CP differs from PC by $(-1)^{2S}$, so the second transformation is (6.138) has phase $\eta_C \eta_P^*(-1)^{2S}$ as compared to $\eta_C \eta_P$ in the first. Thus, in general, $CP \neq PC$. However, for SC particles these phases coincide, giving

$$CP = PC \qquad (SC) \tag{6.139}$$

The squared operators $(CP)^2 = (PC)^2$ are independent of η_C, η_P, however:

$$\begin{aligned} (CP)^2\psi(CP)^{-2} &= (-1)^{2S}\psi \\ (CP)^2 = (PC)^2 &= (-1)^{F_S} \end{aligned} \tag{6.140}$$

For PT the situation is similar:

$$\begin{aligned} PT\psi_{\mu i}^{\alpha}(x)(PT)^{-1} &= \eta_P \eta_T (-1)^{n_S} \psi_{\mu j}^{\alpha}(-x)(PT')_{ji} \\ TP\psi_{\mu i}^{\alpha}(x)(TP)^{-1} &= \eta_P^* \eta_T (-1)^{n_S} \psi_{\mu j}^{\alpha}(-x)(T'P)_{ji} \end{aligned} \tag{6.141}$$

so that, in general, $PT \neq TP$, although for SC multiplets $PT = (-1)^F TP$. η_S is the greatest integer contained in S: $\eta_S = S$ for mesons and $S - \frac{1}{2}$ for Fermions. However, $(PT)^2 = (TP)^2$ is independent of phases

$$\begin{aligned} (PT)^2\psi(TP)^{-2} &= (-1)^{2S}\psi \\ (PT)^2 = (TP)^2 &= (-1)^{F_S} \end{aligned} \tag{6.142}$$

The operations CT and TC give

$$\begin{aligned} CT\psi_{\mu i}^{\alpha}(\mathbf{x}, t)(CT)^{-1} &= \eta_C \eta_T \psi_{\mu j}^{\alpha *}(\mathbf{x}, -t)(CT')_{ji} \\ TC\psi_{\mu i}^{\alpha}(\mathbf{x}, t)(TC)^{-1} &= \eta_C^* \eta_T^* \psi_{\mu j}^{\alpha *}(\mathbf{x}, -t)(T'C)_{ji} \end{aligned} \tag{6.143}$$

so that $CT = TC$ only when $\eta_C \eta_T$ is real. When SC multiplets are under consideration, $\eta_C^* \eta_T^*$ is equal to $(-1)^{2I}\eta_C \eta_T$, so that $CT = (-1)^{F_I} TC$, F_I being the isofermion number.

The corresponding squared operators give

$$\begin{aligned} (CT)^2\psi(CT)^{-2} &= (-1)^{2S}(\eta_C^* \eta_T^*)^2\psi \\ (TC)^2\psi(TC)^{-2} &= (-1)^{2S}(\eta_C \eta_T)^2\psi \end{aligned} \tag{6.144}$$

For SC multiplets $(\eta_C \eta_T)^2$ is $(-1)^{2I}$, giving

$$(CT)^2 = (TC)^2 = (-1)^{F_S + F_I} \tag{6.145}$$

It may be helpful to classify the previous results in order of decreasing generality.

1. Relations true for arbitrary phases:

$$
\begin{aligned}
C^2 &= I \\
P^2 &= \begin{cases} (-1)^{F_S} & (SC) \\ \exp\,[2i\varphi_P(N - \bar{N})] & (PC) \end{cases} \\
T^2 &= (-1)^{F_S} \\
(CP)^2 &= (PC)^2 = (-1)^{F_S} \\
(PT)^2 &= (TP)^2 = (-1)^{F_S + F_I}
\end{aligned}
\tag{6.146}
$$

2. SC multiplets; extra relations:

$$
\begin{aligned}
CP &= PC \\
PT &= (-1)^{F_S} TP \\
CT &= (-1)^{F_S} TC \\
(TC)^2 &= (CT)^2 = (-1)^{F_S + F_I}
\end{aligned}
\tag{6.147}
$$

3. PC multiplets; simple useful phase choices:

$$
\begin{aligned}
\eta_P^2 = 1 &\Rightarrow \begin{cases} P^2 = I \\ CP = (-1)^{F_S} PC \\ PT = TP \end{cases} \\
(\eta_C \eta_T)^2 = 1 &\Rightarrow \begin{matrix} CT = TC \\ (CT)^2 = (TC)^2 = (-1)^{F_S} \end{matrix}
\end{aligned}
\tag{6.148}
$$

Now we discuss the CPT transformation Θ. It is important to consider the six distinct operations Θ_i defined in Eq. (6.67).

Under Θ_1, the general field transforms as

$$
\begin{aligned}
\Theta_1 \psi^{\alpha}_{\mu i}(x) \Theta_1^{-1} &= (-1)^{n_S} \psi^{\alpha *}_{\mu j}(-x)(PCT')_{ji} \eta_C \eta_P \eta_T \\
PCT' &= -i\gamma_0\gamma_1\gamma_2\gamma_3 = \gamma_5, \qquad 2S \text{ odd} \\
&= 1, \qquad 2S \text{ even}
\end{aligned}
\tag{6.149}
$$

Clearly, every Θ transformation can be written as

$$
\Theta_i \psi^{\alpha}_{\mu j}(x) \Theta_i^{-1} = (-1)^{n_S} \omega_i \psi^{\alpha *}_{\mu k}(-x)(\theta_i)_{kj}
\tag{6.150}
$$

where ω_i is a product of CPT phases and θ_i is the matrix product of C, P, and T' in the same order as C, P, and T occur in Θ_i:

$$\begin{aligned} \theta_i &= 1, \qquad 2S = \text{even} \\ \theta_1 &= \theta_2 = -\theta_6 = -\theta_3 = -\theta_4 = -\theta_5 = \gamma_5, \qquad 2S = \text{odd} \end{aligned} \tag{6.151}$$

The phase factors ω_i are the same as in Eq. (6.67). The square of *CPT* results in

$$\Theta_i^2 \psi \Theta_i^{-2} = (\omega_i^*)^2 \psi \tag{6.152}$$

For *SC* multiplets the ω_i^2 are completely fixed by Eqs. (6.134) to be $(-1)^{2I+2S}$ *independent* of i:

$$\begin{aligned} \Theta_i^2 \psi \Theta_i^{-2} &= (-1)^{2I+2S} \psi \\ \Theta_i^2 &= (-1)^{F_S + F_I} \end{aligned} \tag{6.153}$$

Equation (6.153) generalizes the usual result that Θ^2 yields $(-1)^{2S}$ when applied to sets of Hermitian fields which define a *local* field theory. The case of self-conjugate isofermions ($2I$ = odd) is not equivalent to a local linear combination of Hermitian fields. We see that Θ^2 provides a simple way to test a self-conjugate field theory for locality. On the surface it appears that the "global" test of the *field* using Θ^2 might be more general than the local-causal properties of the field *theory*. However, it must be recognized that locality (rather, weak local commutativity) enters the discussion of Θ in a critical way.

Now we turn to the *PC* field. The phase ω_i is quite general, and Eq. (6.152) indicates that the operators are gauge transformations

$$\Theta_j^2 = \exp\,(2i\varphi_j(N - \bar{N})) \tag{6.154}$$

Only by making a special phase choice, or a physical assumption, can we make this phase definite. We stress this point because it is not always realized that the common choice $\Theta^2 = (-1)^F$ for *all* particles relies upon a special phase convention.

PROBLEMS

1. The phase ξ in Eq. (6.16) is not fixed and corresponds to the arbitrariness of the overall phase of the $a_\alpha(k)$. We can change $a_\alpha \to e^{i\varphi} a_\alpha$ (φ independent of α) without changing the basic commutation rules or the form of the isospin operators. The resultant phase

change of φ_α does not affect observable quadratics. Writing the family of fields in (6.19) as $\phi_\alpha[\xi, x]$ prove the following relation:

$$e^{-i\psi N}\varphi_\alpha[\xi, x]e^{i\psi N} = e^{i\psi}\varphi_\alpha[\xi e^{-2i\psi}, x]$$

where N is the number operator. This is not a gauge transformation of the usual type; by it we can adjust the phase of ξ to have any desired value.

2. Show that for free nucleons (ψ) and free pions ($\boldsymbol{\pi}$) the isospin operators have the form

$$\mathbf{I} = \int \psi^\dagger \frac{\boldsymbol{\tau}}{2} \psi d^3x \quad \text{(nucleons)}$$

$$\mathbf{I} = \int \boldsymbol{\pi} \times \dot{\boldsymbol{\pi}}\, d^3x \quad \text{(pions)}$$

and compute the commutators $[I_i, \psi]$, $[I_i, \pi_j]$. Prove that $\pi_1 - i\pi_2$ creates positive pions and destroys negative pions. Show that the operators destroying *states* normalized according to Eq. (6.3) are

$$a_{(\pm 1)} = \mp(a_1 \mp ia_2)/\sqrt{2}$$
$$a_0 \quad = a_3$$

where a_i are the destruction operators in the Cartesian basis. Finally use the canonical commutation relations to verify that the isospin commutation rules are satisfied.

3. Discover a unitary operator C having the properties

$$Ca_\alpha^* C^{-1} = \eta_\alpha b_{-\alpha}^*$$
$$Cb_\alpha^* C^{-1} = \eta_{-\alpha}^* a_{-\alpha}^*$$

Hint: employ the identity

$$e^{\lambda A}Be^{-\lambda A} = B + \lambda[A, B] + \frac{\lambda^2}{2!}[A, [A, B]] + \cdots$$

(λ an ordinary number). We expect C to be of the form exp $(i\theta\Lambda)$, Λ Hermitian. The structure of the problem suggests

$$\Lambda = \Lambda_1 \equiv \sum_{k\alpha} (\eta_\alpha a_\alpha(k) b_{-\alpha}^*(k) + \text{H.c.})$$

Prove that $U(\theta) = \exp(i\theta\Lambda_1)$ gives

$$U_1(\theta)a_\alpha^* U_1(\theta)^{-1} = \cos\theta a_\alpha^* + i\sin\theta\eta_\alpha b_{-\alpha}^*$$
$$U_1(\theta)b_\alpha^* U_1(\theta)^{-1} = i\sin\theta\eta_{-\alpha}^* a_{-\alpha}^* + \cos\theta b_\alpha^*$$

For $\theta = \pi/2$ this is nearly the desired result. The phase i can be transformed away using the number operator in the form $O(\theta) = \exp(i\theta N)$

$$N = \sum_{k\alpha} [a_\alpha^*(k)a_\alpha(k) + b_\alpha^*(k)b_\alpha(k)]$$

$$O(\theta)a_\alpha^* O^{-1}(\theta) = e^{i\theta}a_\alpha^*$$
$$O(\theta)b_\alpha^* O^{-1}(\theta) = e^{i\theta}b_\alpha^*$$

Show that C has the form

$$C = \exp\left[-i\frac{\pi}{2}\sum_{k\alpha}(a_\alpha^*(k) - \eta_\alpha b_{-\alpha}^*(k))(a_\alpha(k) - \eta_\alpha^* b_{-\alpha}(k))\right]$$

REFERENCES

1. L. Michel, in "Group Theoretical Concepts and Methods in Elementary Particle Physics," Ed. F. Gursey (Gordon and Breach, Inc., N.Y., 1964).
2. T. D. Lee and G. C. Wick, *Phys. Rev.* **148,** 1385 (1966).
3. L. C. Biedenharn, J. Nuyts and H. Ruegg, *Commun. Math. Phys.* **2,** 231 (1966).
4. P. Carruthers, *Phys. Rev. Letters* **18,** 353 (1967).
5. P. Carruthers, *J. Math. Phys.* **9,** 928 (1968); *ibid* **9,** 1835 (1968).
6. P. Carruthers, "Introduction to Unitary Symmetry" (John Wiley and Sons, Inc., N.Y., 1966), Chap. 1.
7. G. Feinberg and S. Weinberg, *Nuovo Cimento* **14,** 571 (1959).
8. P. B. Kantor, *Phys. Rev. Letters* **19,** 394 (1967).
9. O. Steinmann, *Phys. Letters* **25B,** 234 (1967).
10. B. Zumino and D. Zwanziger, *Phys. Rev.* **164,** 1959 (1967).
11. P. Carruthers, *Phys. Letters* **26B,** 158 (1968).
12. Huan Lee, *Phys. Rev. Letters* **18,** 1098 (1967).
13. M. Einhorn, (unpublished).
14. G. N. Fleming and E. Kazes, *Phys. Rev. Letters* **18,** 764 (1967).
15. Y. S. Jin, *Phys. Letters* **24B,** 411 (1967).
16. E. P. Wigner, "Group Theory" (Academic Press, N.Y., 1959).
17. E. U. Condon and G. H. Shortley, "The Theory of Atomic Spectra" (Cambridge University Press, London, 1957), Chap. 3.
18. A. R. Edmonds, "Angular Momentum in Quantum Mechanics" (Princeton University Press, Princeton, N.J., 1960).
19. S. Weinberg, *Phys. Rev.* **133B,** 1318 (1964).
20. W. Rarita and J. Schwinger, *Phys. Rev.* **60,** 61 (1941).
21. E. Majorana, Nuovo Cimento **14,** 171 (1937).

7

CROSSING PROPERTIES OF SCATTERING AMPLITUDES AND VERTICES

7.1 INTRODUCTION

One of the most remarkable results of local relativistic quantum field theory is that a given reaction is intimately related to allied reactions obtained by transferring particles from one side to the other, provided that the transferred particles are changed to antiparticles. The relations in question, which are obtained by analytic continuation in the relevant energy variables, were described as "the substitution rule" in the heyday of perturbation theory. We shall refer to the process of obtaining one reaction from another as "crossing." We give a detailed description of this process for two-body reactions and derive the crossing matrices which relate isospin amplitudes in various channels. In this analysis the fundamental role of the local field is especially evident.

We then consider the simpler process of "vertex crossing" wherein only one particle is transferred between initial and final states. Finally in Sec. 7.4 we consider the relations among scattering amplitudes which follow from combined isospin invariance and the discrete symmetry transformations of parity, charge conjugation, time reversal and their combination *TCP*. As in Chapter 6, we keep the various phase factors arbitrary in the general analysis. Our intent is not to create a "cult of the arbitrary phase factor" but rather to analyze one of the most troublesome (although conceptually simple) problems which one encounters in practice.

Consider the scattering of particles a, b, c, d with 4-momenta p_a, p_b, p_c, p_d and masses m_a, m_b, m_c, m_d. These have isospin I_a, I_b, I_c, I_d, I_3 components

α, β, γ, δ, hypercharges Y_i and charges $Q_i = I_{3i} + \frac{1}{2}Y_i$. In this chapter we consider particles without spin.

A basic result of local field theory is that the three distinct types of reactions (see Fig. 7.1)

$$\begin{aligned} &a + b \to c + d \qquad s \\ &a + \bar{d} \to c + \bar{b} \qquad u \\ &a + \bar{c} \to \bar{b} + d \qquad t \end{aligned} \tag{7.1}$$

and all described by the same function.

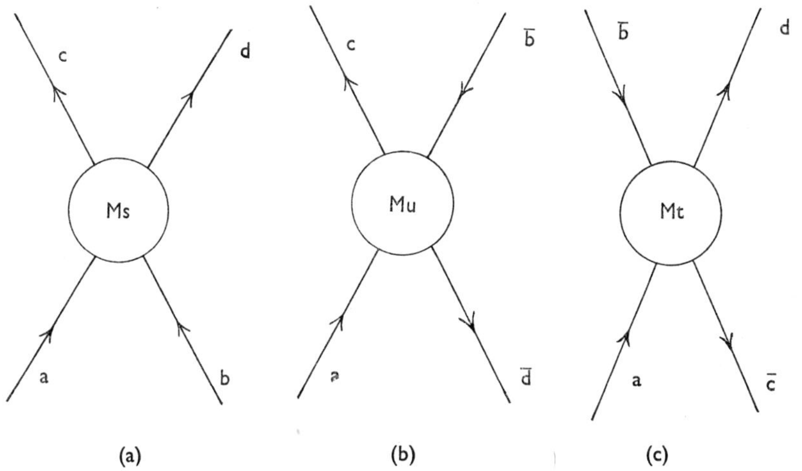

Figure 7.1 The s, t, and u channel reactions are connected by analytic continuation of one basic amplitude.

Each of the above reactions can be written in four equivalent ways; this is only a change in labeling, not of physics. Nevertheless when the order is changed certain phase changes occur that have to be watched.

In addition to equations (7.1) we have the corresponding *antiparticle* reactions

$$\begin{aligned} &\bar{a} + \bar{b} \to \bar{c} + \bar{d} \qquad \bar{s} \\ &\bar{a} + d \to \bar{c} + b \qquad \bar{u} \\ &\bar{a} + c \to b + \bar{d} \qquad \bar{t} \end{aligned} \tag{7.2}$$

and the inverse reactions to each of these:

$$\begin{aligned} c + d &\to a + b \qquad s' \\ c + \bar{b} &\to a + \bar{d} \qquad u' \\ \bar{b} + d &\to a + \bar{c} \qquad t' \end{aligned} \tag{7.3}$$

and

$$\begin{aligned} \bar{c} + \bar{d} &\to \bar{a} + \bar{b} \qquad \bar{s}' \\ \bar{c} + b &\to \bar{a} + d \qquad \bar{u}' \\ b + \bar{d} &\to \bar{a} + c \qquad \bar{t}' \end{aligned} \tag{7.4}$$

We shall find that these four sets of reactions are not independent when the discrete strong interaction symmetries are taken into account. For example under *TCP*, the reactions (7.4) become the same as (7.1); also (7.3) is transformed to (7.2). Similarly antiparticle conjugation has the effect. (7.1) $\leftrightarrow$ (7.2), (7.3) $\leftrightarrow$ (7.4) when C is conserved. Further under *TP* (7.1) $\leftrightarrow$ (7.3) and (7.2) $\leftrightarrow$ (7.4) when *TP* is conserved. If we were to consider the weak interactions, only the *TCP* symmetry relation would survive.

In Sec. 7.2 we study the relations among the reactions (7.1) which are a consequence of crossing. In Sec. 7.3 we describe the simpler case of crossing single particles and apply the result to vertex crossing. In Sec. 7.4 the discrete symmetries are studied.

7.2 CROSSING SYMMETRY FOR TWO-BODY AMPLITUDES

The reactions (7.1) may be regarded as various "channels" of one basic reaction, the distinct physical processes all being connected by analytic continuation of the invariant energy variables from one domain to another. The detailed labeling of the particle states may be found from Fig. (7.1). We follow the discussion of refs. (1–2).

As a technical device we introduce $2I + 1$ component vectors to describe the isospin of the particles and antiparticles. Since we shall use the fields extensively in our discussion of crossing we define the antiparticle wave functions in such a manner as to give the fields a simple form. We begin with the general spin zero pair-conjugate field (6.20) and arrange its components as a $2I + 1$ component column vector ψ. For this purpose we introduce the usual real eigenfunctions of the matrix I_3, χ_α, with components $(\chi_\alpha)_i = \delta_{\alpha i}$.

Then we can write

$$\begin{aligned}\psi(x) &= \sum_{\alpha} \chi_\alpha \psi_\alpha(x) \\ &= \sum_{k\alpha} (a_\alpha(k)\chi_\alpha f_k(x) + b_\alpha^*(k)\chi_\alpha^c f_k^*(x)) \end{aligned} \tag{7.5}$$

The antiparticle isospin wave function χ_α^c is

$$\chi_\alpha^c = \eta_{-\alpha}\chi_{-\alpha} \tag{7.6}$$

The χ_α^c obey the usual orthogonality and completeness relations. The χ_α^c transform exactly as χ_α and hence may be used combined with the latter by means of the usual Clebsch-Gordan coefficients[3,4]. This will be found useful in Chapter VIII when projection operators are derived. χ_α^c may be written as $C\chi_\alpha^*$ (here χ_α was chosen real) where $C_{\alpha\beta} = \xi(-1)^{I+\beta}\,\delta_{\alpha,-\beta}$ is the matrix which transforms the complex conjugate (c.c.) representation basis so that it transforms as χ_α. This is completely analogous to the way one relates the positron and electron wave functions.

We shall often simplify our general formulae by using the phase choice suitable to the eightfold way (Eq. (6.56)) in which case

$$\chi_\alpha^c = (-1)^{Q-\alpha}\chi_{-\alpha} \tag{7.7}$$

For example, the isospin wave functions for nucleons and antinucleons are given by

$$\begin{aligned} \chi(p) &= \begin{pmatrix}1\\0\end{pmatrix} \qquad \chi(n) = \begin{pmatrix}0\\1\end{pmatrix} \\ \chi^c(\bar{n}) &= \begin{pmatrix}0\\1\end{pmatrix} \qquad \chi^c(\bar{p}) = -\begin{pmatrix}1\\0\end{pmatrix} \end{aligned} \tag{7.8}$$

when we use Eq. (7.7).

In order to pick out a particle operator $a_\alpha(p)$ from $\psi(x)$ we first apply $\chi_\alpha^\dagger$ and then take the appropriate Klein-Gordon inner product. A similar construction gives the antiparticle operator.

$$\begin{aligned} a_\alpha(p) &= i\int d^3x f_p^*(x)\overleftrightarrow{\partial}_0 \chi_\alpha^\dagger \psi(x) \\ b_\alpha(p) &= i\int d^3x f_p^*(x)\overleftrightarrow{\partial}_0 (\psi^\dagger(x)\chi_\alpha^c) \end{aligned} \tag{7.9}$$

It is now a simple exercise using the reduction formula technique to derive the following representations for the Lorentz-invariant amplitudes

T defined by $S - 1 = iT/(16\omega_a\omega_b\omega_c\omega_d)^{\frac{1}{2}}$

$$
\begin{aligned}
M^s(p_c\gamma, p_d\delta; p_a\alpha, p_b\beta) &= \chi_\delta^\dagger T^s_{\gamma\alpha}(p_cp_d; p_ap_b)\chi_\beta \\
M^u(p_c\gamma, \bar{p}_b\bar{\beta}; p_a\alpha, \bar{p}_d\bar{\delta}) &= \chi_{\bar{\delta}}^{c\dagger} T^u_{\gamma\alpha}(p_c\bar{p}_b; p_a\bar{p}_d)\chi^c_{\bar{\beta}}
\end{aligned} \tag{7.10}
$$

$$
\begin{aligned}
T^s_{\gamma\alpha}(p_cp_d; p_ap_b) &= i\int d^4x\, d^4y e^{ip_d\cdot x - ip_b\cdot y} K^d_x K^b_y \langle c\gamma|\, T\psi_d(x)\psi_b^\dagger(y)\, |a\alpha\rangle \\
T^u_{\gamma\alpha}(p_c\bar{p}_b; p_a\bar{p}_d) &= i\int d^4x\, d^4y e^{i\bar{p}_b\cdot y - i\bar{p}_d\cdot x} K^d_x K^b_y \langle c\gamma|\, T\psi_d(x)\psi_b^\dagger(y)\, |a\alpha\rangle
\end{aligned} \tag{7.11}
$$

K^d_x is the Klein-Gordon operator for particle d, $K^d_x = \partial^2 + m_d^2$, etc. We assume that the integrand is sufficiently well behaved that we can continue $T^u_{\gamma\alpha}$ to the point $\bar{p}_b = -p_b$, $\bar{p}_d = -p_d$. At this point, unphysical for the u-channel reaction, $T^u_{\gamma\alpha}$ actually coincides with $T^s_{\gamma\alpha}$.

$$
T^u_{\gamma\alpha}(p_c - p_b; p_a - p_d) = T^s_{\gamma\alpha}(p_cp_d; p_ap_b) \tag{7.12}
$$

If we use the wave functions (7.6) we find the s-u crossing condition for the full amplitude

$$
M^u(p_c\gamma, -p_b - \beta; p_a\alpha, -p_d - \delta) = \eta^b_\beta \eta^{d*}_\delta M^s(p_c\gamma, p_d\delta; p_a\alpha, p_b\beta) \tag{7.13}
$$

Next we consider s-t crossing. By relabeling in the above calculation we have

$$
\begin{aligned}
M^s(p_c\gamma, p_d\delta; p_a\alpha, p_b\beta) &= \chi_\gamma^\dagger T^s_{\delta\alpha}(p_cp_d; p_ap_b)\chi_\beta \\
M^t(\bar{p}_b\bar{\beta}, p_d\delta; p_a\alpha\bar{p}_c\bar{\gamma}) &= \chi_{\bar{\gamma}}^{c\dagger} T^t_{\delta\alpha}(\bar{p}_bp_d; p_a\bar{p}_c)\chi^c_{\bar{\beta}}
\end{aligned} \tag{7.14}
$$

where the matrices $T^s_{\delta\alpha}$, $T^t_{\delta\alpha}$ are given by formulas similar to (7.11). The crossing formulas analogous to (7.12)–(7.13) are

$$
\begin{aligned}
T^t_{\delta\alpha}(-p_b, p_d; p_a, -p_c) &= T^s_{\delta\alpha}(p_cp_d; p_ap_b) \\
M^t(-p_b - \beta, p_d\delta; p_a\alpha, -p_c - \gamma) &= \eta^b_\beta \eta^{c*}_\gamma M^s(p_c\gamma, p_d\delta; p_a\alpha, p_b\beta)
\end{aligned} \tag{7.15}
$$

It is clear that the four-momentum-conserving δ-function is common to the amplitudes (7.13) and (7.15) and may be cancelled. This gives physical amplitudes M, which can be expanded in terms of isospin amplitudes by means of Clebsch-Gordan coefficients

$$
\begin{aligned}
(cd|\, M_s\, |ab) &= \sum_{s'} M_s(s')C(abs'; \alpha, \beta)C(cds'; \gamma, \delta) \\
(c\bar{b}|\, M_u\, |a\bar{d}) &= \sum_{u'} M_u(u')C(adu'; \alpha, -\delta)C(cbu'; \gamma, -\beta) \\
(\bar{b}d|\, M_t\, |a\bar{c}) &= \sum_{t'} M_t(t')C(act'; \alpha, -\gamma)C(bdt'; -\beta, \delta)
\end{aligned} \tag{7.16}
$$

In these expressions the amplitude $M_v(v')$ denotes the amplitude for scattering in a state of isospin v' in channel v. The detailed analysis of the expansions (7.16) is of great practical importance and has been described in many textbooks. An equivalent method, using projection operators, is given in Sec. 8.2.

The results (7.13) and (7.15) show that the isospin amplitudes in various channels are related by analytic continuation. Since the relation is important for many dynamical purposes, we explore it in detail. We summarize (7.13) and (7.15) by

$$\begin{aligned} (c\bar{b}|\, M_u\, |a\bar{d})_C &= \zeta_{su}(cd|\, M_s\, |ab), \qquad \zeta_{su} = \eta^b_\beta \eta^{d*}_\delta \\ (\bar{b}d|\, M_t\, |a\bar{c})_C &= \zeta_{st}(cd|\, M_s\, |ab), \qquad \zeta_{st} = \eta^b_\beta \eta^{c*}_\gamma \end{aligned} \tag{7.17}$$

where the subscript C means that the proper continuation has been made. We shall call the ζ's "crossing phases." Using (7.17) we can solve (7.16) for the isospin amplitudes using the identities:

$$\begin{aligned} C(abs';\alpha\beta)C(cds';\gamma\delta) &= \sum_{u'} (2s'+1)(-1)^{\gamma-\alpha+a-c} W(abdc; s'u') \\ &\quad \times C(adu';\alpha,-\delta)C(cbu';\gamma,-\beta) \\ C(abs';\alpha\beta)C(cds';\gamma\delta) &= \sum_{t'} (2s'+1)(-1)^{\delta-\alpha+a+d} W(as't'd; bc) \\ &\quad \times C(bdt';-\beta,\delta)C(act';\alpha,-\gamma) \end{aligned} \tag{7.18}$$

The Racah coefficients $W(abcd; ef)$ and the Clebsch-Gordan coefficients are as defined in ref. 3. In terms of the continued u and t channel amplitudes the crossing relations are

$$\begin{aligned} M_u(u') &= \sum_{s'} X_{u's'} M_s(s') \\ M_t(t') &= \sum_{s'} X_{t's'} M_s(s') \end{aligned} \tag{7.19}$$

where the crossing matrices are given by

$$\begin{aligned} X_{us} &= \zeta_{su}(-1)^{\gamma-\alpha+a-c}(2s+1)W(abdc; su) \\ X_{ts} &= \zeta_{st}(-1)^{\delta-\alpha+a+d}(2s+1)W(astd; bc) \end{aligned} \tag{7.20}$$

In these equations a, b, etc. are an abbreviation for I_a, I_b, etc.

The Racah coefficients are often called "recoupling coefficients" since they relate schemes in which different particles are paired to give a resultant angular momentum.[3] This clearly corresponds to the physical interpretation of crossing as the switching of the legs of the graphs of Fig. (7.1) and the resultant necessity for recoupling.

It is clear that the crossing matrices are independent of the azimuthal components. If we employ the useful special phase convention (7.7) we get the explicit formulas

$$\begin{aligned} X_{us} &= (-1)^{\pi_{su}}(2s+1)W(abdc; su) \\ X_{ts} &= (-1)^{\pi_{st}}(2s+1)W(astd; bc) \\ \pi_{su} &= a - c + \tfrac{1}{2}Y_d - \tfrac{1}{2}Y_b \\ \pi_{st} &= a + d + \tfrac{1}{2}Y_c - \tfrac{1}{2}Y_b \end{aligned} \tag{7.21}$$

One may now employ tables of Racah coefficients (or equivalently, $6j$ symbols) to give explicit crossing matrices. Most of those which arise in practice are given in Table (7.1). Those reactions which involve pions are special cases of more general expressions given in Eqs. (8.90), (8.93), (8.98) and (8.99). When applying these results to Fermions, one has to adhere to a particular ordering convention besides paying attention to the spin dependence of the amplitude. In using these relations one must (at the beginning) decide which are particles and which are antiparticles. Although the names are arbitrary, the distinction cannot be interchanged casually without incurring well-deserved phase errors when isospinors are involved.

The crossing matrices have a number of simple but important properties which we now discuss. These properties follow from the orthogonality property of the Racah coefficients[3]

$$\sum_e (2e+1)(2f+1)W(abdc; ef)W(abdc; eg) = \delta_{fg} \tag{7.22}$$

setting $e = s, f = u, g = u'$ gives on using (7.21) the relation

$$\sum_s \left(\frac{2u+1}{2s+1}\right)(X)_{u's}(X)_{us} = \delta_{uu'} \tag{7.23}$$

for the components of the matrix X_{us}. Hence the inverse to the matrix $(X)_{us}$ has components which we denote by

$$(X^{-1})_{su} = \frac{2u+1}{2s+1} X_{us} \tag{7.24}$$

The same formula is valid for s-t crossing on substituting $u \to t$. The result (7.24) is a special case of a more general result which involves the dimensions of the irreducible representations involved in the crossing (ref. 5).

Table 7.1 Crossing matrices and their inverses are given for common reactions. Each matrix is labeled, left to right, and up to down, by increasing isospins of the allowed values. The isospins of particles N, K, π, Σ, Y_1^*, N^*, Y_2^*, N', are respectively $\frac{1}{2}$, $\frac{1}{2}$, 1, 1, 1, 2, $\frac{5}{2}$. To find the ouput force due to the "exchange" of a particle of isospin I, read down the I column of the appropriate crossing matrix.

Process	X_{us}	$(X^{-1})_{su}$	(X_{ts})	$(X^{-1})_{st}$
$NK \to NK$ $N\bar{K} \to N\bar{K}$ $N\bar{N} \to \bar{K}K$	$\begin{pmatrix} -\frac{1}{2} & \frac{3}{2} \\ \frac{1}{2} & \frac{1}{2} \end{pmatrix}$	X_{su}	$\begin{pmatrix} -\frac{1}{2} & -\frac{3}{2} \\ -\frac{1}{2} & \frac{1}{2} \end{pmatrix}$	X_{ts}
$N\pi \to N$ $N\bar{\pi} \to N\bar{\pi}$ $N\bar{N} \to \bar{\pi}\pi$	$\begin{pmatrix} -\frac{1}{3} & \frac{4}{3} \\ \frac{2}{3} & \frac{1}{3} \end{pmatrix}$	X_{su}	$\begin{pmatrix} \frac{\sqrt{6}}{3} & \frac{2\sqrt{6}}{3} \\ \frac{2}{3} & -\frac{2}{3} \end{pmatrix}$	$\begin{pmatrix} \frac{\sqrt{6}}{6} & 1 \\ \frac{\sqrt{6}}{6} & -\frac{1}{2} \end{pmatrix}$
$N\pi \to N^*\pi$ $N\bar{\pi} \to N^*\bar{\pi}$ $N\bar{N}^* \to \bar{\pi}\pi$	$\begin{pmatrix} \frac{2}{3} & -\frac{\sqrt{10}}{3} \\ -\frac{\sqrt{10}}{6} & -\frac{2}{3} \end{pmatrix}$	X_{su}	$\begin{pmatrix} \frac{1}{3} & \frac{\sqrt{10}}{3} \\ \frac{\sqrt{3}}{3} & -\frac{\sqrt{30}}{15} \end{pmatrix}$	$\begin{pmatrix} \frac{1}{2} & \frac{5\sqrt{3}}{6} \\ \frac{\sqrt{10}}{4} & -\frac{\sqrt{30}}{12} \end{pmatrix}$
$N^*\pi \to N^*\pi$ $N^*\bar{\pi} \to N^*\bar{\pi}$ $N^*\bar{N}^* \to \bar{\pi}\pi$	$\begin{pmatrix} \frac{1}{6} & -\frac{2}{3} & \frac{3}{2} \\ -\frac{1}{3} & \frac{11}{15} & \frac{3}{5} \\ \frac{1}{2} & \frac{2}{5} & \frac{1}{10} \end{pmatrix}$	X_{su}	$\begin{pmatrix} -\frac{\sqrt{3}}{3} & -\frac{2\sqrt{3}}{3} & -\sqrt{3} \\ -\frac{\sqrt{10}}{6} & -\frac{2\sqrt{10}}{15} & \frac{3\sqrt{10}}{10} \\ -\frac{\sqrt{6}}{6} & \frac{4\sqrt{6}}{15} & -\frac{\sqrt{6}}{10} \end{pmatrix}$	$\begin{pmatrix} -\frac{\sqrt{3}}{6} & -\frac{\sqrt{10}}{4} & -\frac{5\sqrt{6}}{12} \\ -\frac{\sqrt{3}}{6} & -\frac{\sqrt{10}}{10} & \frac{\sqrt{6}}{3} \\ -\frac{\sqrt{3}}{6} & \frac{3\sqrt{10}}{20} & -\frac{\sqrt{6}}{12} \end{pmatrix}$

(continued overleaf)

Table 7.1 (*continued*)

Process	X_{us}	$(X^{-1})_{su}$	(X_{ts})	$(X^{-1})_{st}$
$\Sigma\pi \to \Sigma\pi$ $\Sigma\bar{\pi} \to \Sigma\bar{\pi}$ $\Sigma\bar{\Sigma} \to \bar{\pi}\pi$	$\begin{pmatrix} \frac{1}{3} & -1 & \frac{5}{3} \\ -\frac{1}{3} & \frac{1}{2} & \frac{5}{6} \\ \frac{1}{3} & \frac{1}{2} & \frac{1}{6} \end{pmatrix}$	X_{su}	$\begin{pmatrix} \frac{1}{3} & 1 & \frac{5}{3} \\ \frac{1}{3} & \frac{1}{2} & -\frac{5}{6} \\ \frac{1}{3} & -\frac{1}{2} & \frac{1}{6} \end{pmatrix}$	X_{ts}
$N^*K \to N^*K$ $N^*\bar{K} \to N^*\bar{K}$ $N^*\bar{N}^* \to \bar{K}K$	$\begin{pmatrix} -\frac{1}{4} & \frac{5}{4} \\ \frac{3}{4} & \frac{1}{4} \end{pmatrix}$	X_{su}	$\begin{pmatrix} \frac{3\sqrt{2}}{4} & \frac{5\sqrt{2}}{4} \\ \frac{\sqrt{10}}{4} & -\frac{\sqrt{10}}{4} \end{pmatrix}$	$\begin{pmatrix} \frac{\sqrt{2}}{4} & \frac{\sqrt{10}}{4} \\ \frac{\sqrt{2}}{4} & -\frac{3\sqrt{10}}{20} \end{pmatrix}$
$N^*\pi \to N'\pi$ $N^*\bar{\pi} \to N'\bar{\pi}$ $N^*\bar{N}' \to \bar{\pi}\pi$	$\begin{pmatrix} \frac{2}{5} & -\frac{3\sqrt{14}}{10} \\ -\frac{\sqrt{14}}{5} & -\frac{2}{5} \end{pmatrix}$	X_{su}	$\begin{pmatrix} -\frac{\sqrt{10}}{5} & -\frac{\sqrt{35}}{5} \\ -\frac{\sqrt{14}}{5} & \frac{3}{5} \end{pmatrix}$	$\begin{pmatrix} -\frac{3\sqrt{10}}{20} & -\frac{\sqrt{14}}{4} \\ -\frac{\sqrt{35}}{10} & \frac{1}{2} \end{pmatrix}$
$N'\pi \to N'\pi$ $N'\bar{\pi} \to N'\bar{\pi}$ $N'\bar{N}' \to \bar{\pi}\pi$	$\begin{pmatrix} \frac{1}{15} & -\frac{4}{15} & \frac{2}{3} \\ -\frac{2}{5} & \frac{31}{35} & \frac{2}{7} \\ \frac{4}{3} & \frac{8}{21} & \frac{1}{21} \end{pmatrix}$	X_{su}	$\begin{pmatrix} \frac{2\sqrt{2}}{3} & \sqrt{2} & \frac{4\sqrt{2}}{3} \\ \frac{2\sqrt{35}}{15} & \frac{2\sqrt{35}}{35} & -\frac{4\sqrt{35}}{21} \\ \frac{2\sqrt{14}}{15} & -\frac{8\sqrt{14}}{35} & \frac{2\sqrt{14}}{21} \end{pmatrix}$	$\begin{pmatrix} \frac{\sqrt{2}}{6} & \frac{\sqrt{35}}{10} & \frac{\sqrt{14}}{6} \\ \frac{\sqrt{2}}{6} & \frac{\sqrt{35}}{35} & -\frac{4\sqrt{14}}{21} \\ \frac{\sqrt{2}}{6} & -\frac{\sqrt{35}}{14} & \frac{5\sqrt{14}}{84} \end{pmatrix}$

$Y_1^*\pi \to Y_2^*\pi$ $Y_1^*\bar{\pi} \to Y_2^*\bar{\pi}$ $Y_1^*\bar{Y}_2^* \to \bar{\pi}\pi$	$\begin{pmatrix} \frac{1}{2} & -\frac{\sqrt{5}}{2} \\ -\frac{3\sqrt{5}}{10} & -\frac{1}{2} \end{pmatrix}$	X_{us}	$\begin{pmatrix} \frac{1}{2} & \frac{\sqrt{5}}{2} \\ \frac{3\sqrt{5}}{10} & -\frac{1}{2} \end{pmatrix}$	X_{ts}
$N^*\pi \to Y_2^*K$ $N^*\bar{K} \to Y_2^*\bar{\pi}$ $N^*\bar{Y}_2^* \to \bar{\pi}K$	$\begin{pmatrix} -\frac{\sqrt{30}}{15} & \frac{3\sqrt{5}}{5} \\ \frac{3\sqrt{2}}{5} & \frac{\sqrt{3}}{5} \end{pmatrix}$	$\begin{pmatrix} -\frac{\sqrt{30}}{20} & \frac{3\sqrt{50}}{20} \\ \frac{3\sqrt{20}}{20} & \frac{\sqrt{3}}{6} \end{pmatrix}$	$\begin{pmatrix} \frac{\sqrt{10}}{5} & \frac{2\sqrt{15}}{5} \\ \frac{2\sqrt{5}}{5} & -\frac{\sqrt{30}}{10} \end{pmatrix}$	$\begin{pmatrix} \frac{\sqrt{10}}{10} & \frac{2\sqrt{5}}{5} \\ \frac{2\sqrt{15}}{15} & -\frac{\sqrt{30}}{15} \end{pmatrix}$
$Y_2^*\pi \to Y_2^*\pi$ $Y_2^*\bar{\pi} \to Y_2^*\bar{\pi}$ $Y_2^*\bar{Y}_2^* \to \bar{\pi}\pi$	$\begin{pmatrix} \frac{1}{10} & -\frac{1}{2} & \frac{7}{5} \\ -\frac{3}{10} & \frac{5}{6} & \frac{7}{15} \\ \frac{3}{5} & \frac{1}{3} & \frac{1}{15} \end{pmatrix}$	X_{us}	$\begin{pmatrix} -\frac{\sqrt{15}}{5} & -\frac{\sqrt{15}}{3} & -\frac{7\sqrt{15}}{15} \\ -\frac{3\sqrt{5}}{10} & -\frac{\sqrt{5}}{6} & \frac{7\sqrt{5}}{15} \\ -\frac{\sqrt{21}}{10} & \frac{\sqrt{21}}{6} & -\frac{\sqrt{21}}{15} \end{pmatrix}$	$\begin{pmatrix} -\frac{\sqrt{15}}{15} & -\frac{3\sqrt{5}}{10} & -\frac{\sqrt{21}}{6} \\ -\frac{\sqrt{15}}{15} & -\frac{\sqrt{5}}{10} & \frac{\sqrt{21}}{6} \\ -\frac{\sqrt{15}}{15} & \frac{\sqrt{5}}{5} & -\frac{\sqrt{21}}{21} \end{pmatrix}$

Further simplifications occur if the crossed particles have the same isospin, since the isospins in the crossed channels have the same set of values. In this case the crossing matrix obeys $X^2 = 1$, e.g. for s-u crossing

$$\sum_s (X)_{u's}(X)_{su} = \delta_{uu'} \tag{7.25}$$

Here the meaning of the notation should be stressed to avoid possible confusion. In Eq. (7.25) the symbol $(X)_{su}$ means the su component of the crossing matrix X_{us}. Hence a different notation such as $\sum_j (X_{us})_{ij}(X_{us})_{jk} = \delta_{ik}$ might be preferred. Using a symmetry relation of the Racah coefficients then yields ($b = d$)

$$(X)_{us} = \frac{2s + 1}{2u + 1}(X)_{su} \tag{7.26}$$

so that (7.23) simplifies in the special case $b = d$. The result (7.25) provides a useful check on the accuracy of crossing matrices and can also be used as an aid in their construction. The crossing matrix X_{ts} also satisfies $X^2 = 1$ if $b = c$, i.e. if the crossed particles have the same isospin. If all four particles in a reaction have the same isospin then the X_{ts} is the same as X_{us} apart from a sign easily computed using Eqs. (7.21).

Further properties of the isospin crossing matrices are given in refs. 1 and 2. General discussions of isospin crossing matrices have been given in refs. 1, 2, 6–11, and elsewhere. Extensions to the group SU_3 are given in refs. 2, 9, 12, 13, 14. A treatment appropriate to SU_6 is given in ref. 15 and SU_n is considered in ref. 16. It is also interesting to consider crossing properties of amplitudes in which more than four particles are involved (ref. 17). Further interesting problems arise due to modifications necessitated by the existence of spin. Some of these questions are taken up in Chapter 10.

7.3 CROSSING OF A SINGLE PARTICLE: VERTEX CROSSING

From the foregoing discussion one correctly surmises that general rules may be given for crossing of a single particle: we expect that $A + a \to B$ (A, B states of arbitrary composition) can be changed to $A \to B + \bar{a}$ if we change the momentum and internal quantum numbers appropriately. Thus we define

$$
\begin{aligned}
M_1(k, \alpha) &= \sqrt{2\omega}\,\langle B|\, a^*_{\text{in}}(k, \alpha)\,|A\rangle \qquad && (a + A \to B) \\
M_2(\bar{k}, \bar{\alpha}) &= \sqrt{2\bar{\omega}}\,\langle B|\, b_{\text{out}}(\bar{k}, \bar{\alpha})\,|A\rangle \qquad && (A \to \bar{a} + B) \\
M_3(k, \alpha) &= \sqrt{2\omega}\,\langle B|\, b^*_{\text{in}}(k, \alpha)\,|A\rangle \qquad && (\bar{a} + A \to B) \\
M_4(\bar{k}, \bar{\alpha}) &= \sqrt{2\bar{\omega}}\langle B|\, a_{\text{out}}(\bar{k}, \bar{\alpha})\,|A\rangle \qquad && (A \to a + B)
\end{aligned}
\tag{7.27}
$$

We suppose that the constituents of A have "in" boundary conditions and that B is subject to "out" boundary conditions. The amplitudes (7.27) may be related if we employ the usual "LSZ" reduction formulas.[18,19] These are conveniently written as

$$
\begin{aligned}
a_{\text{out}}(k, \alpha) &= a_{\text{in}}(k, \alpha) + i \int d^4x f^*_k(x)\, j_\alpha(x) \\
b_{\text{out}}(k, \alpha) &= b_{\text{in}}(k, \alpha) + i\eta_{-\alpha} \int d^4x f^*_k(x)\, j^*_{-\alpha}(x) \\
a^*_{\text{out}}(k, \alpha) &= a^*_{\text{in}}(k, \alpha) - i \int d^4x f_k(x)\, j^*_\alpha(x) \\
b^*_{\text{out}}(k, \alpha) &= b^*_{\text{in}}(k, \alpha) - i\eta^*_{-\alpha} \int d^4x f_k(x)\, j_{-\alpha}(x)
\end{aligned}
\tag{7.28}
$$

where $j_\alpha \equiv K_\alpha \psi_\alpha$. These are supposed to be valid expressions within matrix elements. As an example consider the amplitude M_1. Recalling that Eq. (7.9) may be used to define a_{in} in terms of $\langle B|\,\psi\,|A\rangle \to \langle B|\,\psi_{\text{in}}\,|A\rangle$ as $t \to -\infty$ we integrate by parts and use $(\partial^2 + m^2) f_k = 0$ to obtain

$$
\begin{aligned}
\langle B|\, a^*_{\text{in}}(k, \alpha)\,|A\rangle &= i \int_{t\to-\infty} d^3x \langle B|\, \psi^*_\alpha(x)\,|A\rangle \overleftrightarrow{\partial}_0 f_k(x) \\
&= \langle B|\, a^*_{\text{out}}(k, \alpha)\,|A\rangle - i \int d^4x\, \partial_0[\langle B|\, \psi^*_\alpha(x)\,|A\rangle \overleftrightarrow{\partial}_0 f_k(x)] \\
&= \langle B|\, a^*_{\text{out}}(k, \alpha)\,|A\rangle + i \int d^4x f_k(x) K^\alpha_\alpha \langle B|\, \psi^*_\alpha(x)\,|A\rangle
\end{aligned}
\tag{7.29}
$$

We see that this gives the third of Eqs. (7.28). The amplitude $\langle B|\, a^*_{\text{out}}\,|A\rangle$ either vanishes or corresponds to a disconnected graph if $|B\rangle$ contains a particle in the same state. Hence the connected parts of the amplitudes M_i are given by

$$\begin{aligned}
M_1(k, \alpha) &= i \int d^4x e^{-ik\cdot x} \langle B|\, j_\alpha^*(x)\, |A\rangle \\
M_2(\bar{k}, \bar{\alpha}) &= i\eta_{-\bar{\alpha}} \int d^4x e^{i\bar{k}\cdot x} \langle B|\, j_{-\alpha}^*(x)\, |A\rangle \\
M_3(k, \alpha) &= i\eta_{-\alpha}^* \int d^4x e^{-ik\cdot x} \langle B|\, j_{-\alpha}(x)\, |A\rangle \\
M_4(\bar{k}, \bar{\alpha}) &= i \int d^4x e^{i\bar{k}\cdot x} \langle B|\, j_{\bar{\alpha}}(x)\, |A\rangle
\end{aligned} \tag{7.30}$$

Hence the following crossing relations hold:

$$\begin{aligned}
M_1(-k, -\alpha) &= \eta_{-\alpha}^* M_2(k, \alpha) \\
M_2(-k, -\alpha) &= \eta_\alpha M_1(k, \alpha) \\
M_3(-k, -\alpha) &= \eta_\alpha^* M_4(k, \alpha) \\
M_4(-k, -\alpha) &= \eta_{-\alpha} M_3(k, \alpha)
\end{aligned} \tag{7.31}$$

Although only half of these relations are independent we find it convenient to display them all. The first relation says that if we have an initial particle ($aA \to B$) and transform (p, α) to $(-p, -\alpha)$ we get the amplitude for $A \to B\bar{a}(p, \alpha)$ multiplied by $\eta_{-\alpha}^*$. The other relations have similar meanings, which are easily discovered from the definitions of Eq. (7.27). We only note one further instance of the need for care with antiparticles. The third of equations (7.31) says that if we transfer an initial antiparticle the phase factor is η_α^*, which differs by $(-1)^{2I}$ as compared to transferring an initial particle. For self-conjugate fields ($b = a$ in Eqs. 7.27) this is consistent since only integral I is allowed by local field theory.

By repeated application of the rules of Eq. (7.31) we can obtain (7.13). A more interesting practical application is to let A and B be various single particle states. To be sure, the physical three-particle vertices of (7.27) do not always satisfy energy-momentum conservation. However the kinematic structure of such vertices would be the same if we were to make the appropriate off-shell definitions, so we henceforth ignore this point.

We begin by considering a basic vertex $a + b \to c$ and compare this with crossed vertices $a \to c + \bar{b}$, $a + \bar{c} \to b$. Clearly these are special cases of the two body reaction, with particle d suppressed. In analogy to Eq. (7.1) it is useful to define three "channels"

$$
\begin{aligned}
&a(p_a\alpha) + b(p_b\beta) \rightarrow c(p_c\gamma) \\
&a(p_a\alpha) \rightarrow c(p_c\gamma) + \bar{b}(\bar{p}_b\bar{\beta}) \\
&a(p_a\alpha) + \bar{c}(\bar{p}_c\bar{\gamma}) \rightarrow \bar{b}(\bar{p}_b\bar{\beta})
\end{aligned} \tag{7.32}
$$

We could also consider various inverse and antiparticle reactions. In the next section we shall give appropriate methods for this task.

From Eq. (7.29) and similar calculations we find

$$
\begin{aligned}
\sqrt{2\omega_b}\,\langle c \mid ab \text{ in}\rangle &= i(2\pi)^4\,\delta(p_a + p_b - p_c)\langle c|\, j_\beta^*(0)\,|a\rangle \\
\sqrt{2\omega_b}\,\langle cb \text{ out} \mid a\rangle &= i(2\pi)^4\,\delta(p_a - \bar{p}_b - p_c)\langle c|\, j_{-\bar{\beta}}^*(0)\,|a\rangle\eta_{-\bar{\beta}} \\
\sqrt{2\omega_b}\,\langle \bar{b} \mid a\bar{c} \text{ in}\rangle &= i(2\pi)^4\,\delta(\bar{p}_b - p_a - \bar{p}_c)\langle 0|\, j_{-\bar{\beta}}^*(0)\,|a\bar{c} \text{ in}\rangle\eta_{-\bar{\beta}}
\end{aligned} \tag{7.33}
$$

Hence we are led to define three covariant vertex functions by

$$
\begin{aligned}
\Gamma^s(ab \rightarrow c) &= \sqrt{4\omega_a\omega_c}\,\langle c|\, j_\beta^*(0)\,|a\rangle \\
\Gamma^u(a \rightarrow c\bar{b}) &= \sqrt{4\omega_a\omega_c}\eta_{-\bar{\beta}}\langle c|\, j_{-\bar{\beta}}^*(0)\,|a\rangle \\
\Gamma^t(a\bar{c} \rightarrow \bar{b}) &= \sqrt{4\omega_a\omega_c}\eta_{-\bar{\beta}}\langle 0|\, j_{-\bar{\beta}}^*(0)\,|a\bar{c} \text{ in}\rangle
\end{aligned} \tag{7.34}
$$

The definitions (7.34) are to be used for particle b off-shell. The crossing conditions follow simply by using the *LSZ* formalism to "pull in" the crossed particles:

$$
\begin{aligned}
\Gamma^s &= i\int d^4x e^{ip_c\cdot x} K_x^c \langle 0|\, T\psi_\gamma^c(x) j_\beta^*(0)\,|a\rangle\sqrt{2\omega_a} \\
\Gamma^t &= i\eta_{-\bar{\beta}}\eta_{-\bar{\gamma}}^*\int d^4x e^{-i\bar{p}_c\cdot x} K_x^c \langle 0|\, T j_{-\bar{\beta}}^*(0)\psi_{-\bar{\gamma}}^c(x)\,|a\rangle\sqrt{2\omega_a}
\end{aligned} \tag{7.35}
$$

Writing $\Gamma^s = \Gamma^s(p_c;\, \alpha\beta\gamma)$, $\Gamma^t = \Gamma^t(\bar{p}_c;\, \alpha\bar{\beta}\bar{\gamma})$, (7.35) gives the crossing condition

$$
\Gamma^t(-p_c;\, \alpha - \beta - \gamma) = \eta_\beta\eta_\gamma^*\Gamma^s(p_c;\, \alpha\beta\gamma) \tag{7.36}
$$

A similar computation gives

$$
\Gamma^u(-p_b;\, \alpha - \beta\gamma) = \eta\ \Gamma^s(p_b,\, \alpha\beta\gamma) \tag{7.37}
$$

These results are as expected from the rules of Eq. (7.31).

Next we define vertex couplings by making use of the Wigner-Eckart theorem to exhibit the dependence on the "magnetic" quantum numbers α, β, γ.

$$\begin{aligned}\Gamma^s(a\alpha b\beta \to c\gamma) &= g(ab \to c)C(abc; \alpha\beta\gamma)\\ \Gamma^u(a\alpha \to c\gamma\bar{b}\bar{\beta}) &= g(a \to c\bar{b})C(cba; \gamma\bar{\beta}\alpha)\\ \Gamma^t(a\alpha\bar{c}\bar{\gamma} \to \bar{b}\bar{\beta}) &= g(a\bar{c} \to \bar{b})C(acb; \alpha\bar{\gamma}\beta)\end{aligned} \tag{7.38}$$

The couplings $g(ab \to c)$ etc. may be related by use of the crossing relations (7.36–7.37), using standard symmetry relations to eliminate the Clebsch-Gordan coefficients. One finds

$$\begin{aligned}g(ab \to c) &= \zeta_1\sqrt{\frac{2a+1}{2c+1}}\, g(a \to c\bar{b})\\ g(ab \to c) &= \zeta_2\sqrt{\frac{2b+1}{2c+1}}\, g(a\bar{c} \to \bar{b})\end{aligned} \tag{7.39}$$

where the factors ζ_i depend on the phase convention for η_α. $\zeta_1 = \eta_\beta^*(-1)^{b-\beta}(-1)^{a+b-c}$ and $\zeta_2 = \eta_\gamma\eta_\beta^*(-1)^{a-\alpha}$. Writing the phase factors η_β (cf. Eq. (6.16)) in such a manner as to cancel the azimuthal dependence gives $\zeta_1 = \xi_b^*(-1)^{a+b-c}$, $\zeta_2 = \xi_b^*\xi_c(-1)^{a+b-c}$. In order to be compatible with the eightfold way phase choice (6.56), one writes $\xi = (-1)^{I+\frac{1}{2}Y}$. Vertex crossing seems to have been first studied in detail in ref. 11.

7.4 SYMMETRIES OF THE SCATTERING AMPLITUDE WHICH FOLLOW FROM *C*, *P*, AND *T* INVARIANCE

We summarize briefly some well-known symmetries which follow from invariance under the usual discrete symmetries. We also assume isospin invariance so that the S matrix satisfies

$$\begin{aligned}CSC^{-1} &= S\\ PSP^{-1} &= S\\ TS^\dagger T^{-1} &= S\\ \Theta S^\dagger\Theta^{-1} &= S\\ [\mathbf{I}, S] &= 0\end{aligned} \tag{7.40}$$

A simple formal proof of these properties follows from the interaction picture form of the S-matrix: $S = T\exp(-i\int \mathscr{H}(x)\,dx)$. We recall that the "anti-unitary" nature of T and Θ is expressed by the inner product relations

$$\begin{aligned} \langle T\Phi \mid T\Psi\rangle &= \langle \Phi \mid \Psi\rangle^* \\ \langle \Theta\Phi \mid \Theta\Psi\rangle &= \langle \Phi \mid \Psi\rangle^* \end{aligned} \tag{7.41}$$

for arbitrary states Φ and Ψ.

Letting $\Psi = S^\dagger\Phi'$ in (7.41) gives

$$\begin{aligned} \langle\Phi|\, S\,|\Phi'\rangle &= \langle T\Phi'|\, S\,|T\Phi\rangle \\ &= \langle\Theta\Phi'|\, S\,|\Theta\Phi\rangle \end{aligned} \tag{7.42}$$

where Φ, Φ' are both *in* or both *out* states. Similarly for the unitary transformations C and P one has

$$\begin{aligned} \langle\Phi|\, S\,|\Phi'\rangle &= \langle C\Phi|\, S\,|C\Phi'\rangle \\ \langle\Phi|\, S\,|\Phi'\rangle &= \langle P\Phi|\, S\,|P\Phi'\rangle \end{aligned} \tag{7.43}$$

The scattering amplitude $ab \to cd$ (Fig. 7.1) may be written as $\langle cd\,\text{out} \mid ab\,\text{in}\rangle$ and the in, out states are constructed as products of appropriate particle operators. For example the s, u and t channel amplitude are expressible as

$$\begin{aligned} \langle cd|\, S\,|ab\rangle &= \langle 0|\, a_{\text{out}}(c)a_{\text{out}}(d)a^*_{\text{in}}(a)a^*_{\text{in}}(b)\,|0\rangle \\ \langle c\bar{b}|\, S\,|a\bar{d}\rangle &= \langle 0|\, a_{\text{out}}(c)b_{\text{out}}(b)a^*_{\text{in}}(a)b^*_{\text{in}}(d)\,|0\rangle \\ \langle b\bar{d}|\, S\,|\bar{a}c\rangle &= \langle 0|\, a_{\text{out}}(b)b_{\text{out}}(d)b^*_{\text{in}}(a)a^*_{\text{in}}(c)\,|0\rangle \end{aligned} \tag{7.44}$$

The transformation laws of the particle operator under C, P, T, and Θ are exactly as given in Sec. (6.5) except that the *in* and *out* boundary conditions are interchanged by T or Θ, e.g.

$$T a_{\text{in}}(\mathbf{p}, \alpha)T^{-1} = \eta_T a_{\text{out}}(-\mathbf{p}, \alpha) \tag{7.45}$$

It is assumed that the vacuum is invariant under T, C, P, Θ, and **I**. We stress this point because there exist interesting examples in which the vacuum lacks the symmetry of the Hamiltonian.

As our first example we apply the charge conjugation transformation (6.51) to the first of equations (7.44), obtaining

$$\begin{aligned} \langle cd\,\text{out} \mid ab\,\text{in}\rangle &= \langle cd\,\text{out}|\, C^\dagger C\,|ab\,\text{in}\rangle \\ &= \langle 0|\, Ca_{\text{out}}(d)C^{-1}Ca_{\text{out}}(c)C^{-1}Ca^*_{\text{in}}(a)C^{-1}Ca^*_{\text{in}}(b)C^{-1}\,|0\rangle; \end{aligned} \tag{7.46}$$

$$\begin{aligned} &S(a\alpha b\beta \to c\gamma d\delta) \\ &\quad = \eta^*_C(a)\eta^*_C(b)\eta_C(c)\eta_C(d)\eta_\alpha\eta_\beta\eta^*_\gamma\eta^*_\delta\, S(\bar{a} - \alpha\bar{b} - \beta \to \bar{c} - \gamma\bar{d} - \delta) \end{aligned}$$

Since $\alpha + \beta = \gamma + \delta$ it is clear that the overall phase factor is independent of $\alpha, \beta, \gamma, \delta$. Similar results obtain for the other reactions (7.44). The assumption that C commutes with S is contained in the assumption that a_{out} transforms as a_{in} (recall that $a_{\text{out}} = S^+ a_{\text{in}} S$). It is also possible to change the isospin components on the right hand side of (7.46) to positive values by applying an isospin rotation $\exp(i\pi I_2)$. (This amounts to using G conjugation rather than C; cf. Eq. (6.56).)

Note that for elastic scattering ($a = c$, $b = d$) the phase factors cancel in (7.46). Hence for elastic scattering C invariance implies

$$\begin{aligned} S(a\alpha b\beta \to a\alpha' b\beta') &= S(\bar{a} - \alpha\bar{b} - \beta \to \bar{a} - \alpha'\bar{b} - \beta') \\ &= S(\bar{a}\alpha\bar{b}\beta \to \bar{a}\alpha'\bar{b}\beta') \end{aligned} \tag{7.47}$$

Even for distinct particles the relation (7.46) is simple for any reasonable phase choice. Only for experiments sensitive to the phase of the amplitude can the factors matter. All of η_α (save the factor ξ) can be removed by using $\exp(i\pi I_2)\,|I\alpha\rangle = (-1)^{I_z+\alpha}\,|I - \alpha\rangle$ and generalizations thereof. Clearly very similar results hold for the other two reactions in (7.44), and for others as well.

The implications of parity conservation are very simple. For the spinless particles considered here, the pertinent symmetry relation is

$$\langle cd|\, S\,|ab\rangle = \eta_P^*(c)\eta_P^*(d)\eta_P(a)\eta_P(b)\langle(cd)'|\, S\,|(ab)'\rangle \tag{7.48}$$

where the prime means that all three-momenta are reversed. The I_3 components are unchanged.

Next consider the operation of time inversion. This also changes $\mathbf{p}$ to $-\mathbf{p}$ but also interchanges initial and final states:

$$\langle cd|\, S\,|ab\rangle = \eta_T(a)\eta_T(b)\eta_T^*(c)\eta_T^*(d)\langle(ab)'|\, S\,|(cd)'\rangle \tag{7.49}$$

As in (7.48) the I_3 components are unchanged. For both (7.48) and (7.49) the phase factors cancel for elastic scattering.

Note that the combined PT operation merely interchanges initial and final states (no inversion of momenta) except for a composite phase change easily found from Eqs. (7.48)–(7.49). Hence the composite operation PT is often very convenient for proving useful symmetry relations.

Finally consider the TCP operation Θ (6.66) (in any one of its six possible orders). In this case the symmetry relation is (here the momenta do not change sign)

$$S(a\alpha b\beta \rightarrow c\gamma d\delta) = \omega_a \omega_b \omega_c^* \omega_d^* \eta_\alpha^* \eta_\beta^* \eta_\gamma \eta_\delta S(\bar{c} - \gamma \bar{d} - \delta \rightarrow \bar{a} - \alpha \bar{b} - \beta) \tag{7.50}$$

Again an isospin rotation can be applied to the right hand side to give

$$S(a\alpha b\beta \rightarrow c\gamma d\delta) = (\omega_a \xi_a^* \omega_b \xi_b^* \omega_c^* \xi_c \omega_d^* \xi_d) S(\bar{c}\gamma \bar{d}\delta \rightarrow \bar{a}\alpha \bar{b}\beta) \tag{7.51}$$

It should be noted that *TCP* does not change the momentum of a state. Since *TCP* is the most general symmetry guaranteed by local relativistic quantum field theory, its predictions deserve the closest scrutiny.

Many variations of the foregoing combinations of particles are possible but the qualitative features of the various transformations should be clear from the analysis already given. The treatment of reactions involving spinning particles is a straightforward application of the results of Sec. (6.9) and we shall not discuss these problems. It should be stated that when Fermions are under consideration it is important to adhere to an ordering convention for the anticommuting operators which create multiparticle states.

PROBLEMS

1. Verify Eqs. (7.18).
2. In the text the crossing from the s to t, and s to u channels was considered. It is also possible to consider tu crossing. Considering the definitions (7.19), show that the matrix $X_{u't'}$ defined by

$$M_u(u') = \sum_{t'} X_{u't'} M_t(t')$$

is given by

$$X_{u't'} = \sum_{s'} X_{u's'}(X^{-1})_{s't'}$$

which may be written in turn as

$$X_{us} = \sum_{t'} X_{ut} X_{ts}$$

3. In the one photon approximation the interaction $-\int d^3x \mathbf{j} \cdot \mathbf{A}$ is the sum of an isospin scalar S and the third component of an isovector V_3 ($Q = I_3 + \frac{1}{2}Y$). Express the matrix elements for the four reactions $\gamma p \rightarrow \pi^0 p$, $\gamma p \rightarrow \pi^+ n$, $\gamma n \rightarrow \pi^- p$, $\gamma n \rightarrow \pi^0 n$ in terms of the reduced matrix elements $(\frac{1}{2}\| S \|\frac{1}{2})$, $(\frac{1}{2}\| V \|\frac{3}{2})$, $(\frac{1}{2}\| V \|\frac{1}{2})$. What is the charge ratio $\sigma(\pi^0 p)/\sigma(\pi^+ n)$ if V_3 dominates? If V_1 dominates?
4. The practical utility of expansions such as (7.16) has been described in many places. Less elementary examples concern "production" amplitudes involving five particles, such as the reactions $N\pi \rightarrow N\pi\pi$ and $N\gamma \rightarrow N\pi\pi$. Writing I_3 eigenvalues and pion

momenta for the first reaction we have

$$\langle \mathbf{p}\alpha\mathbf{q}\beta\tau'|\, S\, |\mathbf{k}\gamma\tau\rangle = \sum_{It} a_{It}(\alpha\beta\gamma;\, \tau'\tau) S_{It}(\mathbf{pq}, \mathbf{k})$$

where t denotes the isospin of the $\pi\pi$ system and I is the total isospin. Show that the expansion coefficients are given by

$$a_{It}(\alpha\beta\gamma;\, \tau'\tau) = C(11t;\, \alpha\beta)C(t\tfrac{1}{2}I;\, \alpha + \beta\tau')C(1\tfrac{1}{2}I;\, \gamma\tau)$$

and derive Table P.1.

The double pion photoproduction reaction is somewhat more complicated since the electromagnetic current is the sum of an isoscalar and isovector. Although isospin is not conserved in this case, the transformation properties of the interaction are known and relations can be obtained by use of the Wigner-Eckart theorem.

Table P.1 The expansion coefficients relating reactions of definite charges to the isospin amplitudes are given.

	$a_{\frac{1}{2}0}$	$a_{\frac{1}{2}1}$	$a_{\frac{3}{2}1}$	$a_{\frac{3}{2}2}$
$\pi^- p \to \pi^- \pi^+ n$	$-\frac{1}{3}\sqrt{2}$	$\frac{1}{3}$	$-\frac{1}{3}$	$\frac{1}{3\sqrt{5}}$
$\pi^- p \to \pi^- \pi^0 p$	0	$-\frac{1}{3}\sqrt{2}$	$-\frac{1}{3\sqrt{2}}$	$-\frac{1}{\sqrt{10}}$
$\pi^- p \to \pi^0 \pi^0 n$	$\frac{1}{3}\sqrt{2}$	0	0	$\frac{2}{3\sqrt{5}}$
$\pi^+ p \to \pi^+ \pi^0 p$	0	0	$\frac{1}{\sqrt{2}}$	$-\frac{1}{\sqrt{10}}$
$\pi^+ p \to \pi^+ \pi^+ n$	0	0	0	$\frac{2}{\sqrt{5}}$

REFERENCES

1. P. Carruthers and J. P. Krisch, *Ann. Phys.* (N.Y.) **33,** 1 (1965).
2. P. Carruthers, "Introduction to Unitary Symmetry" (John Wiley and Sons, N.Y., 1966), Chap. 7.
3. M. E. Rose, "Elementary Theory of Angular Momentum" (John Wiley and Sons, N.Y., 1957).
4. E. U. Condon and G. H. Shortley, "The Theory of Atomic Spectra" (Cambridge University Press, London, 1957), Chap. 3.
5. R. H. Capps, in "Proceedings of the International Conference on High Energy Physics" (Dubna, 1964), Atomizdat, Moscow, 1964.
6. S. Mandelstam, J. E. Paton, R. F. Peierls and A. Q. Sarker, *Ann. Phys.* (N.Y.) **18,** 198 (1962).
7. A. Kotanski, *Acta Phys. Pol.* **27,** 351 (1965).

8. A. O. Barut and B. C. Unal, *Nuovo Cimento* **28,** 112 (1963).
9. C. Rebbi and R. Slansky, *Rev. Mod. Phys.* **42,** 68 (1970).
10. J. R. Taylor, *J. Math. Phys.* **7,** 181 (1966).
11. D. E. Neville, *Phys. Rev.* **160,** 1375 (1967).
12. R. E. Cutkosky, *Ann. Phys.* (N.Y.) **23,** 415 (1963).
13. J. J. deSwart, *Nuovo Cimento* **31,** 420 (1964).
14. M. M. Nieto, *Phys. Rev.* **140,** B434 (1965); **149,** 1294 (1966) E.
15. H. S. Mani, G. Mohan, L. K. Pandit and V. Singh, *Ann. Phys.* (N.Y.) **36,** 285 (1966).
16. R. H. Capps, *Ann. Phys.* (N.Y.) **43,** 428 (1967).
17. A. Kotanski and K. Zalewski, *Acta Phys. Pol.* **26,** 117 (1964).
18. H. Lehmann, K. Symanzik and W. Zimmerman, *Nuovo Cimento* **1,** 205 (1955); *ibid* **6,** 319 (1957).
19. G. Barton, "Dispersion Techniques in Field Theory" (W. A. Benjamin, Inc., New York, 1965).

8

PROJECTION OPERATORS

8.1 INTRODUCTION

In the present chapter we describe some techniques which are useful for the analysis of the isospin properties of scattering amplitudes. In Sec. 8.2 we discuss the very useful isospin projection operators and work out in detail several important reactions involving pions and nucleons. We use these results to illustrate the connection between isospin projection operators and the process of crossing. These examples, especially those involving anti-nucleons, illustrate clearly the necessity for vigilance with regard to phase conventions. Some of the methods used are no more than a crafty application of the rules of vector addition of angular momenta, using familiar Clebsch-Gordan coefficients and their generalizations. In Sec. 8.3 we introduce the concept of generalized isospin matrices, and describe their utility for the description of pion vertices and reactions among pions and particles of arbitrary isospin.

In the "literature" one may find countless illustrations and variants of the methods presented here. Much of our presentation is derived from refs. 1–3.

In Sec. 8.4 similar techniques are applied to the angular dependence of scattering amplitudes. The angular projection operators for pion-nucleon scattering are derived.

8.2 ISOSPIN PROJECTION OPERATORS

For generality consider the S matrix (or equivalently the transition matrix T) for the transition from the initial state (i) having the quantum numbers α to a final state (f) with quantum numbers β. We expand these states in eigen-states of total isotropic spin. (When several particles are involved it is necessary to specify some intermediate quantum numbers. These are denoted by the symbols q_i, or q_f.)

$$|\alpha(i)\rangle = \sum_{II_3q_i} \langle II_3q_i(i) \mid \alpha(i)\rangle \, |II_3q_i(i)\rangle \tag{8.1}$$

A similar expansion holds for $|\beta(f)\rangle$. The transformation coefficients are simply related to the Clebsch-Gordan coefficients. Utilizing invariance under rotations in isospin space one obtains

$$\langle\beta(f)|\, S \,|\alpha(i)\rangle = \sum_{Iq_fq_i} S_{Iq_fq_i} \sum_{I_3} \langle\beta(f) \mid II_3q_f(f)\rangle \langle II_3q_i(i) \mid \alpha(i)\rangle \tag{8.2}$$

The quantity $S_{Iq_fq_i}$ is completely independent of magnetic quantum numbers.

Defining the projection operators $\mathscr{P}^I_{q\,q_i}$ by the outer product

$$\mathscr{P}^I_{q_fq_i} = \sum_{I_3} |II_3q_f(f)\rangle \langle II_3q_i(i)| \tag{8.3}$$

the S-operator may be written in terms of the submatrices S as follows:

$$S = \sum_{Iq_iq_f} S_{Iq_fq_i} \mathscr{P}^I_{q_fq_i} \tag{8.4}$$

The coefficients $S_{Iq_fq_i}$ determine the probability amplitudes for production of a given final state according to

$$S\,|II_3q_i(i)\rangle = \sum_{q_f} S_{Iq_fq_i} \,|II_3q_f(f)\rangle \tag{8.5}$$

The projection operators have the useful general property

$$\mathscr{P}^I_{q_fq_i} \mathscr{P}^{I'}_{q_f'q_i'} = \delta_{II'} \delta_{q_fq_i'} \mathscr{P}^I_{q_fq_i'} \tag{8.6}$$

When the initial and final states have the same particle composition (elastic scattering) one has the useful relations

$$\begin{aligned} \mathscr{P}^I_q \mathscr{P}^{I'}_{q'} &= \delta_{II'} \delta_{qq'} \mathscr{P}^I_q \\ \sum_{Iq} \mathscr{P}^I_q &= 1 \end{aligned} \tag{8.7}$$

since this merely expresses the completeness relation in the subspace of state vectors under consideration.

In performing calculations it is useful to express $\langle\beta(f)|\, S \,|\alpha(i)\rangle$ in terms of the dynamically significant quantities $S_{Iq_fq_i}$. For this purpose it is useful to have explicit formulas for

$$\langle\beta(f)|\, \mathscr{P}_{Iq_fq_i} \,|\alpha(i)\rangle \tag{8.8}$$

For elastic scattering the latter quantity is nothing but the product of Clebsch-Gordan coefficients. For the reaction $a + b \to c + d$, where a has isospin I and third component α, etc., the T matrix may be written as (7.16) where the sum is over allowed values of the total isospin and $\alpha + \beta = \gamma + \delta$. Similar expressions hold for many-particle reactions; in that case one has to specify a particular coupling scheme. For most theoretical purposes the equivalent technique of projection operators is more efficient and elegant. To illustrate these methods we study the five reactions

$$\begin{aligned} \pi\pi &\to \pi\pi \\ N\pi &\to N\pi \\ N\bar{N} &\to \pi\pi \\ NN &\to NN \\ N\pi &\to N\pi\pi \end{aligned} \tag{8.9}$$

First consider $\pi\pi$ scattering. For many considerations involving pions it is useful and convenient to use a Cartesian basis rather than the charge basis (diagonal I_3). Familiar vector manipulations may then be used. In addition, the use of real fields facilitates crossing, since the amplitudes for emission and absorption become the same. The two descriptions are related by the usual transformation

$$\begin{aligned} \pi^{\pm} &= \mp(\pi_1 \pm i\pi_2)/\sqrt{2} \\ \pi^0 &= \pi_3 \end{aligned} \tag{8.10}$$

In $\pi\pi$ scattering the allowed isospins are $t = 0$, 1, and 2. For $t = 0$ it is very easy to compute (8.7) since the sum contains only one term. The initial (final) pions have charge indices $i, j(k, l)$. The $t = 0$ wave function for two pions is

$$\begin{aligned} \psi_0 &= [\pi^+(1)\pi^-(2) + \pi^-(1)\pi^+(2) - \pi^0(1)\pi^0(2)]/\sqrt{3} \\ &= -\boldsymbol{\pi}(1) \cdot \boldsymbol{\pi}(2)/\sqrt{3} \end{aligned} \tag{8.11}$$

which is clearly invariant under rotations in isospace. The projection operator is

$$\mathscr{P}_0 = \psi_0\psi_0^\dagger = \tfrac{1}{3}\sum_{mn} \pi_m(1)\pi_m(2)\pi_n^\dagger(1)\pi_n^\dagger(2) \tag{8.12}$$

Therefore we find the matrix elements

$$\begin{aligned} (\mathscr{P}_0)_{kl,ij} &= \langle\pi_k(1)\pi_l(2)|\, \mathscr{P}_0 \,|\pi_i(1)\pi_j(2)\rangle \\ &= \tfrac{1}{3}\delta_{kl}\delta_{ij} \end{aligned} \tag{8.13}$$

The components of the wave function for isospin 1 must transform like a vector; more precisely, in a representation in which t_3 is diagonal its components transform as a spherical tensor of rank 1. In fact it is easy to see that

$$\psi_{1t_3} = i(\boldsymbol{\pi}(1) \times \boldsymbol{\pi}(2))_{t_3}/\sqrt{2} \tag{8.14}$$

For example the -1 component of $\mathbf{V} = \boldsymbol{\pi}(1) \times \boldsymbol{\pi}(2)$ is $(V_x - iV_y)/\sqrt{2}$. The corresponding projection operator is

$$\mathscr{P}_1 = \sum_{t_3=-1}^{1} \psi_{1t_3}\psi_{1t_3}^{\dagger} \tag{8.15}$$

The same unitary transformation as (8.10) relates the spherical tensor form (8.15) to the components of the vector $\psi_{1\lambda}$, $\lambda = 1, 2, 3$. These components are

$$\psi_{1\lambda} = i(\boldsymbol{\pi}(1) \times \boldsymbol{\pi}(2))_{\lambda}/\sqrt{2} = i\sum_{\alpha\beta} \epsilon_{\lambda\alpha\beta}\pi_{\alpha}(1)\pi_{\beta}(2)/\sqrt{2} \tag{8.16}$$

($\epsilon_{\lambda\alpha\beta}$ is the completely antisymmetric oriented Cartesian tensor; $\epsilon_{123} = 1$.) Using the identity

$$\sum_{\lambda=1}^{3} \epsilon_{\alpha\beta\lambda}\epsilon_{\alpha'\beta'\lambda} = \delta_{\alpha\alpha'}\delta_{\beta\beta'} - \delta_{\alpha\beta'}\delta_{\beta\alpha'} \tag{8.17}$$

(8.15) may be written as

$$\begin{aligned}\mathscr{P}_1 &= \sum_{\lambda=1}^{3} \psi_{1\lambda}\psi_{1\lambda}^{\dagger} \\ &= \sum_{\alpha\alpha'\beta\beta'} \tfrac{1}{2}(\delta_{\alpha\alpha'}\delta_{\beta\beta'} - \delta_{\alpha\beta'}\delta_{\beta\alpha'})\pi_{\alpha}(1)\pi_{\beta}(2)\pi_{\alpha'}^{\dagger}(1)\pi_{\beta'}^{\dagger}(2)\end{aligned} \tag{8.18}$$

The matrix element of $\mathscr{P}_1$ is by inspection

$$(\mathscr{P}_1)_{kl,ij} = \tfrac{1}{2}(\delta_{ki}\delta_{lj} - \delta_{kj}\delta_{li}) \tag{8.19}$$

$\mathscr{P}_2$ need not be calculated directly. Using the completeness relation (8.7) gives

$$(\mathscr{P}_2)_{kl,ij} = \tfrac{1}{2}(\delta_{ki}\delta_{lj} + \delta_{kj}\delta_{li} - \tfrac{2}{3}\delta_{ij}\delta_{kl}) \tag{8.20}$$

Next we study pion–nucleon scattering.[4] In this case it is convenient to regard (8.3) as a matrix in the isospinor space of the nucleon. Letting χ_N, $\chi_{N'}$ denote the initial and final isospinors we write

$$\chi_{N'}^{\dagger}\langle\pi(\beta)|\, \mathscr{P}_I\, |\pi(\alpha)\rangle\chi_N \tag{8.21}$$

Transforming to the Cartesian basis, we write the 2×2 matrix describing

$N\pi_i \to N\pi_j$ as $\mathscr{P}_{ji}$. In order to calculate $\mathscr{P}_{ji}$ we use the wave functions

$$\begin{aligned} \psi_{\frac{1}{2}\frac{1}{2}} &= (\tfrac{2}{3})^{\frac{1}{2}}\pi^+\chi_n - (\tfrac{1}{3})^{\frac{1}{2}}\pi^0\chi_p \\ \psi_{\frac{1}{2}-\frac{1}{2}} &= (\tfrac{1}{3})^{\frac{1}{2}}\pi^0\chi_n - (\tfrac{2}{3})^{\frac{1}{2}}\pi^-\chi_p \end{aligned} \tag{8.22}$$

and the relations

$$\begin{aligned} \chi_p\chi_p^\dagger &= \tfrac{1}{2}(1 + \tau_3) \\ \chi_n\chi_n^\dagger &= \tfrac{1}{2}(1 - \tau_3) \\ \chi_p\chi_n^\dagger &= \tfrac{1}{2}(\tau_1 + i\tau_2) \\ \chi_n\chi_p^\dagger &= \tfrac{1}{2}(\tau_1 - i\tau_2) \end{aligned} \tag{8.23}$$

where $\boldsymbol{\tau}$ is the Pauli isospin matrix vector. This gives the result

$$(\mathscr{P}_{\frac{1}{2}})_{ji} = \tfrac{1}{3}\tau_j\tau_i \tag{8.24}$$

$j(i)$ refers to the final (initial) state. From (8.6) we find the $\mathscr{P}_{\frac{3}{2}}$ projection operator:

$$(\mathscr{P}_{\frac{3}{2}})_{ji} = \delta_{ji} - \tfrac{1}{3}\tau_j\tau_i \tag{8.25}$$

Yet another equivalent form is worthy of mention. We could write $\mathscr{P}_I$ as a direct product in the pion *and* nucleon isospin spaces: $\chi_{N'}^\dagger\chi_{\pi'}^\dagger\mathscr{P}_I'\chi_\pi\chi_N$. The operator $\mathscr{P}_I'$ is easily constructed making use of the vector addition rule:

$$\begin{aligned} \mathbf{t}\cdot\boldsymbol{\tau} &= -2, \qquad I = \tfrac{1}{2} \\ &= 1, \qquad I = \tfrac{3}{2} \end{aligned} \tag{8.26}$$

Here $\mathbf{t}$ is the set of three 3×3 Hermitian spin one matrices in the Cartesian basis, whose components are

$$(t_i)_{jk} = -i\epsilon_{ijk} \tag{8.27}$$

It is clear that $(\mathbf{t}\cdot\boldsymbol{\tau} + 2)$ projects into the $I = \frac{3}{2}$ subspace while $(\mathbf{t}\cdot\boldsymbol{\tau} - 1)$ leads to pure $I = \frac{1}{2}$.

Making use of the relations

$$\begin{aligned} \tau_i\tau_j &= \delta_{ij} + i\epsilon_{ijk}\tau_k \\ [t_i, t_j] &= i\epsilon_{ijk}t_k \end{aligned} \tag{8.28}$$

we find the identity $(\mathbf{t}\cdot\boldsymbol{\tau})^2 = 2 - \mathbf{t}\cdot\boldsymbol{\tau}$. Hence a set of orthogonal normalized projection operators is easily seen to be

$$\begin{aligned} \mathscr{P}_{\frac{1}{2}}' &= \tfrac{1}{3}(1 - \mathbf{t}\cdot\boldsymbol{\tau}) \\ \mathscr{P}_{\frac{3}{2}}' &= \tfrac{1}{3}(2 + \mathbf{t}\cdot\boldsymbol{\tau}) \end{aligned} \tag{8.29}$$

Using (8.27) and (8.28) it is easy to show that the *ji* components of (8.29) are identical with (8.24) and (8.25).

This construction is easily generalized. For nucleon–nucleon scattering ($NN \to NN$) the projection operators are conveniently written as direct products of 2×2 matrices in the isospin spaces of nucleons number 1 and number 2. Since the product $\boldsymbol{\tau}_1 \cdot \boldsymbol{\tau}_2$ is given by

$$\begin{aligned} \boldsymbol{\tau}_1 \cdot \boldsymbol{\tau}_2 &= -3, \qquad I = 0 \\ &= \;\;\, 1, \qquad I = 1 \end{aligned} \tag{8.30}$$

the same argument given above leads to the projection operators

$$\begin{aligned} \mathscr{P}_0 &= \tfrac{1}{4}(1 - \boldsymbol{\tau}_1 \cdot \boldsymbol{\tau}_2) \\ \mathscr{P}_1 &= \tfrac{1}{4}(3 + \boldsymbol{\tau}_1 \cdot \boldsymbol{\tau}_2) \end{aligned} \tag{8.31}$$

For a thorough treatment of nucleon–nucleon scattering one should consult ref. 5.

As a final illustration of this method we consider $\pi\pi$ scattering. This problem shows the general technique which could be used for less trivial problems involving higher values of isospin. Denoting the isospin vectors of pions 1 and 2 by $\mathbf{t}_1$ and $\mathbf{t}_2$ we find

$$\begin{aligned} \mathbf{t}_1 \cdot \mathbf{t}_2 &= -2, \qquad I = 0 \\ &= -1, \qquad I = 1 \\ &= \;\;\, 1, \qquad I = 2 \end{aligned} \tag{8.32}$$

Hence we learn that $\mathbf{t}_1 \cdot \mathbf{t}_2 + 2$ projects *out of* the $I = 0$ subspace; similarly $\mathbf{t}_1 \cdot \mathbf{t}_2 + 1$, $\mathbf{t}_1 \cdot \mathbf{t}_2 - 1$ project out of the $I = 1$, $I = 2$ subspaces. Hence we see that $\mathscr{P}_0 \propto (2 + t_1 t_2)(-1 + t_1 t_2)$ etc. The properly normalized operators are given by

$$\begin{aligned} \mathscr{P}_0' &= \tfrac{1}{3}[(\mathbf{t}_1 \cdot \mathbf{t}_2)^2 - 1] \\ \mathscr{P}_1' &= \tfrac{1}{2}[2 - (\mathbf{t}_1 \cdot \mathbf{t}_2) - (\mathbf{t}_1 \cdot \mathbf{t}_2)^2] \\ \mathscr{P}_2' &= \tfrac{1}{6}[2 + 3(\mathbf{t}_1 \cdot \mathbf{t}_2) + (\mathbf{t}_1 \cdot \mathbf{t}_2)^2] \end{aligned} \tag{8.33}$$

Remembering that initial and final pions 1 have labels i and k, while initial and final pions 2 have labels j and l, one may use (8.28) to establish the results

$$\begin{aligned} (\mathbf{t}_1 \cdot \mathbf{t}_2)_{kl,ij} &= \delta_{kj}\delta_{li} - \delta_{kl}\delta_{ij} \\ [(\mathbf{t}_1 \cdot \mathbf{t}_2)^2]_{kl,ij} &= \delta_{ij}\delta_{kl} + \delta_{ik}\delta_{jl} \\ (1)_{kl,ij} &= \delta_{ik}\delta_{jl} \end{aligned} \tag{8.34}$$

which proves the equivalence of (8.33) with the results previously obtained.

For general isospin, we note that $\mathbf{I}_1 \cdot \mathbf{I}_2$ has a definite (distinct) value for each total I; call this ξ_I. Then the quantities $Q_I = \mathbf{I}_1 \cdot \mathbf{I}_2 - \xi_I$ project out of the subspace of isospin I. The product $\Pi Q_{I'}$ with $I' \neq (I, I'')$ is then proportional to $\mathscr{P}_{I''}$. Similar methods to those used above will determine the proper normalization factors. Subsequently we shall give general formulas for a special class of high isospin projection operators. The method is clearly applicable to more general groups, for example[6] to the internal symmetry group SU_3. Again the method is simple for those representations having relatively low dimensions.

As an example of the application of the method to non-elastic reactions we consider the projection operators for the reaction $N\pi \rightarrow N\pi\pi$. The interesting new feature of this problem arises from the distinct ways the various isospins can be combined to form the total isospin I. We can first combine $N\pi$ for a selected pion and add the result to the other pion, or first combine the isospins of the two pions. The two coupling schemes may be labeled as $((N\pi)\pi)$ and $((\pi\pi)N)$ respectively. The relation between the two schemes is given by the Racah coefficients (or equivalently the $6j$ symbols) and has been described in ref. 1 for the problem at hand.

Calculations of this sort are usually tedious and we give only the results for the $((\pi\pi)N)$ coupling scheme. Denoting the $\pi\pi$ isospin by t and the total isospin by I, one may insert explicit wave functions in

$$\mathscr{P}_{It} = \sum_{I_3} |II_3t(f)\rangle\langle II_3(i)| \tag{8.35}$$

Working this out in a Cartesian basis gives the result

$$\begin{aligned}
(\mathscr{P}_{\frac{1}{2}0})_{ij,k} &= \tfrac{1}{3}\delta_{ij}\tau_k \\
(\mathscr{P}_{\frac{1}{2}1})_{ij,k} &= (\tau_i\tau_j - \delta_{ij})\tau_k/3\sqrt{2} \\
(\mathscr{P}_{\frac{3}{2}1})_{ij,k} &= [\tau_j\delta_{ki} - \tau_i\delta_{kj} + \tfrac{2}{3}(\tau_i\tau_j - \delta_{ij})\tau_k]/\sqrt{2} \\
(\mathscr{P}_{\frac{1}{2}2})_{ij,k} &= -[\tau_i\delta_{kj} + \tau_j\delta_{ki} - \tfrac{2}{3}\delta_{ij}\tau_k]/\sqrt{10}
\end{aligned} \tag{8.36}$$

It will be noted that the even t amplitudes are symmetric in i and j, while the odd t amplitudes are antisymmetric, as required by Bose symmetry.

We note a few general relations useful for computing products of the various transition amplitudes involving pions and nucleons. In Eq. (8.37)

$\mathscr{P}_I$ denotes the projection operator for $N\pi \to N\pi$.

$$\begin{aligned}
\mathscr{P}_I^\dagger &= \mathscr{P}_I \\
\mathscr{P}_I \mathscr{P}_{I'} &= \delta_{II'} \mathscr{P}_I \\
\mathscr{P}_{It} \mathscr{P}_{I'} &= \delta_{II'} \mathscr{P}_{It} \\
\mathscr{P}_{It}^\dagger \mathscr{P}_{I'} &= \mathscr{P}_{I'} \mathscr{P}_{It} = 0 \\
\mathscr{P}_{It}^\dagger \mathscr{P}_{I't'} &= \delta_{II'} \delta_{tt'} \mathscr{P}_I \\
\mathscr{P}_{I't'} \mathscr{P}_{It}^\dagger &= \delta_{II'} \delta_{tt'} \mathscr{P}_{It}^{2\pi N}
\end{aligned} \tag{8.37}$$

where $\mathscr{P}_{It}^{2\pi N}$ is the projection operator for the scattering process

$$2\pi + N \to 2\pi + N \tag{8.38}$$

Incidentally the last of the relations (8.37) provides an easy way to compute some of the projection operators for (8.38). Since these are easily found, and since the results are rather lengthy, we do not present them here. (Note that the $I = \frac{5}{2}$ state cannot be found in this way—but we do not need this state to discuss the processes resulting from a single pion incident on a nucleon.)

As an illustration of the utility of these formulas we express the single pion exchange production matrix in terms of the $\pi\pi$ isospin vertex functions. The π–N vertex carries a factor τ_i, so the S matrix is

$$S_{ij,k} = \sum_l \tau_l [S_0(\mathscr{P}_0)_{ij,lk} + S_1(\mathscr{P}_1)_{ij,lk} + S_2(\mathscr{P}_2)_{ij,lk}] \tag{8.39}$$

The functions S_t contain a common factor describing the π–N vertex and the pion propagation. They differ in the structure of the $\pi\pi$ interaction in the three isospin states.

From Eqs. (8.13), (8.19), (8.20), and (8.36) we see at once that

$$\begin{aligned}
\sum_i \tau_i (\mathscr{P}_0)_{pq,ik} &= 3(\mathscr{P}_{\frac{1}{2}0})_{pq,k} \\
\sum_i \tau_i (\mathscr{P}_1)_{pq,ik} &= \tfrac{1}{2}(\tau_p \delta_{kq} - \tau_q \delta_{kp}) \\
\sum_i \tau_i (\mathscr{P}_2)_{pq,ik} &= -\tfrac{1}{2}\sqrt{10}(\mathscr{P}_{\frac{3}{2}2})_{pq,k}
\end{aligned} \tag{8.40}$$

Hence we can write the single pion exchange contribution to the production process as follows:

$$S_{pq,k} = 3S_0(\mathscr{P}_{\frac{1}{2}0})_{pq,k} + \sqrt{2}S_1(\mathscr{P}_{\frac{1}{2}1})_{pq,k} - \frac{1}{\sqrt{2}} S_1(\mathscr{P}_{\frac{3}{2}0})_{pq,k} - \tfrac{1}{2}\sqrt{10}S_2(\mathscr{P}_{\frac{3}{2}2})_{pq,k} \tag{8.41}$$

The analysis of the nucleon annihilation reaction $N_\alpha(1)\bar{N}_{\bar{\alpha}'}(2) \rightarrow \pi_i(1)\pi_j(2)$ reveals some interesting and delicate phase questions. Here α, $\bar{\alpha}'$ denote isospin components I_3, and i, j are Cartesian pion indices. The projection operator

$$\mathscr{P}^I(N\bar{N} \rightarrow \pi\pi) = \sum_{I_3} |\pi_1\pi_2\rangle\langle N_1\bar{N}_2| \tag{8.42}$$

has the properties

$$\begin{aligned}\mathscr{P}^I(N\bar{N} \rightarrow \pi\pi)\mathscr{P}_I^\dagger(N\bar{N} \rightarrow \pi\pi) &= \mathscr{P}^I(\pi\pi \rightarrow \pi\pi)\\ \mathscr{P}^{I^\dagger}(N\bar{N} \rightarrow \pi\pi)\mathscr{P}^I(N\bar{N} \rightarrow \pi\pi) &= \mathscr{P}^I(N\bar{N} \rightarrow N\bar{N})\end{aligned} \tag{8.43}$$

Clearly only $I = 0, 1$ are allowed in (8.42).

To obtain a matrix representation in the nucleon–antinucleon isospin space, we first project $\mathscr{P}^I$ onto definite pion coordinates using the explicit wave functions of Eqs. (8.11) and (8.16). For $I = 0$ we find

$$\begin{aligned}\mathscr{P}^0_{ij} &= -\frac{1}{\sqrt{3}}\,\delta_{ij}\Psi_0^\dagger(N_1\bar{N}_2)\\ \Psi_0(N_1\bar{N}_2) &= \frac{1}{\sqrt{2}}(\chi(p)\chi^c(\bar{p}) - \chi(n)\chi^c(n))\end{aligned} \tag{8.44}$$

The operator (8.44) may be regarded as the direct product of two row spinors, whose purpose is to act on a linear combination of the nucleon and antinucleon spinors. It is possible and useful (even natural from the field-theoretic standpoint of Chapter 7) to transfer the antinucleon spinor to the left and reinterpret (8.44) as a 2×2 matrix standing between $\chi_{\bar{N}}^\dagger$ and χ_N. Using the reality of the χ's we use Eq. (7.8) to obtain

$$\Psi_0^\dagger(N\bar{N}) = -\frac{1}{\sqrt{2}}(\alpha(2)\alpha^\dagger(1) + \beta(2)\beta^\dagger(1)) \tag{8.45}$$

where the label 1 refers to N, 2 to $\bar{N}$. α and β are identical to $\chi(p)$, $\chi(n)$ in Eq. (7.8). Using Eqs. (8.23) we recognize that (8.45) is nothing but the unit matrix. Hence we may write

$$\mathscr{P}^0_{ij} = \delta_{ij}/\sqrt{6} \tag{8.46}$$

In order to capture $\mathscr{P}^1_{ij}$ we first write the $I = 1$, I_3-diagonal $\pi\pi$ wave functions in the form

$$\begin{aligned}\Psi_{I_3=1} &= -(\psi_1 + i\psi_2)/\sqrt{2}\\ \Psi_{I_3=0} &= \psi_3\\ \Psi_{I_3=-1} &= (\psi_1 - i\psi_2)/\sqrt{2}\end{aligned} \tag{8.47}$$

where ψ_α is defined by (8.16). Writing the $I = 1$ wave functions in analogy to (8.45) then gives directly

$$\mathscr{P}^1_{ij} = -\tfrac{1}{2}i\epsilon_{ijk}\tau_k = \tfrac{1}{4}[\tau_j, \tau_i] \tag{8.48}$$

The symmetry of $\mathscr{P}^0_{ij}$, $\mathscr{P}^1_{ij}$ under the interchange $i \leftrightarrow j$ is expected on general principles. For the inverse reaction $\pi\pi \to N\bar{N}$ the projection operators are found by taking the adjoint of (8.46) and (8.48). This agrees with the results first given in ref. 7.

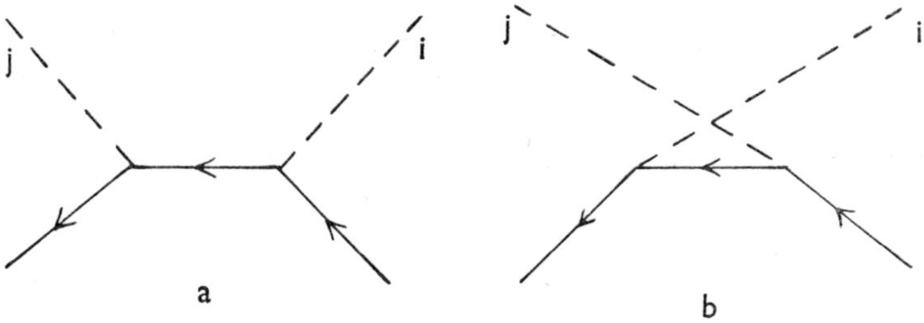

Figure 8.1 The Born terms contributing to scattering of pions from spinless nucleons are illustrated.

The projection operators $\mathscr{P}^0$, $\mathscr{P}^1$ for $N\bar{N} \to \pi\pi$ may be expressed in terms of the $\mathscr{P}^{\frac{1}{2}}$, $\mathscr{P}^{\frac{3}{2}}$ for the reaction $N\pi \to N\pi$. These relations are easily found to be

$$\begin{pmatrix} \mathscr{P}_0 \\ \mathscr{P}_1 \end{pmatrix}_{ij} = X_{st}^{-1} \begin{pmatrix} \mathscr{P}_{\frac{1}{2}} \\ \mathscr{P}_{\frac{3}{2}} \end{pmatrix}_{ji}$$
$$\begin{pmatrix} \mathscr{P}^{\frac{1}{2}} \\ \mathscr{P}^{\frac{3}{2}} \end{pmatrix}_{ji} = X_{ts} \begin{pmatrix} \mathscr{P}_0 \\ \mathscr{P}_{\frac{1}{2}} \end{pmatrix}_{ij} \tag{8.49}$$

where the relevant *crossing matrices* are given in Table 7.1.

As a simple illustration we analyze the isospin content of the lowest order Born terms for the scattering of pions off spinless nucleons. The interaction Lagrangian density is

$$\mathscr{L} = g\bar{N}\boldsymbol{\tau}N \cdot \boldsymbol{\pi} \tag{8.50}$$

The S matrix is related to the invariant transition matrix T by $S - 1 = i(2\pi)^4\delta(p' + q' - p - q)T/(16EE'\omega\omega')^{\frac{1}{2}}$ where the variables are identified by $N_\alpha(p) + \pi_i(q) \to N_{\alpha'}(p') + \pi_j(q')$, $E = p^0$, $\omega = q_0$, etc. We find (cf. Fig. 8.1)

$$T_s = g^2\chi^\dagger_{\alpha'}\left[\frac{\tau_j\tau_i}{m^2 - s} + \frac{\tau_i\tau_j}{m^2 - u}\right]\chi_\alpha \tag{8.51}$$

where $s = (p + q)^2$, $u = (p - q')^2$. If we expand this amplitude in terms of the projection operators (8.24)–(8.25) we find the pure isospin amplitudes

$$T_s = \sum_{I_s} \mathscr{P}^{I_s}_{ji} T_{I_s}$$
$$T_{\frac{1}{2}} = g^2\left(\frac{3}{m^2 - s} - \frac{1}{m^2 - u}\right) \tag{8.52}$$
$$T_{\frac{3}{2}} = 2g^2/(m^2 - u)$$

The same type of calculation may be done for the annihilation reaction

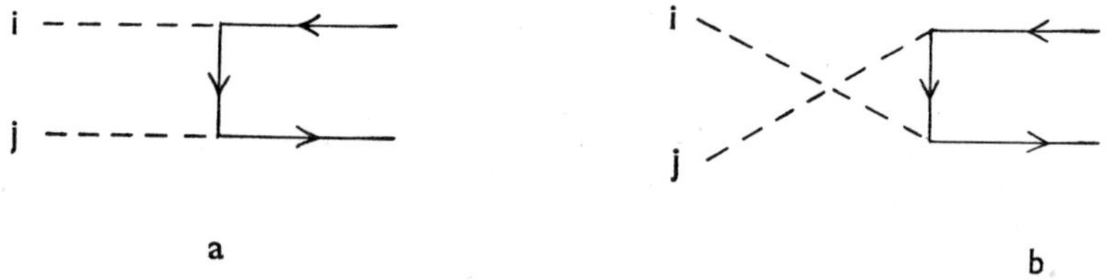

Figure 8.2 The Born terms for the annihilation of spinless nucleons into pions are illustrated.

$N_\alpha(p) + \bar{N}_{\bar{\alpha}'}(\bar{p}') \rightarrow \pi_i(\bar{q}) + \pi_j(q')$, yielding (Fig. 8.2)

$$T_t = g^2 \chi^{c\dagger}_{\alpha'} \left[\frac{\tau_j \tau_i}{m^2 - \bar{s}} + \frac{\tau_i \tau_j}{m^2 - \bar{u}}\right] \chi_\alpha \tag{8.53}$$

where $\bar{s} = (p - \bar{q})^2$, $\bar{u} = (p - q')^2 = u$. If we evaluate T_t at the unphysical point $\bar{p}' = -p'$, $\bar{q} = -q$ and set $\bar{\alpha}' = -\alpha$ we obtain $(-1)^{\frac{1}{2}+\alpha'} T_s$, in agreement with the results of the preceding chapter. Expanding T_t in isospin amplitudes gives

$$T_t = \sum_{I_t} \mathscr{P}^{I_t}_{ij} T_{I_t}$$
$$T_0 = \sqrt{6} g^2 \left(\frac{1}{m^2 - \bar{s}} + \frac{1}{m^2 - \bar{u}}\right) \tag{8.54}$$
$$T_1 = 2g^2 \left(\frac{1}{m^2 - \bar{s}} - \frac{1}{m^2 - \bar{u}}\right)$$

(In Chapter 10 we give the complete results for the scattering of pions off nucleons in the case that the nucleons have the correct spin.) Comparing the results of Eqs. (8.49), (8.52), and (8.54) we find the crossing relation connecting the invariant isospin amplitudes in the s and t channels to be

$$T_t(I_t) = \sum_{I_s} X_{ts} T^c_s(\bar{I}_s) \tag{8.55}$$

where c denotes that the proper analytic continuation $s \to \bar{s}$, $u \to \bar{u}$ has been performed. This specific result is in agreement with the general considerations of the preceding chapter.

Next consider the reaction $\pi\pi \to \pi\pi$ in the s, t, and u channels. The s-channel is defined by

$$\pi_i(p_1) + \pi_j(p_2) \to \pi_k(p_3) + \pi_l(p_4)$$

$$M^s_{kl,ij}(s, t, u) = \sum_{I_s} \mathscr{P}^{I_s}_{kl,ij} M^{I_s}(s, t, u) \tag{8.56}$$

$$s = (p_1 + p_2)^2, \qquad t = (p_1 - p_3)^2, \qquad u = (p_1 - p_4)^2$$

The u-channel is described by

$$\pi_i(p_1) + \pi_l(\bar{p}_4) \to \pi_k(p_3) + \pi_j(\bar{p}_2)$$

$$M^u_{kj,il}(s', t', u') = \sum_{I_u} \mathscr{P}^{I_u}_{kj,il} M^{I_u}(s', t', u') \tag{8.57}$$

$$s' = (p_1 + \bar{p}_4)^2, \qquad u' = (p_1 - \bar{p}_2)^2, \qquad t' = (p_1 - p_3)^2$$

and finally the t channel by

$$\pi_i(p_1) + \pi_k(\bar{p}_3) \to \pi_j(\bar{p}_2) + \pi_l(p_4)$$

$$M^t_{jl,ik}(s'', t'', u'') = \sum_{I_t} \mathscr{P}^{I_t}_{jl,ik} M^{I_t}(s'', t'', u'') \tag{8.58}$$

$$s'' = (p_1 + \bar{p}_3)^2, \qquad t'' = (p_1 - \bar{p}_2)^2, \qquad u'' = (p_1 - p_4)^2$$

In Eqs. (8.56–8.58) we have written $M^s_{kl,ij}(s, t, u)$ in place of $M^s_{kl,ij}(p_3 p_4; p_1 p_2)$, $M^u_{kj,il}(s', t', u')$ in place of $M^u_{kj,il}(p_3 \bar{p}_2; p_1 \bar{p}_4)$ and $M^t_{jl,ik}(s'', t'', u'')$ in place of $M^t_{jl,ik}(\bar{p}_2 p_4; p_1 \bar{p}_3)$.

The present reaction is very simple in that all three channels describe the same physical process. Hence we may erase the superscript labeling the various channels. The crossing conditions are very simple in terms of the amplitudes defined above. For example, when $\bar{p}_2$ and $\bar{p}_4$ are continued to the values $\bar{p}_2 = -p_2$, $\bar{p}_4 = -p_4$ the t-channel amplitude coincides with the s-channel amplitude and $(s'', t'', u'') \to (t, s, u)$. Thus we have the crossing conditions

$$\begin{aligned} M^t_{jl,ik}(t, s, u) &= M^s_{kl,ij}(s, t, u) \\ M^u_{kj,il}(u, t, s) &= M^s_{kl,ij}(s, t, u) \end{aligned} \tag{8.59}$$

Here we have retained the channel label to emphasize the physical significance of the relation, e.g. when the invariants (s'', t'', u'') are continued to the values (t, s, u), which are *unphysical* in the t-channel, one finds the indicated s-channel amplitude. Combining Eqs. (8.56–8.59) allows the derivation of crossing relations for the invariant amplitudes. The key step is to express projection operators in one channel in terms of those in another channel. For example, we find $(\mathscr{P}^t_0)_{jl,ik} = \frac{1}{3}(\mathscr{P}^s_0 + \mathscr{P}^s_1 + \mathscr{P}^s_2)_{kl,ij}$, $(\mathscr{P}^t_1)_{jl,ik} = (\mathscr{P}^s_0 + \frac{1}{2}(\mathscr{P}^s_1 - \mathscr{P}^s_2))_{kl,ij}$, etc. which results may be summarized as

$$\begin{aligned}\mathscr{P}^{I_s}_{kl,ij} &= \sum_{I_t} \mathscr{P}^{I_t}_{jl,ik} X(I_t, I_s)\\ \mathscr{P}^{I_s}_{kl,ij} &= \sum_{I_u} \mathscr{P}^{I_u}_{jl,ik} X(I_u, I_s)\end{aligned} \tag{8.60}$$

where the crossing matrices were given in Chapter 7. The isospin amplitudes then obey the familiar rules

$$\begin{aligned}M_{I_u}(u, t, s) &= \sum_{I_s} X(I_u, I_s) M^c_{I_s}(s, t, u)\\ M_{I_t}(t, s, u) &= \sum_{I_s} X(I_t, I_s) M^c_{I_s}(s, t, u)\end{aligned} \tag{8.61}$$

The basic dynamical information about $\pi\pi$ scattering is seen to be contained in three Lorentz-invariant isospin amplitudes. The use of such invariant amplitudes is also useful for reactions involving spinning particles and is considered in Chapter 11.

For some purposes it is useful to consider a different set of invariant amplitudes which do not have definite isospin but which have very simple properties under crossing.[8] In the s-channel we write

$$M^s_{kl,ij}(s, t, u) = \delta_{kl}\delta_{ij}A(s, t, u) + \delta_{ki}\delta_{lj}B(s, t, u) + \delta_{kj}\delta_{li}C(s, t, u) \tag{8.62}$$

The A, B, C amplitudes are related to the isospin amplitudes by the formulas

$$\begin{aligned}M_{I_s=0} &= 3A + B + C\\ M_{I_s=1} &= B - C\\ M_{I_s=2} &= B + C\end{aligned} \tag{8.63}$$

Writing similar expansions in the u and t-channels and imposing the crossing relations (8.59) yields the crossing conditions for A, B, C: under *su* crossing $A \leftrightarrow C$, $B \leftrightarrow B$; under *st* crossing $A \leftrightarrow B$, $C \leftrightarrow C$. We could also consider *tu* crossing which gives the same result as the Bose symmetry condition: $A \leftrightarrow A$, $B \leftrightarrow C$.

Equations (8.60) are easily extended to arbitrary reactions. We define a set of projection operators labeled by the pairs of isospin indices involved in *su* crossing:

$$T_s = \sum_{I_s} \chi_{\alpha'}^{\dagger} \mathscr{P}^s_{\beta'\beta}(I_s)\chi_\alpha M_s(I_s)$$

$$T_u = \sum_{I_u} \chi_{\gamma'} \mathscr{P}^u_{\bar{\beta}\bar{\delta}}(I_u)\chi_\alpha M_u(I_u) \tag{8.64}$$

$$T_t = \sum_{I_t} \chi_{\gamma}^{c\dagger} \mathscr{P}^t_{\bar{\delta}\beta'}(I_t)\chi_\alpha M_t(I_t)$$

The I_3 components refer to the spherical basis, although conversion to a Cartesian basis is simple (when appropriate).

We recall that the crossing matrices are defined by

$$M_u(I_u) = \sum_{I_s} X(I_u, I_s)M_s^c(I_s)$$

$$M_t(I_t) = \sum_{I_s} X(I_t, I_s)M_s^c(I_s) \tag{8.65}$$

From the crossing conditions (7.13)–(7.15) one can derive

$$\mathscr{P}^s_{\beta'\beta}(I_s) = \sum_{I_u} \mathscr{P}^u_{\beta\bar{\beta}'}(I_u)X(I_u, I_s)$$

$$\mathscr{P}^s_{\beta'\beta}(I_s) = \sum_{I_t} \mathscr{P}^t_{\bar{\beta}\beta'}(I_t)X(I_t, I_s) \tag{8.66}$$

where $\bar{\beta} = -\beta$, $\bar{\beta}' = -\beta'$.

The contrast in structure of Eqs. (8.65–8.66) should be noted. In matrix form the first of Eqs. (8.65) is $M_u = X_{us}M_s^c$, while the first of Eqs. (8.66) is $\mathscr{P}^u_{\beta'\beta} = (X_{us}^T)^{-1}\mathscr{P}^s_{\beta\beta'}$, where T denotes transpose.

When two pions are involved it is convenient to call particles b and d pions, and to convert (8.66) to a Cartesian basis. The relations remain true if the bars are erased and β, δ regarded as Cartesian indices 1, 2, 3. Further, one conventionally removes the isospin wave functions in (8.64), so that the invariant amplitudes are matrices in the isospin space of particles a and c. All these remarks are simple extensions of the treatment of pion-nucleon scattering.

Finally we should draw attention to the fact that many projection operators can be found most easily by construction of simple Feynman graphs having intermediate states of known isospin content.[9] For example the graph of Fig. (8.1a) has an isospin factor $\tau_j\tau_i$ associated with the $I = \frac{1}{2}$ nucleon intermediate state (Eq. 8.51). Hence we expect $\mathscr{P}^{\frac{1}{2}}_{ji} = k\tau_j\tau_i$. Requiring $\mathscr{P}^2 = \mathscr{P}$ gives $k = \frac{1}{3}$. Then the completeness relation determines $\mathscr{P}^{\frac{3}{2}}$. It is worth looking somewhat closer at this method, since its naive application leads to errors when applied to spin-parity. For example in the c.m. frame one might suppose that the πN scattering pole term of Fig. (8.1a) would have $J^P = \frac{1}{2}^+$. However this graph actually has a $\frac{1}{2}^-$ component, as is easily seen by recalling that the Feynman graph (8.1a) actually corresponds to two (separately non-covariant) time ordered graphs depicted in Fig. 8.3. The nucleon pair has negative intrinsic parity so that coupling to a 0^- pion in an s state is possible. However the isospin couplings are not changed in any basic way on going from (8.3a) to (8.3b).

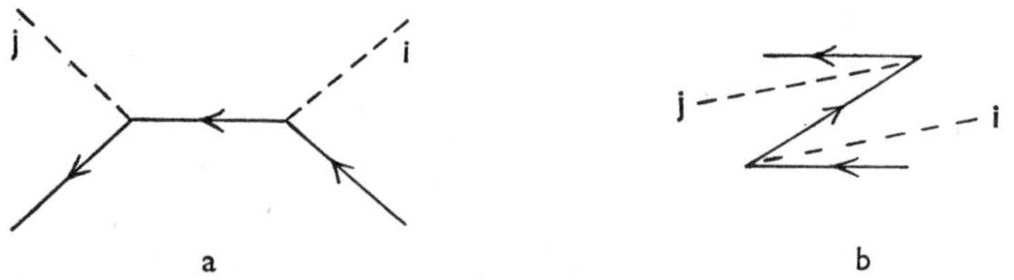

Figure 8.3 The two time orderings contained in the Feynman graph (8.1a) exhibit the spin-parity content of the pole term.

8.3 GENERALIZED ISOSPIN MATRICES

Consider the problem of constructing trilinear couplings of two pions to particles of arbitrary isospin. Equation (8.50) is the simplest example of such a coupling. In this case the Pauli isospin matrices τ_i act as Clebsch-Gordan coefficients combining the components of N and N^* to make an isovector. From the work of Chapter 5 it is clear that the three quantities $N_I^\dagger \mathbf{t} N_I$ also comprise an isovector. As there, the t_i are standard Hermitian I-dimensional isospin matrices and the fields N_I transform as φ_α in Eq. (6.19). Thus the Yukawa couplings of pions to "nucleons" of isospin I may be written as $g\bar{N}_I \mathbf{t} N_I \cdot \boldsymbol{\pi}$ apart from the spin-parity complications treated in Chapter 9. This scarcely exhausts the possibilities since the pion can induce changes of isospin $\Delta I = \pm 1, 0$. Hence we expect to have couplings of the form

$$\mathscr{L}_{\pi NN'} = g\bar{N}'\mathbf{T}(I', I)N \cdot \boldsymbol{\pi} + \text{H.c.} \tag{8.67}$$

where N has isospin I and $N_{I'}$ has isospin I' ($I' = I \pm 1$ or I). The "generalized isospin matrices" $\mathbf{T}(I', I)$ not only make possible a systematic treatment of the $NN'\pi$ vertex but allow a simple construction of projection operators for the reactions $N\pi \to N'\pi$, $N\bar{N}' \to \pi\pi$ and the corresponding crossing matrices. Special cases were introduced in refs. 10 and 11, and the general case treated in ref. 2.

Consider the isospin-conserving vertex formed from fields $\psi_{I'}^{(\mu)}$, $\psi_{I'}^{(\nu)}$, and $\pi^{(\rho)}$, where the superscripts denote the decrease in I_z resulting from application of the field operator in question. These fields transform according to the representations $D^{(I')*}$, $D^{(I)*}$, and $D^{(1)*}$, respectively. The Fourier components of the conjugate operators create isomultiplets related by the standard phase convention. The only tricky point is that the complex conjugate representations have to be converted to standard form in order to use the conventional Clebsch-Gordan coefficients. Thus we construct the isovector $V^{(\rho)}$ from the baryon fields

$$V^{(\rho)} = \sum_{\nu\mu} C(I'I1;\, \nu\mu\rho)(-1)^{I-\mu}\psi_{I'}^{(\nu)*}\psi_{I}^{(-\mu)} \tag{8.68}$$

$V^{(\rho)}$ transforms as $D^{(1)}$. The notation of Rose[12] is used for the Clebsch-Gordan coefficients. Recalling that $\pi^{(\rho)*} = (-1)^{\rho}\pi^{(-\rho)}$, we can form two equivalent isoscalars:

$$\boldsymbol{\pi} \cdot \mathbf{V} = \sum_{\rho} (-1)^{\rho}\pi^{(\rho)*}V^{(-\rho)} = \sum_{\rho} \pi^{(\rho)}V^{(\rho)} \tag{8.69}$$

Here we note that the fields $\pi^{(\rho)}$ are related to the Cartesian fields π_1, π_2, π_3 by

$$\begin{aligned} \pi^{(1)*} &= -(\pi_1 + i\pi_2)/\sqrt{2} \\ \pi^{(0)*} &= \pi_3 \\ \pi^{(-1)*} &= (\pi_1 - i\pi_2)/\sqrt{2} \end{aligned} \tag{8.70}$$

i.e. the operators $\pi^{(\rho)*}$ transform exactly as the *states* Ψ_α of Eq. (8.47). We emphasize this because the literature is full of ambiguity on this point. Clearly the matrices **T** are the Cartesian analogues of the quantities $(-1)^{I-\mu}C(I'I1;\, \nu\mu\rho)$, where the latter define matrices labeled by (ν, μ). For purposes of normalization we introduce an extra factor.

In the spherical basis the generalized isospin matrices are defined in terms of the Clebsch-Gordon coefficients.

$$T_{\nu\mu}^{(\rho)}(I', I) = \left(\frac{(2I'+1)(2I+1)}{3}\right)^{\frac{1}{2}}(-1)^{I-\mu}C(I'I1;\, \nu - \mu\rho) \tag{8.71}$$

As usual the Cartesian matrices are related to $T^{(\rho)}$ by

$$\mathbf{T}^{(\pm 1)} = \mp(\mathbf{T}_1 \pm i\mathbf{T}_2)/\sqrt{2}, \qquad \mathbf{T}^{(0)} = \mathbf{T}_3 \tag{8.72}$$

The indices ν, μ which label rows and columns refer to I_3 eigenvalues, and $T(I', I)$ is a $(2I' + 1) \times (2I + 1)$ matrix.

The isospin-invariant coupling is then given by

$$\mathscr{L} = g \sum_{\rho} \psi_{I'}^{\dagger} T^{(\rho)} \psi_I \pi^{(\rho)} + \text{H.c.} \tag{8.73}$$

Using Eqs. (8.70) and (8.72), this yields the generalized isospin matrices defined in Eq. (8.67) when we write $\psi_I = N$, $\psi_{I'} = N'$.

Using symmetry and orthonormality of the Clebsch-Gordan coefficients we get

$$\begin{aligned} \mathbf{T}^{\dagger} \cdot \mathbf{T} &= \sum_i T_i^{\dagger} T_i = 2I' + 1 \qquad (\text{dimension } (2I + 1) \times (2I + 1)) \\ \mathbf{T} \cdot \mathbf{T}^{\dagger} &= \sum_i T_i T_i^{\dagger} = 2I + 1 \qquad (\text{dimension } (2I' + 1) \times (2I' + 1)) \end{aligned} \tag{8.74}$$

From (8.74) we see that

$$\text{Tr } \mathbf{T}^{\dagger} \cdot \mathbf{T} = \text{Tr } \mathbf{T} \cdot \mathbf{T}^{\dagger} = (2I + 1)(2I' + 1) \tag{8.75}$$

This result can be generalized by writing the Cartesian T's in terms of spherical components and using orthogonality properties. We obtain

$$\begin{aligned} \text{Tr } T_i T_j^{\dagger} &= \tfrac{1}{3}\delta_{ij}(2I + 1)(2I' + 1) \\ &= \text{Tr } T_i^{\dagger} T_j \end{aligned} \tag{8.76}$$

For $I' = I$, $\mathbf{T}(I, I)$ is proportional to the usual isospin matrix

$$\mathbf{T}(I, I) = \left(\frac{2I + 1}{I(I + 1)}\right)^{\frac{1}{2}} \mathbf{t} \tag{8.77}$$

The quantities $\mathbf{T}^{\rho}$ in (8.71) are real. For $I' = I + 1$ we find the explicit formulas

$$\begin{aligned} T_{\nu\mu}^{(1)}(I + 1, I) &= \left[\frac{(I + \mu + 1)(I + \mu + 2)}{2(I + 1)}\right]^{\frac{1}{2}} \delta_{\mu,\nu-1} \\ T_{\nu\mu}^{(0)}(I + 1, I) &= \left[\frac{(I + 1)^2 - \mu^2}{I + 1}\right]^{\frac{1}{2}} \delta_{\nu\mu} \\ T_{\nu\mu}^{(-1)}(I + 1, I) &= \left[\frac{(I - \mu + 1)(I - \mu + 2)}{2(I + 1)}\right]^{\frac{1}{2}} \delta_{\mu,\nu+1} \end{aligned} \tag{8.78}$$

Table 8.1 gives explicit matrices for $I = \frac{1}{2}, \frac{3}{2}, \frac{5}{2}$.

Equation (8.67) indicates that the adjoint of $\mathbf{T}(I', I)$ is proportional to $\mathbf{T}(I, I')$. The relation

$$[\mathbf{T}^{\dagger}(I', I)]_{\mu\nu} = [\mathbf{T}(I', I)]_{\nu\mu}^{*} \tag{8.79}$$

Table 8.1 The generalized isospin matrices are given in spherical bases for $(I', I) = (\frac{3}{2}, \frac{1}{2}), (\frac{5}{2}, \frac{3}{2})$, and $(\frac{7}{2}, \frac{5}{2})$.

$$T^{(1)}(\tfrac{3}{2}, \tfrac{1}{2}) = \frac{1}{\sqrt{3}} \begin{bmatrix} \sqrt{6} & 0 \\ 0 & \sqrt{2} \\ 0 & 0 \\ 0 & 0 \end{bmatrix} \qquad T^{(-1)}(\tfrac{3}{2}, \tfrac{1}{2}) = \frac{1}{\sqrt{3}} \begin{bmatrix} 0 & 0 \\ 0 & 0 \\ \sqrt{2} & 0 \\ 0 & \sqrt{6} \end{bmatrix} \qquad T^{(0)}(\tfrac{3}{2}, \tfrac{1}{2}) = \frac{1}{\sqrt{3}} \begin{bmatrix} 0 & 0 \\ 2 & 0 \\ 0 & 2 \\ 0 & 0 \end{bmatrix}$$

$$T^{(1)}(\tfrac{5}{2}, \tfrac{3}{2}) = \frac{1}{\sqrt{5}} \begin{bmatrix} \sqrt{20} & 0 & 0 & 0 \\ 0 & \sqrt{12} & 0 & 0 \\ 0 & 0 & \sqrt{6} & 0 \\ 0 & 0 & 0 & \sqrt{2} \\ 0 & 0 & 0 & 0 \\ 0 & 0 & 0 & 0 \end{bmatrix} \qquad T^{(-1)}(\tfrac{5}{2}, \tfrac{3}{2}) = \frac{1}{\sqrt{5}} \begin{bmatrix} 0 & 0 & 0 & 0 \\ 0 & 0 & 0 & 0 \\ \sqrt{2} & 0 & 0 & 0 \\ 0 & \sqrt{6} & 0 & 0 \\ 0 & 0 & \sqrt{12} & 0 \\ 0 & 0 & 0 & \sqrt{20} \end{bmatrix} \qquad T^{(0)}(\tfrac{5}{2}, \tfrac{3}{2}) = \frac{2}{\sqrt{5}} \begin{bmatrix} 0 & 0 & 0 & 0 \\ \sqrt{2} & 0 & 0 & 0 \\ 0 & \sqrt{3} & 0 & 0 \\ 0 & 0 & \sqrt{3} & 0 \\ 0 & 0 & 0 & \sqrt{2} \\ 0 & 0 & 0 & 0 \end{bmatrix}$$

$$T^{(1)}(\tfrac{7}{2}, \tfrac{5}{2}) = \frac{1}{\sqrt{7}} \begin{bmatrix} \sqrt{42} & 0 & 0 & 0 & 0 & 0 \\ 0 & \sqrt{30} & 0 & 0 & 0 & 0 \\ 0 & 0 & \sqrt{20} & 0 & 0 & 0 \\ 0 & 0 & 0 & \sqrt{12} & 0 & 0 \\ 0 & 0 & 0 & 0 & \sqrt{6} & 0 \\ 0 & 0 & 0 & 0 & 0 & \sqrt{2} \\ 0 & 0 & 0 & 0 & 0 & 0 \\ 0 & 0 & 0 & 0 & 0 & 0 \end{bmatrix} \qquad T^{(-1)}(\tfrac{7}{2}, \tfrac{5}{2}) = \frac{1}{\sqrt{7}} \begin{bmatrix} 0 & 0 & 0 & 0 & 0 & 0 \\ 0 & 0 & 0 & 0 & 0 & 0 \\ \sqrt{2} & 0 & 0 & 0 & 0 & 0 \\ 0 & \sqrt{6} & 0 & 0 & 0 & 0 \\ 0 & 0 & \sqrt{12} & 0 & 0 & 0 \\ 0 & 0 & 0 & \sqrt{20} & 0 & 0 \\ 0 & 0 & 0 & 0 & \sqrt{30} & 0 \\ 0 & 0 & 0 & 0 & 0 & \sqrt{42} \end{bmatrix} \qquad T^{(0)}(\tfrac{7}{2}, \tfrac{5}{2}) = \frac{1}{\sqrt{7}} \begin{bmatrix} 0 & 0 & 0 & 0 & 0 & 0 \\ \sqrt{12} & 0 & 0 & 0 & 0 & 0 \\ 0 & \sqrt{20} & 0 & 0 & 0 & 0 \\ 0 & 0 & \sqrt{24} & 0 & 0 & 0 \\ 0 & 0 & 0 & \sqrt{24} & 0 & 0 \\ 0 & 0 & 0 & 0 & \sqrt{20} & 0 \\ 0 & 0 & 0 & 0 & 0 & \sqrt{12} \\ 0 & 0 & 0 & 0 & 0 & 0 \end{bmatrix}$$

becomes, in the spherical basis,

$$[\mathbf{T}^{\dagger}(I', I)]^{\rho}_{\nu\mu} = (-1)^{\rho}[\mathbf{T}(I', I)]^{(-\rho)}_{\mu\nu} \tag{8.80}$$

with $(T^{\dagger})^{(\pm 1)} = \mp(T^{\dagger}_x \pm iT^{\dagger}_y)$, $(T^{\dagger})^0 = T_z$. Using simple symmetry properties of the Clebsch-Gordan coefficients gives*

$$\mathbf{T}^{\dagger}(I', I) = (-1)^{I'-I}\mathbf{T}(I, I') \tag{8.81}$$

Writing the fields in terms of isospin wave functions as in Eq. (7.5), we see that the vertex (8.67) contains four types of terms. The isospin parts of the vertices associated with emission (or absorption) of Cartesian pions π_i are

$$\begin{array}{ll} N(\alpha) \to \pi_i + N'(\alpha') & \chi^{\dagger}_{\alpha'} T_i \chi_{\alpha} \\ \bar{N}'(\alpha') \to \pi_i + \bar{N}(\alpha) & \chi^{c\dagger}_{\alpha'} T_i \chi^{c}_{\alpha} \\ N(\alpha) + \bar{N}'(\alpha') \to \pi_i & \chi^{c\dagger}_{\alpha'} T_i \chi_{\alpha} \\ \pi_i \to N'(\alpha') + \bar{N}(\alpha) & \chi^{\dagger}_{\alpha'} T_i \chi^{c}_{\alpha} \end{array} \tag{8.82}$$

where T_i is $T_i(I', I)$. The use of this formalism easily avoids tricky sign errors in antiparticle problems.

Further relations among the generalized isospin matrices follow from a study of the crossing conditions.

In order to illustrate the foregoing techniques we consider the $N\Delta\pi$ vertex, where N is the nucleon ($I = \frac{1}{2}, J^P = \frac{1}{2}^+$) and the $(I, J^P) = (\frac{3}{2}, \frac{3}{2}^+)$ resonance at 1238 MeV. We treat this resonant state as a particle and introduce the Rarita-Schwinger field Δ_μ to describe it. The effective $N\Delta\pi$ coupling is then

$$\mathscr{L}_{N\Delta\pi} = g\bar{\Delta}_\mu \mathbf{T} N \cdot \partial^\mu \boldsymbol{\pi} + \text{H.c.} \tag{8.83}$$

where the matrices $\mathbf{T} = \mathbf{T}(\frac{3}{2}, \frac{1}{2})$ are

$$T_1 = \begin{bmatrix} -1 & 0 \\ 0 & -\frac{1}{\sqrt{3}} \\ \frac{1}{\sqrt{3}} & 0 \\ 0 & 1 \end{bmatrix}, \quad T_2 = i\begin{bmatrix} 1 & 0 \\ 0 & \frac{1}{\sqrt{3}} \\ \frac{1}{\sqrt{3}} & 0 \\ 0 & 1 \end{bmatrix}, \quad T_3 = \frac{2\sqrt{3}}{3}\begin{bmatrix} 0 & 0 \\ 1 & 0 \\ 0 & 1 \\ 0 & 0 \end{bmatrix} \tag{8.84}$$

* There are two equivalent ways to write the $NN'\pi$ vertex:

$$H_1 = g_1 N' \mathbf{T}(I', I) N \cdot \boldsymbol{\pi} + \text{H.c.},$$

$$H_2 = g_2 N \mathbf{T}(I, I') N' \cdot \boldsymbol{\pi} + \text{H.c.}$$

These are identical provided $g_2 = (-1)^{I'-I} g_1^*$.

To exhibit the charge structure we rewrite (8.83) in the spherical basis

$$\mathscr{L} = g \sum_{\rho} \partial_\mu \pi^{(\rho)*} \bar{N} T^{(\rho)\dagger} \Delta^\mu + \text{H.c.} \tag{8.85}$$

where $\pi^{(\rho)*}$ is given in Eq. (8.70). Writing $N = \text{col}\,(p, n)$,

$$\Delta = \text{col}\,(\Delta^{++}, \Delta^{+}, \Delta^{0}, \Delta^{-})$$

gives, omitting the Lorentz space structure,

$$\mathscr{L}_{N\Delta\pi}/g = (\pi^+)^*[\sqrt{2}\bar{p}\,\Delta^{++} + \sqrt{\tfrac{2}{3}}\bar{n}\,\Delta^{+}] + (\pi^-)^*[\sqrt{\tfrac{2}{3}}\bar{p}\,\Delta^{0} + \sqrt{2}\bar{n}\,\Delta^{-}]$$
$$+ \frac{2}{\sqrt{3}}(\pi^0)^*[\bar{p}\,\Delta^{+} + \bar{n}\,\Delta^{0}] + \text{H.c.} \tag{8.86}$$

The Δ width is given by

$$\Gamma(\Delta \to N\pi) = \frac{2}{3}\left(\frac{g^2}{4\pi}\right)\left(\frac{p}{\mu}\right)^3 \frac{E_N + m_N}{m_\Delta} \tag{8.87}$$

As another illustration we use the **T** matrices to construct the projection operators for the reactions $N_I + \pi_i \to N_I + \pi_j$, $N_I + \bar{N}_{I'} \to \pi_i + \pi_j$. Paying attention to the rules (8.82) and considering the contribution of $N_{I'}$ to the intermediate state of Fig. (8.1a) gives directly

$$\begin{aligned}
\mathscr{P}_{ji}(I+1) &= T_j^\dagger(I+1, I)T_i(I+1, I)/(2I+1) \\
\mathscr{P}_{ji}(I) &= T_j^\dagger(I, I)T_i(I, I)/(2I+1) = t_j t_i/I(I+1) \\
\mathscr{P}_{ji}(I-1) &= T_j^\dagger(I-1, I)T_i(I-1, I)/(2I+1)
\end{aligned} \tag{8.88}$$

where the normalization factor is determined by $\mathscr{P}^2 = \mathscr{P}$. The further relation $\sum_I \mathscr{P}_I = 1$ gives another result for T_i

$$\sum_I T_j^\dagger(I', I)T_i(I', I) = (2I+1)\delta_{ij} \tag{8.89}$$

Although the matrices (8.88) are explicitly known it is convenient to express them in terms of the three independent quantities δ_{ij}, $t_i t_j$, $t_j t_i$ or equivalently the symmetrical and antisymmetrical objects δ_{ij}, A_{ij}^+, A_{ji}^-:

$$\begin{aligned}
&\delta_{ij} \\
2A_{ij}^- &\equiv [t_i, t_j] = i\epsilon_{ijk}t_k = -2A_{ji}^- \\
2A_{ij}^+ &\equiv \{t_i, t_j\} = 2A_{ji}^+
\end{aligned} \tag{8.90}$$

These quantities appear naturally in the projection operators for the annihilation reaction $N_I + \bar{N}_I \to \pi_j + \pi_i$ because of the symmetry of the two-pion wave function.

The $\mathscr{P}_{ji}(I')$ can be written as a linear combination of this set. By definition we know that

$$\mathscr{P}_{ji}(I) = (A_{ji}^{(+)} + A_{ji}^{(-)})/I(I+1) \tag{8.91}$$

In order to express $\mathscr{P}_{ji}(I-1)$ in terms of $A^{\pm}$, 1, we regard (8.91) as referring to the s channel and expand it in terms of u-channel $\mathscr{P}$'s using Eq. (8.66) in a Cartesian basis. Using explicit tables of Racah coefficients one finds the relevant crossing matrix[13]

$$X(I_u, I_s) = \begin{bmatrix} \dfrac{1}{2I+1} & -\dfrac{1}{I} & \dfrac{2I+3}{2I+1} \\[2ex] -\dfrac{(2I-1)}{I(2I+1)} & \dfrac{I^2+I-1}{I(I+1)} & \dfrac{2I+3}{(I+1)(2I+1)} \\[2ex] \dfrac{2I-1}{2I+1} & \dfrac{1}{I+1} & \dfrac{1}{(I+1)(2I+1)} \end{bmatrix} \tag{8.92}$$

Since the s and u channels are identical we find

$$\mathscr{P}_{ji}^{I} = -\frac{1}{I}\mathscr{P}_{ij}^{I-1} + \frac{I^2+I-1}{I(I+1)}\mathscr{P}_{ij}^{I} + \frac{1}{I+1}(\delta_{ij} - \mathscr{P}_{ij}^{I} - \mathscr{P}_{ij}^{I-1}) \tag{8.93}$$

We have used the completeness relation to eliminate $\mathscr{P}^{I+1}$. Now $\mathscr{P}^{I-1}$ may be calculated in terms of known quantities. The results of this calculation and a similar one for $\mathscr{P}^{I+1}$ are given by:

$$\begin{aligned} \mathscr{P}_{ji}^{s}(I-1) &= \frac{I}{2I+1}\delta_{ji} - \frac{1}{I(2I+1)}A_{ji}^{+} + \frac{2I-1}{I(2I+1)}A_{ji}^{-} \\ \mathscr{P}_{ji}^{s}(I+1) &= \frac{I+1}{2I+1}\delta_{ji} - \frac{1}{(I+1)(2I+1)}A_{ji}^{+} - \frac{2I+3}{(I+1)(2I+1)}A_{ji}^{-} \end{aligned} \tag{8.94}$$

It is not difficult to use these projection operators to rederive the crossing matrix (8.92).

Next consider the t-channel reaction $N_I\bar{N}_I' \rightarrow \pi_i\pi_j$. To compute the projection operators we proceed as before, using the relation

$$\mathscr{P}_{ij}^{t}(I_t) = \sum_{I_s} X^{-1}(I_s, I_t)\mathscr{P}_{ji}^{s}(I_s) \tag{8.95}$$

The phase of the crossing matrix, and of the t-channel projection operators, depends on the phase factors η_α. Making the standard choice $\eta_\alpha = (-1)Q_\alpha$

we obtain the explicit crossing matrix

$$X^{-1}(I_s, I_t) = \frac{(-1)^\sigma}{\sqrt{3(2I+1)}} \begin{bmatrix} 1 & 3\sqrt{\dfrac{I+1}{2I}} & \sqrt{\dfrac{5(I+1)(2I+3)}{2I(2I-1)}} \\ 1 & \dfrac{3}{\sqrt{2I(I+1)}} & -\sqrt{\dfrac{5(2I-1)(2I+3)}{2I(I+1)}} \\ 1 & -3\sqrt{\dfrac{I}{2(I+1)}} & \sqrt{\dfrac{5I(2I-1)}{2(I+1)(2I+3)}} \end{bmatrix} \tag{8.96}$$

where $\sigma = I + 1 + \frac{1}{2}Y$ where Y is the hypercharge of N_I. The inverse matrix $X(I_t, I_s)$ is given by Eq. (7.24) and was given explicitly in ref. 3.

Many of the explicit crossing matrices of Chapter 7 are special cases of Eqs. (8.96).

The projection operators for $I_t = 0, 1, 2$ given by Eqs. (8.95) and (8.96) are found to be

$$\begin{aligned} \mathscr{P}_{ij}^{I_t=0} &= \frac{(-1)^\sigma \delta_{ji}}{\sqrt{3(2I+1)}} \\ \mathscr{P}_{ij}^{I_t=1} &= \frac{(-1)^\sigma \sqrt{6}}{\sqrt{I(I+1)(2I+1)}} A_{ji}^{-} \\ \mathscr{P}_{ij}^{I_t=2} &= -(-1)^\sigma \left(\frac{30}{I(I+1)(4I^2-1)(2I+3)}\right)^{\frac{1}{2}} [A_{ji}^{+} - \tfrac{1}{3}I(I+1)\delta_{ji}] \end{aligned} \tag{8.97}$$

The symmetry of $\mathscr{P}(I_t)$ in the pion labels is obvious from Bose symmetry.

For pions and nucleons $I_t = 0$ and 1 only. $\sigma = 2$, $I = \frac{1}{2}$, and $A_{ji} = (i/2)\epsilon_{jik}\tau_k$ give $\mathscr{P}_{ij}^0 = \delta_{ij}/\sqrt{6}$, $\mathscr{P}_{ij}^1 = i\epsilon_{jik}\tau_k 2$ in agreement with Sec. 8.2. For $\bar{\Delta}\Delta \to \pi\pi$ (Δ has $I = \frac{3}{2}$), one finds

$$\begin{aligned} \mathscr{P}_{ij}^{I_t=0} &= \sqrt{3}\,\delta_{ji}/6 \\ \mathscr{P}_{ij}^{I_t=1} &= i\epsilon_{jik}t_k/\sqrt{10} \\ \mathscr{P}_{ij}^{I_t=2} &= -\frac{\sqrt{6}}{12}(t_i t_j + t_j t_i) - \tfrac{2}{3}I(I+1)\delta_{ij} \end{aligned} \tag{8.98}$$

As an example consider the isospin content of the pole terms (Fig. 8.1) in πN_I scattering due to the N_I, intermediate state. Neglecting spin, we find the invariant transition matrix

$$T_{ji} = g^2\left[\frac{T_j^\dagger(I', I)T_i(I', I)}{(m')^2 - s} + \frac{T_i^\dagger(I', I)T_j(I', I)}{(m')^2 - u}\right] \tag{8.99}$$

where m' is the mass of N_I. From our previous results we find

$$T_{I_s} = \frac{(2I + 1)g^2}{(m')^2 - s}\delta_{I'I_s} + \frac{(2I + 1)X_{us}(I_s, I')g^2}{(m')^2 - u} \tag{8.100}$$

with X_{us} given by Eq. (8.92).

A similar analysis may be done for the inelastic processes $N_I + \pi_i \rightarrow N_{I+1} + \pi_j$, $N_I + \bar{N}_{I+1} \rightarrow \pi_i + \pi_j$. Details may be found in ref. 3. For completeness we give the crossing matrices appropriate to these reactions.

$$X(I_u, I_s) = \frac{1}{I+1}\begin{bmatrix} 1 & -\sqrt{\dfrac{I(I+2)(2I+3)}{2I+1}} \\ -\sqrt{\dfrac{I(I+2)(2I+1)}{2I+3}} & -1 \end{bmatrix} \tag{8.101}$$

$$X^{-1}(I_s, I_t) = \frac{(-1)^\sigma}{\sqrt{2(I+1)}}\begin{bmatrix} \sqrt{\dfrac{3I}{2I+1}} & \sqrt{\dfrac{5(I+2)}{2I+1}} \\ \sqrt{\dfrac{3(I+2)}{2I+3}} & -\sqrt{\dfrac{5I}{2I+3}} \end{bmatrix} \tag{8.102}$$

8.4 ANGULAR MOMENTUM PROJECTION OPERATORS

The expansion of scattering amplitudes in terms of components having definite values of angular momenta is an old procedure, going back to Lord Rayleigh.[14] Detailed treatments of such decompositions have been given in recent textbooks on scattering theory.[15–17] For many purposes the helicity partial wave expansion is to be preferred in applications to particle physics. However for several important reactions the older descriptions are well entrenched and provide a familiar reference point for modern applications. Hence we give here a very brief analysis of the method of angular projection operators.[18–19] Recent general and systematic analyses of the "non-dynamical" structure of particle reactions will not be considered here in view of the excellent treatments already available (see ref. 20 and the literature cited therein).

It is clear that the method of isospin projection operators developed in Sec. 8.2 can be extended to the case of angular momentum projection operators. We work in the center of mass frame and assume the validity of parity and time-reversal invariance. For two particle reactions we can expand the scattering matrix in analogy to Eq. (8.4):

$$\begin{aligned} \omega_{fi} &= \sum_{J\pi} S_J^\pi \mathscr{P}_J^\pi \\ \mathscr{P}_J^\pi &= \sum_m |Jm\pi f\rangle\langle Jm\pi i| \end{aligned} \tag{8.103}$$

where m is the eigenvalue of J_z and π denotes the parity. If more than two particles are involved it is necessary to specify certain "intermediate" quantum numbers.

As in the case of isospin, there are various ways of obtaining the projection operators. We give two methods for the reactions of the type spin 0 + spin $\frac{1}{2} \to$ spin 0 + spin $\frac{1}{2}$. The most important example of such a reaction is pion-nucleon scattering. Although the resulting formulas are non-covariant and appear nonrelativistic they actually hold in the relativistic domain although the labels "spin" and orbital momentum no longer have a clearly defined meaning. It is instructive to compare these results with the helicity partial wave expansion for the same reaction (Sec. 10.5).

The free wave function of a spin $\frac{1}{2}$ particle has the asymptotic form

$$e^{ikr\cos\theta}\chi_\sigma \sim \sum_{l=0}^{\infty} \frac{e^{ikr} - (-1)^l e^{-ikr}}{2ikr}(2l+1)P_l(\cos\theta)\chi_\sigma \tag{8.104}$$

the coordinate $z = r\cos\theta$ describes the relative coordinate of the spin 0, spin $\frac{1}{2}$ pair, or the distance from a fixed spherically symmetric potential. $\sigma = \pm\frac{1}{2}$ gives the spin projection on the z axis. The wave (8.104) may be further decomposed into components having definite angular momentum $j = l \pm \frac{1}{2}$ and parity $(-1)^{l+1}$. (Here the extra -1 in the parity factor is due to the odd intrinsic parity of the pion.) The projection operators are

$$\begin{aligned} j = l + \tfrac{1}{2}: &\qquad \mathscr{P}_{l+} = \frac{l + 1 + \mathbf{l}\cdot\boldsymbol{\sigma}}{2l+1} \\ j = l - \tfrac{1}{2}: &\qquad \mathscr{P}_{l-} = \frac{l - \boldsymbol{\sigma}\cdot\mathbf{l}}{2l+1} \end{aligned} \tag{8.105}$$

where $\mathbf{l} = -i\mathbf{r} \times \nabla$ (cf. Eq. (8.29)). These operators obey the relations

$$\begin{aligned} \mathscr{P}_{l\pm}^2 &= \mathscr{P}_{l\pm} \\ \mathscr{P}_{l\pm}\mathscr{P}_{l\mp} &= 0 \\ \mathscr{P}_{l+} + \mathscr{P}_{l-} &= 1 \\ \mathbf{J}^2\mathscr{P}_{l\pm} &= j(j+1)\mathscr{P}_{l\pm}, \qquad j = l \pm \tfrac{1}{2} \end{aligned} \tag{8.106}$$

The effect of interaction is to modify the phase of the outgoing waves of distinct j and parity; hence we find

$$\psi_\sigma^{(+)} \sim \sum_{l=0}^{\infty} \frac{2l+1}{2ikr} [(e^{2i\delta_{l+}}\mathscr{P}_{l+} + e^{2i\delta_{l-}}\mathscr{P}_{l-})e^{ikr} - (-1)^l(\mathscr{P}_{l+} + \mathscr{P}_{l-})e^{-ikr}]P_l(\cos\theta)\chi_\sigma \tag{8.107}$$

and the scattering amplitude $f_{\sigma'\sigma}$ for scattering from spin state σ to σ' is defined as usual by

$$\chi_{\sigma'}^\dagger(\psi_\sigma^{(+)} - e^{ikz}\chi_\sigma) \sim f_{\sigma'\sigma}e^{ikr}/r \tag{8.108}$$

Writing $f_{\sigma'\sigma} = \chi_{\sigma'}^\dagger f \chi_\sigma$, where f is a 2×2 matrix, we find

$$\begin{aligned} f &= \sum_{l=0}^{\infty}(2l+1)(f_{l+}\mathscr{P}_{l+} + f_{l-}\mathscr{P}_{l-})P_l(\cos\theta) \\ f_{l\pm} &= e^{i\delta_{l\pm}}\sin\delta_{l\pm}/k \end{aligned} \tag{8.109}$$

It is easy to express (8.109) in terms of the direction of momenta $\hat{k}$ and $\hat{k}'$ using

$$\mathbf{r} \times \nabla P_l(\cos\theta) = \left[\mathbf{r} \times \nabla\left(\frac{z}{r}\right)\right]P_l'(\cos\theta) = \hat{k}' \times \hat{k}P_l'(\hat{k}\cdot\hat{k}') \tag{8.110}$$

Hence the spin dependent term is

$$\begin{aligned} \boldsymbol{\sigma}\cdot\mathbf{l}P_l(\cos\theta) &= i\boldsymbol{\sigma}\cdot\hat{k} \times \hat{k}'P_l'(\cos\theta) \\ &= i\boldsymbol{\sigma}\cdot\hat{n}\sin\theta P_l'(\cos\theta) \end{aligned} \tag{8.111}$$

where $\hat{n} = \hat{k} \times \hat{k}'/|\hat{k} \times \hat{k}'|$ is the conventional normal to the scattering plane. Sometimes one uses the identity $P_l^1(\cos\theta) = \sin\theta P_l'(\cos\theta)$. The scattering amplitude then has the form

$$\begin{aligned} f &= f_d + i\boldsymbol{\sigma}\cdot\hat{n}f_{sp} \\ f_d &= \sum_{l=0}^{\infty}[(l+1)f_{l+} + lf_{l-}]P_l(\cos\theta) \\ f_{sp} &= \sum_{l=1}^{\infty}(f_{l+} - f_{l-})P_l^1(\cos\theta) \end{aligned} \tag{8.112}$$

Note that if there is no "spin-orbit force" $f_{l+} = f_{l-}$ and one obtains the standard result for spinless particles. f_d is sometimes called the "direct" amplitude and f_{sp} the "spin-flip" amplitude.

The unpolarized differential cross section is

$$d\sigma/d\Omega = \tfrac{1}{2}\,\mathrm{Tr}\, f^{+}f = |f_d|^2 + |f_{sp}|^2 \tag{8.113}$$

If the initial state is unpolarized the final polarization is defined by

$$\mathbf{P} = \langle \boldsymbol{\sigma} \rangle = \frac{\mathrm{Tr}\, f^{\dagger}\boldsymbol{\sigma} f}{\mathrm{Tr}\, f^{\dagger} f} \tag{8.114}$$

which may be cast in the form

$$\mathbf{P}\frac{d\sigma}{d\Omega} = 2\hat{n}\ \mathrm{Im}\,(f_{sp}^{*} f_d) \tag{8.115}$$

Another useful and common form of f follows from the identities

$$\begin{aligned} \boldsymbol{\sigma}\cdot\hat{k}'\boldsymbol{\sigma}\cdot\hat{k} &= \hat{k}\cdot\hat{k}' + i\boldsymbol{\sigma}\cdot\hat{k}'\times\hat{k} \\ zP_l'(z) &= lP_l(z) + P_{l-1}'(z) \\ (2l+1)P_l(z) &= P_{l+1}'(z) - P_{l-1}'(z) \end{aligned} \tag{8.116}$$

This form is

$$\begin{aligned} f &= f_1 + \sigma\cdot\hat{k}'\sigma\cdot\hat{k}f_2 \\ f_1 &= \sum_{l=0}^{\infty}[P_{l+1}'(z)f_{l+} - P_{l-1}'(z)f_{l-}] \\ f_2 &= \sum_{l=1}^{\infty}(f_{l-} - f_{l+})P_l'(z) \end{aligned} \tag{8.117}$$

Finally one may write (8.112) in the form

$$f = \sum_{l=0}^{\infty}(f_{l+}\mathscr{I}_{l+} + f_{l-}\mathscr{I}_{l-}) \tag{8.118}$$

where the angular projection operators are

$$\begin{aligned} \mathscr{I}_{l+}(\hat{k}',\hat{k}) &= (l+1)P_l(z) - i\boldsymbol{\sigma}\cdot\hat{k}'\times\hat{k}P_l'(z) \\ \mathscr{I}_{l-}(\hat{k}',\hat{k}) &= lP_l(z) + i\boldsymbol{\sigma}\cdot\hat{k}'\times\hat{k}P_l'(z) \end{aligned} \tag{8.119}$$

Letting $\alpha, \beta = 1, 2$ corresponding to $j = l - \tfrac{1}{2}$ and $j = l + \tfrac{1}{2}$ respectively we find the crossing relation[21]

$$\begin{aligned} \mathscr{I}_{\alpha}(\hat{k},\hat{k}') &= \sum_{\beta}\mathscr{I}_{\beta}(\hat{k}',\hat{k})N_{\beta\alpha} \\ N &= \frac{1}{2l+1}\begin{pmatrix} -1 & 2l+2 \\ 2l & 1 \end{pmatrix} \end{aligned} \tag{8.120}$$

We now indicate[18] how to derive the result (8.112) using explicit wave functions in the formula (8.103). Apart from a normalization constant the projection operator is formed from the wave functions

$$\Psi_{JMl} = \sum_{\sigma} C(l\tfrac{1}{2}j;\, M-\sigma, \sigma) Y_{l,M-\sigma}(\hat{k})\chi_{\sigma} \tag{8.121}$$

where $Y_{lm}(\hat{k})$ is the usual spherical harmonic. Using Eqs. (8.23) we may write $\mathscr{P}_{jl}$ as a matrix in spin space

$$\mathscr{P}_{jl}(\hat{k}', \hat{k}) = Q^{jl}_{\frac{1}{2}\frac{1}{2}}\left(\frac{1+\sigma_z}{2}\right) + Q^{jl}_{\frac{1}{2}-\frac{1}{2}}\left(\frac{\sigma_x + i\sigma_y}{2}\right) + Q^{jl}_{-\frac{1}{2}\frac{1}{2}}\left(\frac{\sigma_x - i\sigma_y}{2}\right) + Q^{jl}_{-\frac{1}{2}-\frac{1}{2}}\left(\frac{1-\sigma_z}{2}\right) \tag{8.122}$$

where the $Q_{\sigma'\sigma}$ are given by

$$Q^{jl}_{\sigma'\sigma} = \sum_{M} C(l\tfrac{1}{2}j;\, M-\sigma', \sigma')C(l\tfrac{1}{2}j;\, M-\sigma, \sigma)Y^{*}_{lM-\sigma'}(\hat{k}')Y_{l,M-\sigma}(\hat{k}) \tag{8.123}$$

The sum may be reduced to one term by choosing $\hat{k}$ along the z axis, so that $Y_{l,M=\sigma}(\hat{k}) = \delta_{M\sigma}[(2l+1)/4\pi]^{\frac{1}{2}}$. Choosing $\hat{k}'$ in the x-z plane and using explicit forms for the Clebsch-Gordan coefficients yields for $j = l \pm \frac{1}{2}$

$$\begin{aligned} 4\pi\mathscr{P}_{l+\frac{1}{2},l} &= (l+1)P_l + i \sin\theta\sigma_y P'_l \\ 4\pi\mathscr{P}_{l-\frac{1}{2},l} &= lP_l - i \sin\theta\sigma_y P'_l \end{aligned} \tag{8.124}$$

Except for the normalization factors these are seen to be identical with the expressions (8.119).

This method has been applied[18] to other reactions such as pion photoproduction from nucleons, Compton scattering from nucleons, and nucleon-nucleon scattering as well as production reactions[19] such as $\pi N \to 2\pi N$, etc. Further extensions and generalizations are given in refs. 20 and 22.

PROBLEMS

1. Work out the projection operators for the reaction $N + \pi_k \to N + \pi_i + \pi_j$ (i, j, k are Cartesian indices) using
   (a) $(N(\pi\pi))$ coupling
   (b) $((N\pi)\pi)$ coupling
   What is the connection between these coupling schemes (i.e. express one set of projection operators in terms of the other). The solution to problem (1a) is given in Eq. (8.36).

2. Use the projection operators (8.24)–(8.25) for πN scattering to derive the s-u crossing matrix

$$\mathscr{P}_{ji}^{I_s} = \sum_{I_u} \mathscr{P}_{ij}^{I_u} X(I_u, I_s)$$

$$X_{su} = \begin{pmatrix} -\frac{1}{3} & \frac{4}{3} \\ \frac{2}{3} & \frac{1}{3} \end{pmatrix}$$

cf. with Eq. (8.66) and with Table 8.1.

3. Derive the crossing matrices X_{ts} and $(X^{-1})_{st}$ connecting the reactions $N\bar{N} \to \pi_i \pi_j$ and $N\pi_i \to N\pi_j$. Verify that the results agree with Table 7.1.

$$\mathscr{P}_{ji}^{I_s} = \sum_{I_t} \mathscr{P}_{ij}^{I_t} X(I_t, I_s)$$

$$X_{ts} = \begin{pmatrix} \frac{\sqrt{6}}{3} & \frac{2\sqrt{6}}{3} \\ \frac{2}{3} & -\frac{1}{3} \end{pmatrix}$$

4. Verify that Eqs. (8.66) follow from (8.64) and (8.65). When the indices $\beta_1 \beta'$ refer to pions, transform to a Cartesian basis and show that Eq. (8.64) is valid with β, β' regarded as Cartesian indices.

5. Use the explicit matrices $T_i(\frac{3}{2}, \frac{1}{2})$ of Eq. (8.84) to check the known πN projection operators against (8.88).

$$P_{ji}^{\frac{3}{2}} = \tfrac{1}{2} T_j^\dagger(\tfrac{3}{2}, \tfrac{1}{2}) T_i(\tfrac{3}{2}, \tfrac{1}{2}) = \delta_{ji} - \tfrac{1}{3}\tau_j \tau_i$$

$$P_{ji}^{\frac{1}{2}} = \tfrac{1}{2} T_j^\dagger(\tfrac{1}{2}, \tfrac{1}{2}) T_i(\tfrac{1}{2}, \tfrac{1}{2}) = \tfrac{1}{3}\tau_j \tau_i$$

In the second relation recall Eq. (8.77). Verify the "commutator"

$$T_z(\tfrac{3}{2}, \tfrac{1}{2}) T_x(\tfrac{1}{2}, \tfrac{1}{2}) - T_x(\tfrac{3}{2}, \tfrac{1}{2}) T_z(\tfrac{1}{2}, \tfrac{1}{2}) = -i(\tfrac{2}{3})^{\frac{1}{2}} T_y(\tfrac{3}{2}, \tfrac{1}{2})$$

The generalized commutation rule

$$T_i(I+1, I) T_j(I, I) - T_j(I+1, I) T_i(I, I) = -i\left(\frac{I(2I+1)}{I+1}\right)^{\frac{1}{2}} \epsilon_{ijk} T_k(I+1, I)$$

is easily verified for particular components using the formulas (8.78).

REFERENCES

1. P. Carruthers, *Ann. Phys.* (N.Y.) **14,** 229 (1961), Appendix A.
2. P. Carruthers, *Phys. Rev.* **152,** 1345 (1966); *ibid.* **169,** 1398 (1968) E.
3. P. Carruthers and M. M. Nieto, *Ann. Phys.* (N.Y.) **51,** 359 (1969).
4. G. F. Chew and F. E. Low, *Phys. Rev.* **101,** 1570 (1956).
5. M. L. Goldberger, M. T. Grisaru, S. W. MacDowell, and D. Wong, *Phys. Rev.* **120,** 2250 (1960).
6. R. E. Cutkosky, *Ann. Phys.* (N.Y.) **23,** 415 (1963).
7. W. R. Frazer and J. R. Fulco, *Phys. Rev.* **117,** 1603 (1960).

8. G. F. Chew and S. Mandelstam, *Phys. Rev.* **119,** 467 (1960).
9. F. J. Dyson, *Phys. Rev.* **100,** 344 (1955).
10. S. Ciulli and J. Fischer, *Nuovo Cimento* **12,** 264 (1959).
11. C. Ryan, *Ann. Phys.* (N.Y.) **38,** 1 (1966).
12. M. E. Rose, "Elementary Theory of Angular Momentum," Wiley, New York, 1957.
13. S. Mandelstam, J. E. Paton, R. F. Peierls, and A. Q. Sarker, *Ann. Phys.* (N.Y.) **18,** 198 (1962).
14. Lord Rayleigh, "Theory of Sound," 2nd Ed. (The Macmillan Co., London, 1894), p. 323.
15. M. L. Goldberger and K. M. Watson, "Collision Theory" (John Wiley and Sons, Inc., New York, 1964).
16. R. G. Newton, "Scattering Theory of Waves and Particles" (McGraw-Hill Book Co., New York, 1966).
17. N. F. Mott and H. S. W. Massey, "The Theory of Atomic Collisions" (Oxford University Press, Fair Lawn, N.J., 1949).
18. V. I. Ritus, *J. Exptl. Theoret. Phys.* (U.S.S.R.) **32,** 1536 (1957) (English translation: *Soviet Physics J.E.T.P.* **5,** 1249 (1957)).
19. S. Ciulli and J. Fischer, *Nuovo Cimento* **12,** 264 (1959).
20. M. J. Moravcsik in "Recent Developments in Particle Physics," ed. M. J. Moravcsik (Gordon and Breach Science Publishers, New York, 1966).
21. P. Carruthers, *Phys. Rev.* **133B,** 497 (1964).
22. J. Fischer and S. Ciulli, *J. Exptl. Theoret. Phys.* (U.S.S.R.) **38,** 1740 (1960); *ibid*, **39,** 1349 (1960) (English translations: *Soviet Physics J.E.T.P.* **11,** 1256 (1960); *ibid*, **12,** 941 (1961).

9

SPIN STRUCTURE OF VERTEX FUNCTIONS

9.1 INTRODUCTION

The vertex function occupies a central position in elementary particle physics. Its purest form occurs in field theory, in which it occurs as the Fourier transform of a time-ordered product of three field operators at distinct space-time points. Essentially the same construct describes the decay of an unstable particle into two other particles. Strictly speaking, the three particle vertex does not describe transitions between asymptotic states and so is not an observable quantity from the point of view of S-matrix theory.

The remarkable disparity in strength of the gravitational, weak, electromagnetic and strong interactions provides a strong basis for the observability of the vertex, especially for the hadronic matrix elements of the "currents" $\mathcal{J}$ of the gravitational, weak and electromagnetic interactions. In this picture the stress-energy-momentum tensor, the vector and axial vector currents and the electromagnetic current act as probes of hadron structure through their measurement of hadron matrix elements $\langle B|\,\mathcal{J}\,|A\rangle$. The simplest case, in which A and B are single particle states and $\mathcal{J}$ one of the aforementioned currents. leads to the most useful vertices. It is often useful to consider resonating pairs of particles as a single particle state; for example the matrix element $\langle \pi N|\,J_\mu\,|N\rangle$ has in the $(I, J^P) = (\frac{3}{2}, \frac{3}{2}^+)$ final state a resonance (Δ). For many purposes it is adequate to treat this resonant state as a stable particle and to describe the foregoing vertex as $\langle\Delta|\,J_\mu\,|N\rangle$. It is also useful to define other currents which do not correspond to weak probes. These "strong" probes are easy to define theoretically but often difficult to extract from data.

In Sec. 9.2 we review the kinematic structure of the nucleon electromagnetic form factor. We emphasize the helicity description of nucleon spin states, the isospin structure and the crossing properties. The simple properties of the "brick-wall frame" and the $N\bar{N} \to \gamma$ c.m. frame is noted.

Section 9.3 gives an illustration of the use of effective couplings in coordinate space for the construction of vertices having the correct spin and isospin structure. We emphasize couplings of the type $BB'\pi$ where B and B' are baryons. The matrix element in the B' rest frame can be explicitly evaluated for any spin by the use of recursion relations. The results are applied to the practical problem of evaluating coupling strengths from decay rates for $B' \to B\pi$.

In Sec. 9.4 the kinematic structure of more general vertices is analyzed using the Lorentz transformation properties of helicity states. Multipole expansions in the brick-wall frame are also given. This discussion follows the treatment of ref. 1.

9.2 ELECTROMAGNETIC FORM FACTORS OF THE NUCLEON

Consider the process of electron-proton scattering in the one-photon exchange approximation. The lower part of Fig. (9.1) involves the vertex I_μ, defined by

$$I_\mu(p'\lambda', p\lambda) = \left(\frac{EE'}{m^2}\right)^{\frac{1}{2}} \langle p'\lambda'| \, J_\mu(0) \, |p\lambda\rangle \tag{9.1}$$

p, λ are the four-momentum and helicity of the initial proton, etc. The

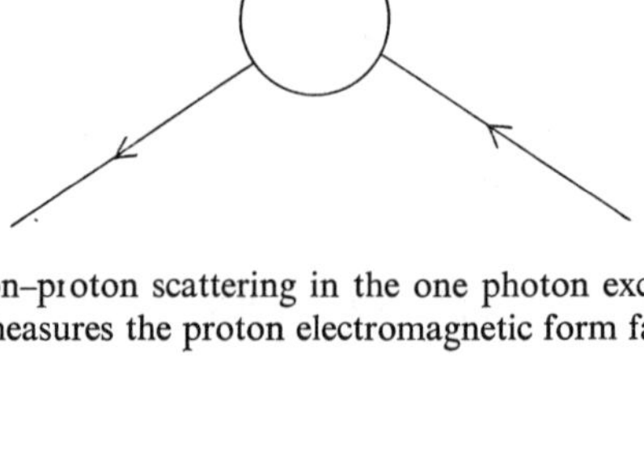

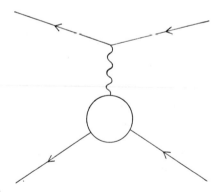

Figure 9.1 Electron–proton scattering in the one photon exchange approximation measures the proton electromagnetic form factor.

quantity I_μ, which transforms as a four-vector may be written in the form

$$I_\mu \equiv \bar{u}(p'\lambda')\Gamma_\mu(p', p)u(p, \lambda) \tag{9.2}$$

The properties of the 4 × 4 matrix Γ_μ are constrained by the transformations by the *symmetry* operations ($\Theta = CPT$)

$$\begin{aligned} CJ_\mu(x)C^{-1} &= -J_\mu \\ PJ_\mu(x)P^{-1} &= g_{\mu\mu}J_\mu(x') \\ TJ_\mu(x)T^{-1} &= g_{\mu\mu}J_\mu(-x') \\ \Theta J_\mu(x)\Theta^{-1} &= -J_\mu(-x) \end{aligned} \tag{9.3}$$

where $x'_\mu = g_{\mu\mu}x_\mu$. The conditions for P and T invariance lead to the following conditions on Γ_μ

$$\Gamma_\mu(\mathbf{p}', \mathbf{p}) = g_{\mu\mu}\gamma_0\Gamma_\mu(-\mathbf{p}', -\mathbf{p})\gamma_0 \tag{9.4}$$

$$\Gamma_\mu(\mathbf{p}', \mathbf{p}) = g_{\mu\mu}T^\dagger\Gamma_\mu^*(-\mathbf{p}', -\mathbf{p})T \tag{9.5}$$

Equation (9.4) simply states that Γ_μ acts as a vector under space inversions. $T = \gamma_3\gamma_1$ with the phase conventions of Chapter 3; Eq. (9.5) leads to reality restrictions on the coefficients A_i in an expansion of Γ_μ in a basis of independent quantities Γ_μ^i:

$$\Gamma_\mu(p', p) = \sum_i A_i(t)\Gamma_\mu^i \tag{9.6}$$

The A_i can only depend on the invariant $t = q^2$.

There are six "independent" Γ_μ^i compatible with P invariance:

$$\begin{aligned} &\gamma_\mu && \sigma_{\mu\nu}P^\nu \\ &P_\mu = (p + p')_\mu && i\sigma_{\mu\nu}q^\nu \\ &q_\mu = (p' - p)_\mu && r_\mu = -i\epsilon_{\mu\nu\lambda\sigma}P^\nu q^\lambda\gamma^\sigma\gamma_5 \\ & && \quad\; = \tfrac{1}{2}[\gamma_\mu \not{P}\not{q} - \not{q}\not{P}\gamma_\mu] \end{aligned} \tag{9.7}$$

γ_5 is given by $\gamma_5 = -i\gamma_0\gamma_1\gamma_2\gamma_3$ and $\epsilon_{0123} = 1$. When Γ_μ is inserted between nucleon spinors, the Gordon identities

$$\begin{aligned} P_\mu &= 2m\gamma_\mu - i\sigma_{\mu\nu}q^\nu \\ iq_\mu &= \sigma_{\mu\nu}P^\nu \\ r_\mu &= -t\gamma_\mu + 2mi\sigma_{\mu\nu}q^\nu \end{aligned} \tag{9.8}$$

may be used to reduce Γ_μ to the form

$$\Gamma_\mu(p', p) = A(t)\gamma_\mu + B(t)P_\mu + C(t)q_\mu \tag{9.9}$$

The T-invariance condition shows that A, B, C are real ($t \leqslant 0$).

Current conservation implies that $q^\mu\Gamma_\mu = 0$. For equal masses, $\bar{u}'\not{q}u = 0$ and $tC(t) = 0$. One can also note that $\langle p'\lambda'| J_\mu |p\lambda\rangle^* = \langle p\lambda| J_\mu |p'\lambda'\rangle$ implies $C^* = -C$, whereas T-invariance requires $C^* = C$. Hence $C = 0$.

Using the Gordon identities (9.8) we may rewrite Γ_μ in a standard form

$$\Gamma_\mu(p', p) = \gamma_\mu F_1(t) + i\frac{\sigma_{\mu\nu}q^\nu}{2m} F_2(t) \tag{9.10}$$

F_1 and F_2 are, respectively, the Dirac and Pauli form factors. They correspond, at low momentum transfer, to the electric and magnetic coupling.

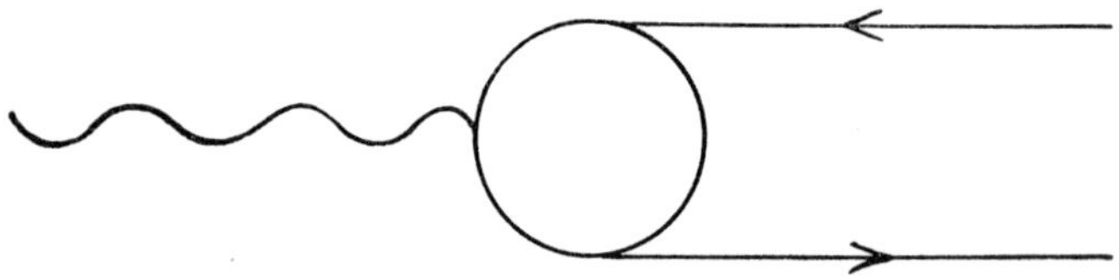

Figure 9.2 The process of nucleon annihilation into a (virtual) photon is related to the vertex of Eq. (9.2) by crossing.

One may also write

$$\begin{aligned} \Gamma_\mu(p', p) &= \frac{P_\mu}{2m} F_1 + \frac{F_1 + F_2}{2m} i\sigma_{\mu\nu}q^\nu \\ F_1(0) &= 1 \\ F_2(0) &= \kappa' \end{aligned} \tag{9.11}$$

κ' is the anomalous magnetic moment of the proton. The separation (9.11) exhibits the contribution of the "ordinary" moment (contained in γ_μ) to the magnetic coupling. It is worth noting that (9.10) results from the lowest order vertex calculated from

$$\mathscr{L} = e\bar{\psi}\gamma_\mu\psi A^\mu + \frac{e}{2m}\kappa'\bar{\psi}\sigma_{\mu\nu}\psi F^{\mu\nu} \tag{9.12}$$

where $F_{\mu\nu} = \partial_\mu A_\nu - \partial_\nu A_\mu$, with $F_1 = 1$, $F_2 = \kappa'$. The deviation from constant values is attributable to structure in the proton.

Another useful set of form factors is $G_{E'}$, $G_{M'}$, the so-called "electric" and "magnetic" form factors. These appear naturally in the c.m. frame for $N\bar{N} \to \gamma$ (Fig. 9.2) or in its crossed-channel equivalent, the "brick-wall" frame. We write

$$\Gamma_\mu(p', p) = \frac{1}{1 - t/4m^2}\left[G_E(t)\frac{P_\mu}{2m} + G_M(t)\frac{r_\mu}{4m^2}\right] \tag{9.13}$$

$$G_E(t) = F_1(t) + \frac{t}{4m^2}F_2$$

$$G_M(t) = F_1(t) + F_2(t) \tag{9.14}$$

$$G_E(0) = 1$$

$$G_M(0) = 1 + \kappa'$$

$$F_1(t) = \left(1 - \frac{t}{4m^2}\right)^{-1}\left(G_E - \frac{t}{4m^2}G_M\right)$$

$$F_2(t) = \left(1 - \frac{t}{4m^2}\right)^{-1}(G_M - G_E) \tag{9.15}$$

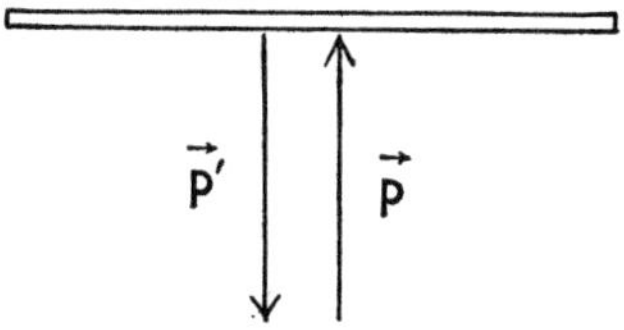

Figure 9.3 The brick-wall frame has especially simple kinematical properties for the analysis of vertices. In this frame the initial and final momenta are opposite and equal in magnitude.

Next we evaluate the vertex in the brick-wall frame, where $\mathbf{p}' = -\mathbf{p}$ (Fig. 9.3)

$$q^\mu = (0, -2\mathbf{p})$$

$$P^\mu = (2E, \mathbf{0}) \tag{9.16}$$

$$t = -4\mathbf{p}^2$$

$\mathbf{p}$ defines the z axis. In terms of the relativistic spin matrix S

$$S_x = \gamma_0\sigma_x, \qquad S_y = \gamma_0\sigma_y, \qquad S_z \tag{9.17}$$

we find $[(1 - v^2)^{\frac{1}{2}} = m/E]$

$$\Gamma^0 = (1 - v^2)^{\frac{1}{2}}G_E$$

$$\mathbf{\Gamma} = (1 - v^2)^{\frac{1}{2}} i\mathbf{S} \times \mathbf{q}\frac{G_M}{2m} \tag{9.18}$$

In order to understand the physical significance of these form factors, we consider scattering in a *time-independent* external field $A_\mu(\mathbf{x})$, with interaction energy $\mathscr{H} = ej_\mu A^\mu$. The S matrix is $S_{p'p} = \delta_{p'p} - 2\pi i\delta(E - E') \times \langle p'|\, H\, |p\rangle$, where $\langle p'|\, H\,|p\rangle = ej_\mu(q)A^\mu(q)$, $j_\mu(q) = \langle p'|\, j_\mu(0)\,|p\rangle$ and $A_\mu(q) = \int d^3x e^{-i\mathbf{q}\cdot\mathbf{x}} A_\mu(x)$. In the brick-wall frame the current is

$$j^\mu(q) = \frac{1}{1 + \mathbf{p}^2/m^2}\left(G_E,\ i\mathbf{S} \times \mathbf{q}\,\frac{G_M}{2m}\right) \tag{9.19}$$

Since $\mathbf{B} = \nabla \times \mathbf{A}$, $\mathbf{B}(\mathbf{q}) = i\mathbf{q} \times \mathbf{A}(\mathbf{q})$ and the transition energy is

$$\langle p'|\, H\,|p\rangle = \frac{e}{1 + \mathbf{p}^2/m^2}\,[G_E\phi(q) - G_M\mathbf{S}\cdot\mathbf{B}(\mathbf{q})] \tag{9.20}$$

As $\mathbf{q} \to 0$ this has the expected form for a particle of charge eG_E and moment $eG_M\mathbf{S}$.

It is useful to define an amplitude for the annihilation reactions $N(p, \lambda) + \bar{N}(\bar{p}', \bar{\lambda}') \to \gamma$:

$$J_\mu(p\lambda, \bar{p}'\lambda') \equiv \left(\frac{E\bar{E}'}{m^2}\right)^{\frac{1}{2}} \langle 0|\, J_\mu(0)\,|p\lambda\bar{p}'\bar{\lambda}' in\rangle \tag{9.21}$$

$$\equiv \bar{v}(\bar{p}'\lambda')\tilde{\Gamma}_\mu(\bar{p}', p)u(p, \lambda) \tag{9.22}$$

To exhibit the connection of the annihilation vertex $\tilde{\Gamma}_\mu$ to Γ_μ we write integral representations for I_μ and J_μ using the *LSZ* formalism:

$$\begin{aligned} I_\mu(p'\lambda', p\lambda) &= -i\left(\frac{E}{m}\right)^{\frac{1}{2}} \int dx e^{ip'\cdot x}\mathscr{K}_x \bar{u}(p', \lambda')\langle 0|\, \theta(x)[\psi(x), J_\mu(0)]\,|p\lambda\rangle \\ J_\mu(p\lambda, \bar{p}'\lambda') &= i\left(\frac{E}{m}\right)^{\frac{1}{2}} \int dx e^{-i\bar{p}'\cdot x}\mathscr{K}_x \bar{v}(\bar{p}', \lambda')\langle 0|\, \theta(-x)[\psi(x), J_\mu(0)]\,|p\lambda\rangle \end{aligned} \tag{9.23}$$

where $\mathscr{K}_x = i\gamma^\mu\,\partial_\mu - m$.

For helicity spinors the standard continuation of the antiparticle wave function gives* (note that helicity doesn't change sign)

$$\bar{v}(-p', \lambda') = i(-1)^{\frac{1}{2}-\lambda'}\bar{u}(p', \lambda') \tag{9.24}$$

so that the crossing relation is

$$J_\mu(p\lambda, -p'\lambda') = i(-1)^{\frac{1}{2}-\lambda'} I_\mu(p'\lambda', p\lambda) \tag{9.25}$$

Hence the vertex matrices satisfy

$$\tilde{\Gamma}_\mu(-p', p) = \Gamma_\mu(p', p) \tag{9.26}$$

---

* Details are given on p. 239.

We may also write the physical vertex appearing in (9.22) as

$$\tilde{\Gamma}_\mu(\bar{p}', p) = \gamma_\mu F_1(t) - i\,\frac{\sigma_{\mu\nu}(p + \bar{p}')^\nu}{2m}\,F_2(t) \tag{9.27}$$

where $t = (p + \bar{p}')^2$ is the square of the c.m. total $N\bar{N}$ energy. The physical region is $t \geqslant 4m^2$, but ImF_i is nonvanishing down to the 2π threshold $4\mu^2$. Equation (9.27) should be compared with (9.10).

We now use the explicit form of the nucleon helicity spinors to evaluate the vertex $\bar{u}'\Gamma_\mu u = I_\mu$ in the brick-wall frame (Fig. 9.3). As a preliminary step we write $\Gamma_\mu(p', p) = \gamma_\mu(F_1 + F_2) - (p + p')_\mu F_2/2m$. $(p + p')^\mu/2m$ is of the form $(E/m, 0)$ in this frame. In addition we use

$$\begin{aligned}
\bar{u}(-\mathbf{p}, \lambda')u(\mathbf{p}, \lambda) &= 2\lambda\,\frac{E}{m}\,\delta_{\lambda,-\lambda'}\\
\bar{u}(-\mathbf{p}, \lambda')\gamma_0 u(\mathbf{p}, \lambda) &= 2\lambda\delta_{\lambda,-\lambda'}\\
\bar{u}(-\mathbf{p}, \lambda')\gamma_3 u(\mathbf{p}, \lambda) &= 0\\
\bar{u}(-\mathbf{p}, \mp\tfrac{1}{2})(\gamma^1 \pm i\gamma^2)u(\mathbf{p}, \pm\tfrac{1}{2}) &= -2p/m\\
\bar{u}(-\mathbf{p}, \pm\tfrac{1}{2})(\gamma^1 \pm i\gamma^2)u(\mathbf{p}, \pm\tfrac{1}{2}) &= 0
\end{aligned} \tag{9.28}$$

The values of I^μ for various helicity states are given in Table (9.1). p is related to the invariant t by $p/m = \left(\dfrac{-t}{4m^2}\right)^{\frac{1}{2}}$ $(t \leqslant 0)$.

Table 9.1 Matrix elements of the nucleon electromagnetic vertex $\langle N'|\, J_\mu\, |N\rangle$ are given in the brick-wall frame for various helicities

(λ', λ)	$(\frac{1}{2}, \frac{1}{2})$	$(-\frac{1}{2}, -\frac{1}{2})$	$(-\frac{1}{2}, \frac{1}{2})$	$(\frac{1}{2}, -\frac{1}{2})$
I^0	0	0	G_E	$-G_E$
I^3	0	0	0	0
$I^1 + iI^2$	0	$-\dfrac{2p}{m}G_M$	0	0
$I^1 - iI^2$	$-\dfrac{2p}{m}G_M$	0	0	0

The nature of the zeros in the table follows from elementary considerations. In this frame current conservation reduces $pI_z = 0$. Hence the space part of I_μ is purely transverse. J^0 is a scalar and so cannot change the angular momentum **J**. Therefore the helicity must flip for nonzero matrix elements I_0. $I^1 \pm iI^2$ correspond to the spherical vector $j^1 \pm ij^2$ and must increase, decrease J_z respectively. Thus G_E and G_M are the helicity flip, nonflip amplitudes in the brick wall frame.

The situation is similar in the annihilation frame, except that now the roles of time-like and longitudinal are interchanged. Again take $p^\mu = (E, 0, 0, p)$, $\bar{p}'^\mu = (E, 0, 0, -p)$. The amplitude $J_\mu = \bar{v}'\tilde{\Gamma}_\mu u$ is evaluated as above. $\tilde{\Gamma}_\mu$ can be written as $\tilde{\Gamma}_\mu(\bar{p}', p) = \gamma_\mu(F_1 + F_2) - \frac{(p - \bar{p}')\mu}{2m} F_2$. Current conservation: $0 = (p + \bar{p}')^\mu J_\mu = 2EJ_0 = 0$ implies that $J_0 = 0$. Furthermore

$$\begin{aligned} \bar{v}(-\mathbf{p}, \lambda')u(\mathbf{p}, \lambda) &= -(p/m)2\lambda\delta_{\lambda\lambda'} \\ \bar{v}(-\mathbf{p}, \lambda')\gamma^3 u(\mathbf{p}, \lambda) &= 2\lambda\delta_{\lambda\lambda'} \\ \bar{v}(-\mathbf{p}, \lambda')\gamma^1 u(\mathbf{p}, \lambda) &= (E/m)\delta_{\lambda',-\lambda} \\ \bar{v}(-\mathbf{p}, \lambda')\gamma^2 u(\mathbf{p}, \lambda) &= i(E/m)2\lambda\delta_{\lambda',-\lambda} \end{aligned} \tag{9.29}$$

The values of J^μ for various helicity states is given in Table 9.2. Again,

Table 9.2 Matrix elements for the annihilation matrix element $\langle 0| J_\mu |N\bar{N}\rangle$ are given in the c.m. frame for various helicities

(λ', λ)	$(\frac{1}{2}, \frac{1}{2})$	$(-\frac{1}{2}, -\frac{1}{2})$	$(-\frac{1}{2}, \frac{1}{2})$	$(\frac{1}{2}, -\frac{1}{2})$
J^0	0	0	0	0
J^3	G_E	$-G_E$	0	0
$J^1 + iJ_2$	0	0	0	$\frac{2\bar{E}}{m} G_M$
$J^1 - iJ_2$	0	0	$\frac{2\bar{E}}{m} G_M$	0

the non-zero entries can easily be understood on the basis of angular momentum conservation. $\bar{E}/m = \left(\frac{t}{4m^2}\right)$ for $t \geqslant 4m^2$. The kinematical situation is quite similar in the two cases.

If we wish to deal simultaneously with protons and neutrons it is useful to define isoscalar and isovector form factors. These correspond to the decomposition of the electromagnetic current into isoscalar and isovector parts

$$J_\mu = J_\mu^S + J_\mu^V \tag{9.30}$$

Since J_μ^V transforms as the third component of an isovector, the Wigner-Eckart theorem may be employed to write Γ_μ as a matrix in the nucleon isospin space:

$$\Gamma_\mu = \tfrac{1}{2}(\Gamma_\mu^S + \tau_3\Gamma_\mu^V) \tag{9.31}$$

The form factors F_1 and F_2 have the same decomposition. The relation to physical quantities is

$$\begin{aligned} F_i^s &= F_i^p + F_i^n & \qquad F_i^v &= F_i^p - F_i^n \\ F_i^p &= \tfrac{1}{2}(F_i^s + F_i^v) & \qquad F_i^n &= \tfrac{1}{2}(F_i^s - F_i^v) \end{aligned} \tag{9.32}$$

With our normalization conventions the $t = 0$ values are

$$\begin{aligned} F_1^p(0) &= 1 & \qquad F_1^n(0) &= 0 \\ F_2^p(0) &= \kappa_p' & \qquad F_1^n(0) &= \kappa_n' \\ F_1^s(0) &= F_1^v(0) = 1 & & \\ F_2^s(0) &= \kappa_p' + \kappa_n' & \qquad F_2^v(0) &= \kappa_p' - \kappa_n' \\ \kappa_p' &= 1.79 & \qquad \kappa_n' &= -1.91 \end{aligned} \tag{9.33}$$

Note that the anomalous moment is almost entirely isovector.

It is not our purpose to review the present theoretical picture of the momentum transfer dependence of the nucleon electromagnetic form factors. We only note that it is reasonable to assume that vector mesons dominate the matrix elements of J_μ at moderate values of q^2. Specifically, the ω, ϕ mesons ($G = -1$) determine the isoscalar form factors while ρ^0 ($G = +1$) dominates the isovector form factors.

9.3 EFFECTIVE YUKAWA COUPLINGS OF PIONS WITH BARYONS OF ANY SPIN

The conventional way to introduce "interactions" in a field theory is to multiply together distinct fields at the same space-time point. The occurrence of such products in a Lagrangian results in the equations of motion of one field depending on the value of other fields; the resulting "coupling" presumably describes interactions. The aforementioned couplings, which describe various vertices of the theory, are Lorentz scalars formed from the fields of the theory. It is often useful to introduce effective couplings among unstable or high spin particles even though the latter may not correspond to fundamental fields of a field theory. Such effective couplings, used in lowest order perturbation theory, reproduce the appropriate spin structure and provide a convenient parametrization of electromagnetic and decay properties. This abstraction may be justified and refined for the case of unstable particles by a careful study of the corresponding resonance poles which occur in the scattering matrix.

An interesting and important example is provided by the effective vertices $NN'\pi$ where N and N' have arbitrary spin and isospin. Since the isospin structure is easily treated by means of the generalized isospin matrices of Sec. 8.3, we ignore isospin at first. The spin structure is easily analyzed by means of the Rarita-Schwinger formalism (Sec. 3.5). (A detailed treatment of high spin vertices using Weinberg fields is given in ref. 11.) Typical early treatments of problems of this type were given in refs. 12 and 13. General analyses were first presented in refs. 14–17. Here we shall follow the method of ref. 16.

For orientation we consider the well-known case of pions and spin $\frac{1}{2}$ nucleons, generalizing slightly to allow for two distinct nucleons N_1, N_2 of masses m_1, m_2. Keeping the isospin structure, there are two parity-conserving couplings

$$\begin{aligned} \mathscr{L}_{ps} &= g_{ps}\bar{N}_2 i\gamma_5 \boldsymbol{\tau} N_1 \cdot \boldsymbol{\pi} + \text{H.c.} \\ \mathscr{L}_{pv} &= (g_{pv}/\mu)\bar{N}_2\gamma_5\gamma_\mu \boldsymbol{\tau} N_1 \cdot \partial^\mu \boldsymbol{\pi} + \text{H.c.} \end{aligned} \tag{9.34}$$

μ is the pion mass, inserted to make g_{pv} dimensionless. As is well known,[18] the two Lagrangians of Eq. (9.34) lead to identical lowest-order effective vertices, with all particles on the mass shell. To verify this consider the decay $N_1(p) \to N_2(p') + \pi(q)$; the factor ∂_μ becomes $iq_\mu = i(p - p')_\mu$ so that the Dirac equations for N_1, N_2 may be used to reduce $\mathscr{L}_{pv}$ to $\mathscr{L}_{ps}$ provided the couplings are related by

$$g_{pv} = \mu g_{ps}/(m_1 + m_2) \tag{9.35}$$

Next we generalize to the case for which the nucleons N_1, N_2 have arbitrary spins and parities. The general vertex has the form

$$g\bar{N}_2^{\boldsymbol{\mu}}\Gamma_{\boldsymbol{\mu}'}N_1^{\boldsymbol{\mu}''}\left(\prod_i \partial_\mu\right)\pi + \text{H.c.} \tag{9.36}$$

where all the $\boldsymbol{\mu}$ indices must be contracted. The structure of $\Gamma_{\boldsymbol{\mu}}$ is extremely simple, since (9.36) is to be used for mass shell particles. First we note that $\Gamma_{\boldsymbol{\mu}}$ cannot contain γ_μ or ∂_μ factors acting on the nucleon fields because of the subsidiary conditions (3.90). Further, if $\Gamma_{\boldsymbol{\mu}}$ contains terms matching those acting on π, the mass-shell condition reduces such contributions to a constant times a vertex of the same structure lacking the two gradients. Hence one may completely omit gradient terms from $\Gamma_{\boldsymbol{\mu}}$. Moreover, the argument leading to (9.35) shows that any γ_μ factor accompanied by a matching $\partial^\mu\pi$ factor may be eliminated. Hence Γ_μ can only be I or $i\gamma_5$ depending on the relative parity of the particles in the vertex.

It is now simple to write down the general vertex. First consider the nucleons to have the same spin $s = k + \frac{1}{2}$. The number of gradients acting on π depends on how many indices of the Rarita-Schwinger fields are contracted, as shown by the listing

$$\begin{gathered}
\bar{\psi}_{\mu_1\cdots\mu_k}\Gamma\psi^{\mu_1\cdots\mu_k}\pi \\
\bar{\psi}_{\mu_1\mu_2\cdots\mu_k}\Gamma\psi^{\mu_1\cdots\mu_2}_{\nu_1}\partial^{\mu_k}\partial^{\nu_1}\pi \\
\cdot \\
\cdot \\
\cdot \\
\bar{\psi}_{\mu_1\cdots\mu_k}\Gamma\psi_{\nu_1\cdots\nu_k}\partial^{\mu_1}\cdots\partial^{\mu_k}\partial^{\nu_1}\cdots\partial^{\nu_k}\pi
\end{gathered} \tag{9.37}$$

It will be noted that there are $k + 1 = S + \frac{1}{2}$ distinct couplings. A slight extension of the above argument handles the case of unequal spins $S = k + \frac{1}{2}$, $S' = k' + \frac{1}{2}$, $k'k$. In this case we need at least $k' - k$ gradients on the pion field. The total number of couplings is again $s + \frac{1}{2}$, so that the minimum baryon spin delimits the number of independent vertices. The allowed couplings have the structure

$$\begin{gathered}
\bar{\psi}_{\mu_1\cdots\mu_k\mu_{1+k}\cdots\mu_{k'}}\Gamma\psi^{\mu_1\cdots\mu_k}\partial^{\mu_{k+1}}\cdots\partial^{\mu_{k'}}\pi + \text{H.c.} \\
\cdot \\
\cdot \\
\cdot \\
\bar{\psi}_{\mu_1\cdots\mu_{k'}}\Gamma\psi_{\nu_1\cdots\nu_k}\partial^{\mu_1}\cdots\partial^{\mu_{k'}}\partial^{\nu_1}\cdots\partial^{\nu}\,\pi + \text{H.c.}
\end{gathered} \tag{9.38}$$

At very low pion momenta it is expected that the coupling containing the smallest number of gradients will dominate. We shall solve this case completely and explicitly. The relation to the multipole decomposition of the vertex will be considered in the following section.

Next we consider the parity structure of the vertices, employing the parity transformation law (6.135). For orientation we first consider the coupling of a $\frac{1}{2}^+$ nucleon to "nucleons" of any spin parity S^P. The phenomenological coupling is

$$\mathscr{L} = (g/\mu^k)\bar{\psi}_{\mu_1\cdots\mu_k}(x)\Gamma\psi(x)\partial^{\mu_1}\cdots\partial^{\mu_k}\pi + \text{H.c.} \tag{9.39}$$

Under the parity transformation, (9.39) goes into

$$\eta_P(-1)^{k+1}\bar{\psi}_{\mu_1}(x')\gamma_0\Gamma\gamma_0\psi(x')\partial'^{\mu_1}\cdots\partial'^{\mu_k}\pi(x') + \text{H.c.} \tag{9.40}$$

where $x_\mu = g_{\mu\mu}x_\mu$. Parity conservation therefore requires

$$\gamma_0\Gamma\gamma_0 = \eta_P(-1)^{s+\frac{1}{2}} \tag{9.41}$$

which implies

$$\begin{aligned} \Gamma &= i\gamma_5, \quad s^P = \tfrac{1}{2}^+, \tfrac{3}{2}^-, \tfrac{5}{2}^+, \tfrac{7}{2}^-, \ldots \\ \Gamma &= 1, \quad\;\; s^P = \tfrac{1}{2}^-, \tfrac{3}{2}^+, \tfrac{5}{2}^-, \tfrac{7}{2}^+, \ldots \end{aligned} \tag{9.42}$$

The classification of baryons according to the two series in (9.42) actually determines the structure of the more general vertex and is important for other considerations (cf. Sec. 10.2). To codify this distinction we define a quantity ν called *normality* by

$$\nu \equiv P(-1)^{s-\frac{1}{2}} \tag{9.43}$$

where $\nu = +1$ for the sequence $\frac{1}{2}^+, \frac{3}{2}^-, \ldots$ and -1 for $\frac{1}{2}^-, \frac{3}{2}^+, \ldots$. The corresponding quantity for mesons is $\nu = P(-1)^s$ so that mesons in the sequence $0^+, 1^-, 2^+, \ldots$ are "normal" and those in the sequence $0^-, 1^+, 2^-, \ldots$ are "abnormal."

Other meson-baryon couplings can be constructed in a similar way. For a scalar meson, one interchanges $i\gamma_5$ and 1 in the above considerations. The simplest vertex for a vector meson V follows on replacing the axial vector $\partial_\mu\pi$ by V_μ and interchanging $i\gamma_5$ and 1. Thus, the (magnetic dipole) coupling for ρNN^* is $\bar{\psi}_\mu i\gamma_5\psi\rho^\mu$ + H.c.

We now construct the general $\pi N_a N_b$ vertex of lowest multipolarity. It will become apparent that the normality is a key quantity. From Eq. (9.38) we see that the condition for parity invariance is

$$\begin{aligned} \Gamma &= -\eta_P^a\eta_P^b(-1)^{k'-k}\gamma_0\Gamma\gamma_0, \\ \Gamma &= -\nu_a\nu_b\gamma_0\Gamma\gamma_0 \end{aligned} \tag{9.44}$$

Thus we find the result

$$\begin{aligned} \nu_R &\equiv \nu_a\nu_b = 1 \Rightarrow \Gamma = i\gamma_5 \\ \nu_R &= -1 \Rightarrow \Gamma = 1 \end{aligned} \tag{9.45}$$

Hence the relative normality determines the matrix structure of the vertex function. The invariant vertex is momentum space for $N_a + \pi \rightarrow N_b$ is ($\partial_\mu \rightarrow iq_\mu$ is equivalent to $+ip_\mu$)

$$\Gamma(s_{k'}, s_k; p'\lambda', p\lambda) = (i)^{k'-k}\bar{u}^{\mu_1\cdots\mu_{k'}}(p', \lambda')(i\gamma_5)^{\frac{1}{2}(1+\nu_R)}u_{\mu_1\cdots\mu_k}(p, \lambda)p_{\mu_{k+1}}\cdots p_{\mu_{k'}} \tag{9.46}$$

The recursive construction of wave functions given in Eq. (3.109) may be used to derive recursion relations for the vertices Γ. The simplest relation is

$$\Gamma(s_{k'} + 1, s_k; p'\lambda', p\lambda) = i \sum_{\lambda_1\lambda_2} e^*(p', \lambda_2) \cdot p C(s_{k'} 1 s_{k'} + 1; \lambda_1\lambda_2\lambda')\Gamma(s_{k'}, s_k; p'\lambda_1 p\lambda) \quad (9.47)$$

This process can be continued until $s_{k'} = s_k$ appears. Then all the p_μ factors have been used up. If desired the process can be continued using a second recursion relation

$$\Gamma(s_k + 1, s_k + 1; p'\lambda' p\lambda) = \sum_{\lambda_1\lambda_2\lambda_1'\lambda_2'} e^*(p'\lambda_2') \cdot e(p\lambda_2) C(s_k 1 s_k + 1; \lambda_1'\lambda_2') \times C(s_k 1 s_k + 1; \lambda_1\lambda_2)\Gamma(s_k s_k; p'\lambda_1' p\lambda_1) \quad (9.48)$$

It will be noted that only vertices in which the relative normality of the two baryons is the same are connected by the recursion relations.

Equations (9.47)–(9.48) show how the basic spin $\frac{1}{2}$ vertices $\Gamma_s \equiv u(p', \lambda')u(p, \lambda)$ and $\Gamma_p \equiv u(p', \lambda')i\gamma_5 u(p, \lambda)$ may be used to build up the arbitrary vertex of Eq. (9.47). Explicit forms for Γ_s and Γ_p are given by (see Eq. (3.101) for the explicit helicity spinors)

$$\begin{aligned} \Gamma_s = \bar{u}(p', \lambda')u(p, \lambda) &= \delta_{\lambda\lambda'} \cos\frac{\theta}{2} \cosh \tfrac{1}{2}(\zeta' - \zeta) \\ &\quad - 2\lambda\delta_{\lambda'-\lambda} \sin\frac{\theta}{2} \cosh \tfrac{1}{2}(\zeta' + \zeta) \end{aligned} \quad (9.49)$$

$$\begin{aligned} -i\Gamma_p = \bar{u}(p', \lambda')\gamma_5 u(p, \lambda) &= -2\lambda\delta_{\lambda\lambda'} \cos \tfrac{1}{2}\theta \sinh \tfrac{1}{2}(\zeta - \zeta') \\ &\quad + \delta_{\lambda'-\lambda} \sin \tfrac{1}{2}\theta \sinh (\zeta' + \zeta) \end{aligned}$$

Note that for $|\mathbf{p}| = |\mathbf{p}'|$ only the helicity flip contributes. Regarded as matrices in helicity space, the vertices in the brick-wall frame ($\theta = \pi, p = p'$) reduce to $\Gamma_s = \sigma_x \sinh \zeta$, $\Gamma_p = i\sigma_x \cosh \zeta$. The reader is urged to use the standard form of the Dirac spinors to evaluate the left-hand sides of Eqs. (9.49) in order to appreciate the compactness of using the parameter ζ. For reference note the relations $\cosh \frac{1}{2}\zeta = [(E + M)/2M]^{\frac{1}{2}}$, $\sinh \frac{1}{2}\zeta = [(E - M)/2M]^{\frac{1}{2}}$, and $\tanh \frac{1}{2}\zeta = p/(E + M)$.

In order to work out the expressions (9.47) and (9.48) we choose the z axis along $\mathbf{p}$ and make $\mathbf{p}'$ lie in the x-z plane at an angle θ relative to $\mathbf{p}$. The vector wave functions are given in Eq. (3.105), with ϕ set equal to zero. It should be especially noted that the e vectors depend on the mass of the baryon whose wave function they belong to. We use the abbreviations

$c = \cosh \zeta$, $c' = \cosh \zeta'$, $s = \sinh \zeta$, $s' = \sinh \zeta'$.

$$\begin{aligned} e^*(p', \pm 1) \cdot e(p, \pm 1) &= -\cos^2(\theta/2) \\ e^*(p', \pm 1) \cdot e(p, 0) &= \mp c \sin \theta/\sqrt{2} \\ e^*(p', 0) \cdot e(p, 0) &= ss' - cc' \cos \theta \\ p \cdot e^*(p', \pm 1) = p \cdot e(p', \pm 1) &= \mp p \sin \theta/\sqrt{2} \\ p \cdot e^*(p', 0) = p \cdot e(p', 0) &= -pc' \cos \theta + Es' \\ e^*(p', \pm 1) \cdot e(p, \mp 1) &= -\sin^2(\theta/2) \\ e^*(p', 0) \cdot e(p, \pm 1) &= \pm c' \sin \theta/\sqrt{2} \end{aligned} \tag{9.50}$$

Next we consider the simplifications which occur when one of the baryons is at rest. In that case Γ_s and Γ_p simplify to

$$\begin{aligned} \Gamma_s(\lambda', \lambda) &= (-1)^{\lambda'-\lambda} \cosh \tfrac{1}{2}\zeta d^{\frac{1}{2}}_{\lambda'\lambda}(\theta) \\ \Gamma_p(\lambda', \lambda) &= i(-1)^{\lambda'+\frac{1}{2}} \sinh \tfrac{1}{2}\zeta d^{\frac{1}{2}}_{\lambda'\lambda}(\theta) \end{aligned} \tag{9.51}$$

where the Jacobi polynomials are given in Appendix B. Moreover, (9.50) can be simply expressed in terms of Jacobi polynomials:

$$\begin{aligned} p \cdot e(p', \lambda) &\rightarrow -p d^1_{0\lambda}(\theta) \\ e^*(p', \lambda') \cdot e(p, \lambda) &\rightarrow -\xi_\lambda d^1_{\lambda\lambda'}(\theta) \\ \xi_{\pm 1} = 1, \qquad \xi_0 &= \cosh \zeta = c \end{aligned} \tag{9.52}$$

From (9.48) we find

$$\begin{aligned} \Gamma(\tfrac{3}{2}^\pm, \tfrac{1}{2}^\pm; p_0'\lambda' p\lambda) &= -ip \sum_{\lambda_1\lambda_2} (-1)^{\lambda_1+\lambda_2-\lambda} C(\tfrac{1}{2}1\tfrac{3}{2}; \lambda_1\lambda_2\lambda') d^{\frac{1}{2}}_{\lambda_1\lambda}(\theta) d^1_{\lambda_2 0}(\theta) \cosh \tfrac{1}{2}\zeta \\ &= -i(\tfrac{2}{3})^{\frac{1}{2}}(-1)^{\lambda'-\lambda} p \cosh \tfrac{1}{2}\zeta d^{\frac{3}{2}}_{\lambda'\lambda}(\theta) \end{aligned} \tag{9.53}$$

In the same way one finds

$$\Gamma(\tfrac{3}{2}^\pm, \tfrac{1}{2}^\mp; p_0'\lambda' p\lambda) = (\tfrac{2}{3})^{\frac{1}{2}}(-1)^{\lambda' \frac{1}{2}} p \sinh \tfrac{1}{2}\zeta d^{\frac{3}{2}}_{\lambda'\lambda}(\theta) \tag{9.54}$$

Here we have used the Clebsch-Gordan series[19] to sum up the Jacobi polynomials

$$d^j_{\mu m}(\theta) = \sum_{\mu_1 m_1} C(j_1 j_2 j; \mu, \mu - \mu_1) C(j_1 j_2 j; m_1, m - m_1) d^{j_1}_{\mu_1 m_1}(\theta) d^{j_1}_{\mu-\mu_1, m-m_1}(\theta) \tag{9.55}$$

These results suggest that the general $\pi N_a N_b$ vertex is proportional to $d^{s_b}_{\lambda'\lambda}$ when nucleon b is at rest. The truth of this follows from consideration of the $\pi N_a N_b$ vertex function

$$\langle N_b p_0' \lambda' | j(0) | N_a p \lambda \rangle \propto d^{s_b}_{\lambda'\lambda}(\theta) \tag{9.56}$$

(here $j(0)$ is the pion current) when baryon b is at rest. The moving particle N_a has all angular momenta, of which only $J = S_b$ survives when projected onto the rotational scalar $j(0)\,|N_b p_0' \lambda'\rangle$. The surviving part can be referred to states with definite helicity along the $\mathbf{p}'$ axis by the operator $\exp(-i\theta J_y)$, whose matrix elements then yield the result (9.56).

The simplicity of the method should now be apparent. Next consider the pion vertex of a baryon of spin $S = k + \frac{1}{2}$ and another with spin $\frac{1}{2}$. According to whether the relative normality ν_R is even or odd, we have $(S_k = k + \frac{1}{2})$

$$\begin{aligned} \Gamma_{\nu_R}(k + \tfrac{1}{2}, \tfrac{1}{2};\, p_0'\lambda' p\lambda) &= (-ip)^k N_k F^{\nu_R}_{\lambda'\lambda}(\zeta) d^{k+\frac{1}{2}}_{\lambda'\lambda}(\theta) \\ N_k &= \left[\frac{(k+1)!}{(2k+1)!!}\right]^{\frac{1}{2}} \\ F_{\lambda'\lambda} &= i(-1)^{\lambda'+\frac{1}{2}} \sinh \tfrac{1}{2}\zeta, \qquad \nu_R = +1 \\ &= (-1)^{\lambda'-\lambda} \cosh \tfrac{1}{2}\zeta, \qquad \nu_R = -1 \end{aligned} \tag{9.57}$$

This vertex is appropriate to the coupling of Eq. (9.39) except now we allow $\psi(x)$ to represent a particle of either parity. Multiplying (9.57) by g/μ^n, where μ is the π mass, to obtain a dimensionless coupling constant g, leads quickly to the decay rate

$$\Gamma(N_{k+\frac{1}{2}} \to N\pi) = \frac{g^2}{4\pi} \frac{2^k (k!)^2}{(2k+1)!} \left(\frac{p}{\mu}\right)^{2k+1} \left(\frac{E \pm m}{m_R}\right) \tag{9.58}$$

where the $\pm$ sign is to be chosen according to whether $\nu_R = \mp 1$ for the two baryons. m_R is the mass of the decaying state.

The remaining vertices are constructed as follows. The vertex $\Gamma(s, s)$ can be obtained from $\Gamma(\frac{1}{2}, \frac{1}{2})$ by repeated application of Eq. (9.48). Then Eq. (9.47) gives any desired $\Gamma(s', s)$.

The $(\frac{3}{2}\frac{3}{2}\pi)$ vertices follow immediately from (9.47):

$$\begin{aligned} \Gamma_+(\tfrac{3}{2}^\pm, \tfrac{3}{2}^\pm;\, p_0'\lambda' p\lambda) &= (-1)^{\lambda'+\frac{1}{2}} \epsilon^+_{\frac{3}{2},\lambda} \sinh \tfrac{1}{2}\zeta\, d^{\frac{3}{2}}_{\lambda'\lambda}(\theta) \\ \epsilon^+_{\frac{3}{2},\pm\frac{3}{2}} &= 1 \\ \epsilon^+_{\frac{3}{2},\pm\frac{1}{2}} &= \tfrac{1}{3}(1 - 2c) \\ \Gamma_-(\tfrac{3}{2}^\pm, \tfrac{3}{2}^\mp;\, p_0'\lambda' p\lambda) &= (-1)^{\lambda'-\lambda} \epsilon^-_{\frac{3}{2},\lambda} \cosh \tfrac{1}{2}\zeta\, d^{\frac{3}{2}}_{\lambda'\lambda}(\theta) \\ \epsilon^-_{\frac{3}{2},\pm\frac{3}{2}} &= -1 \\ \epsilon^-_{\frac{3}{2},\pm\frac{1}{2}} &= -\tfrac{1}{3}(1 + 2c) \end{aligned} \tag{9.59}$$

The superscript on the ϵ symbols denotes the relative normality.

The general spin vertex must satisfy

$$\begin{aligned} \Gamma_\pm(s+1, s+1;\, p_0'\lambda' p\lambda) = -\sum_{\lambda_1\lambda_2\lambda_1'\lambda_2'} &(-1)^{\lambda_2'-\lambda_2} \xi_{\lambda_2} C(s1s+1;\, \lambda_1\lambda_2) \\ &\times C(s1s+1;\, \lambda_1'\lambda_2') d^1_{\lambda_2'\lambda_2}(\theta) \Gamma_\pm(s, s;\, p_0'\lambda_1' p\lambda_1) \end{aligned} \tag{9.60}$$

The calculations leading to Eqs. (9.58) and (9.59) make it clear that the general vertices have the form

$$\Gamma_+(s, s; p_0'\lambda' p\lambda) = i(-1)^{\lambda'+\frac{1}{2}} \sinh \tfrac{1}{2}\zeta \epsilon_{s\lambda}^{+} d_{\lambda'\lambda}^{s}(\theta) \tag{9.61}$$

for positive relative normality and

$$\Gamma_-(s, s; p_0'\lambda' p\lambda) = (-1)^{\lambda'-\lambda} \cosh \tfrac{1}{2}\zeta \epsilon_{s\lambda}^{-} d_{\lambda'\lambda}^{s}(\theta) \tag{9.62}$$

for odd relative normality. The coefficients $\epsilon_{s\lambda}^{\pm}$ are defined by the recursion relations obtained by substituting the latter equations into (9.60):

$$\begin{aligned}\epsilon_{s+1,\lambda}^{+} &= \sum_{\lambda_1\lambda_2} (-1)^{\lambda_2+1} \xi_{\lambda_2} C^2(s1s+1; \lambda_1\lambda_2\lambda)\epsilon_{s\lambda_1}^{+}\\ \epsilon_{s+1,\lambda}^{-} &= -\sum_{\lambda_1\lambda_2} \xi_{\lambda_2} C^2(s1s+1; \lambda_1\lambda_2\lambda)\epsilon_{s\lambda_1}^{-}\end{aligned} \tag{9.63}$$

By induction one sees that the $\epsilon_{s\lambda}^{\pm}$ depend only on λ^2 since the coefficients in (9.63) are unchanged when $(\lambda_1, \lambda_2) \to (-\lambda_1, -\lambda_2)$. These equations are easy to use becasuse only three terms occur in each of the sums. ($\epsilon_{\frac{1}{2}\frac{1}{2}}^{\pm}$ are defined to be unity.) In the limit $c \to 1$ the $\epsilon_{s\lambda}^{\pm}$ are $3-j$ symbols.[1]

The energy dependence of the $N_s N_s \pi$ vertex is thus given by the functions $\epsilon_{s\lambda}^{\pm}(c)$, except for the additional factor $\sinh \frac{1}{2}\zeta$ or $\cosh \frac{1}{2}\zeta$. The $\epsilon_{s\lambda}^{\pm}$ are polynomials in $c = E/m$, of degree $|s-\lambda|$. Rather than study these functions in generality, we shall give them for the lower spin values of greatest practical interest.

$$\begin{aligned}\epsilon_{\frac{1}{2}\frac{1}{2}}^{\pm} &= 1\\ \epsilon_{\frac{3}{2}\frac{3}{2}}^{\pm} &= \pm 1\\ \epsilon_{\frac{3}{2}\frac{1}{2}}^{\pm} &= \pm\tfrac{1}{3}(1 \mp 2c)\\ \epsilon_{\frac{5}{2}\frac{5}{2}}^{\pm} &= 1\\ \epsilon_{\frac{5}{2}\frac{3}{2}}^{\pm} &= \tfrac{1}{5}(1 \mp 4c)\\ \epsilon_{\frac{5}{2}\frac{1}{2}}^{\pm} &= \tfrac{1}{5}(1 \mp 2c + 2c^2)\\ \epsilon_{\frac{7}{2}\frac{7}{2}}^{\pm} &= \pm 1\\ \epsilon_{\frac{7}{2}\frac{5}{2}}^{\pm} &= \pm(\tfrac{1}{7})(1 \mp 6c)\\ \epsilon_{\frac{7}{2}\frac{3}{2}}^{\pm} &= \pm(\tfrac{1}{7})(1 \mp 2c + 4c^2)\\ \epsilon_{\frac{7}{2}\frac{1}{2}}^{\pm} &= \pm(\tfrac{1}{35})(3 \mp 12c + 12c^2 \mp 8c^3)\end{aligned} \tag{9.64}$$

It follows easily by induction that

$$\epsilon_{s\lambda}^{-}(c) = (-1)^{s-\frac{1}{2}}\epsilon_{s\lambda}^{+}(-c) \tag{9.65}$$

(It is also convenient to write ξ_λ as $1 + (c - 1)\,\delta_{\lambda 0}$ and $(-1)^\lambda \xi_\lambda = -1 + (c + 1)\,\delta_{\lambda 0}$, cf. Eq. (9.63).)

Referring to (9.61) and (9.62), we can establish an interesting general relation:

$$\Gamma_+(ss;\, \lambda'\lambda;\, E) = (-1)^{s+\lambda}\Gamma_-(ss;\, \lambda'\lambda;\, -E) \tag{9.66}$$

noting that $c = E/M$, where E is the energy of the baryon with four-momentum p. The vertex of Eq. (9.59) obeys the same relation. Below we establish its general validity for any pair of spins.

The construction of the general vertex is now quite simple. For $k' = k$, the recursion relation (9.47) gives (in the rest frame of the final particle)

$$\Gamma_+(s + 1, s;\, p_0'\lambda' p\lambda) = (-1)^{\lambda'+\frac{1}{2}} p \sinh \tfrac{1}{2}\zeta\epsilon_{s\lambda}^{+} C(s1s + 1;\, \lambda 0) d_{\lambda'\lambda}^{s+1}(\theta) \tag{9.67}$$

on using the explicit form (9.61). Iteration gives the general vertex for positive relative normality

$$\Gamma_+(s + n, s;\, p_0'\lambda' p\lambda) = (-ip)^n i \sinh \tfrac{1}{2}\zeta(-1)^{\lambda'+\frac{1}{2}}\epsilon_{s\lambda}^{+} d_{\lambda'\lambda}^{s+n}(\theta) \times \prod_{k=1}^{n} C(s + k - 1, 1, s + k;\, \lambda 0) \tag{9.68}$$

A similar calculation yields for the odd relative normality vertex

$$\Gamma_-(s + n, s;\, p_0'\lambda' p\lambda) = (-ip)^n \cosh \tfrac{1}{2}\zeta(-1)^{\lambda'-\lambda}\epsilon_{s\lambda}^{-} d_{\lambda'\lambda}^{s+n}(\theta) \times \prod_{k=1}^{n} C(s + k - 1, 1, s + k;\, \lambda 0) \tag{9.69}$$

Since p is an even function of E, we deduce from the latter equations the general vertex symmetry

$$\Gamma_+(s + n, s;\, \lambda'\lambda;\, E) = (-1)^{\lambda+s}\Gamma_-(s + n, s;\, \lambda'\lambda;\, -E) \tag{9.70}$$

We are now in a position to write down the decay width for the process $B_{k'} \to B_k + \pi$ where $B_{k'}$ has spin $s' = k' + \frac{1}{2}$, isospin I', while B_k has spin $s = k + \frac{1}{2}$ and isospin I. The relative normality has to be specified as well. The coupling constant is a function of all the quantities necessary to label these particles uniquely. Generally we shall not write such labels explicitly.

First consider the space-time factors. According to Eqs. (9.68) and (9.69) the invariant vertex has the form

$$\Gamma_\pm(s', s;\, p_0'\lambda' p\lambda) = gE_{s's\lambda}^{\pm} d_{\lambda'\lambda}^{s'}(\theta) \tag{9.71}$$

This equation defines the (momentum-dependent) functions $E^{\pm}_{s's\lambda}$. Inserting normalization constants and evaluating the trivial phase space integral gives immediately the width

$$\Gamma(B_{k'} \rightarrow B_k + \pi) = \frac{g^2}{4\pi} \frac{2M_s p}{(2s'+1)M_{s'}} \sum_{\lambda} |E^{\pm}_{s's\lambda}|^2 \tag{9.72}$$

Next we put in the isospin factors. To avoid ambiguity we give the explicit form of the interaction vertex

$$\mathscr{L} = g\bar{\psi}_{I'}^{\mu_1\cdots\mu_{k'}} \Gamma \mathbf{T} \psi_{I\mu_1\cdots\mu_k} \partial_{\mu_{k+1}} \cdots \partial_{\mu_{k'}} \boldsymbol{\pi} + \text{H.c.} \tag{9.73}$$

Summing on the I_3 states of the final particles gives

$$\sum_{i,I_3} |\chi^+(I_3) T_i^\dagger \chi(I_3')|^2 = \chi^+(I_3')\mathbf{T} \cdot \mathbf{T}^\dagger \chi(I_3')$$
$$= 2I + 1 \tag{9.74}$$

The matrices $\mathbf{T}$ have been defined in a symmetrical way, so that if we have $B_I \rightarrow B_{I'} + \pi$, the corresponding isospin factor is $2I' + 1$.

We now give the final formulas. A baryon of mass $M_{s'}$, spin s', isospin I' decays into a baryon of mass $M_{s'}$ spin s, isospin I, energy E_s, momentum p and a pion of momentum p. If the relative normality of the baryons if even the decay width is

$$\Gamma_+(B_{s'} \rightarrow B_s + \pi) = \frac{g^2}{4\pi} \frac{2I+1}{2s'+1} \frac{p^{2(s'-s)+3} \sum_\lambda (\epsilon^+_{s\lambda})^2}{M_{s'}(E_s + M_s)}$$
$$\times \left(\prod_{k=1}^{n} C(s+k-1, 1, s+k; \lambda 0) \right)^2 \tag{9.75}$$

If the relative normality is odd the width is

$$\Gamma_-(B_{s'} \rightarrow B_s + \pi) = \frac{g^2}{4\pi} \frac{2I+1}{2s'+1} \frac{p^{2(s'-s)+1}(E_s + M_s)}{M_{s'}}$$
$$\times \left(\prod_{k=1}^{n} C(s+k-1, 1, s+k; \lambda 0) \right)^2 \tag{9.76}$$

The main energy dependence of the decay widths is given by the p^{2L+1} factor where L is the orbital momentum of the emitted meson. In addition the polynomials $\epsilon^{\pm}_{s\lambda}$ are mildly energy-dependent. However, from resonance theory in potential scattering[20] one knows of significant modifications due to the structure of the decaying state. Clearly our effective interaction is a point interaction.

The experimental evidence on the energy dependence of resonance widths indicates that the essentially p^{2L+1} dependence of Γ is considerably too rapidly varying.[21] The reader should bear in mind that results may depend sensitively on the "barrier penetration factor" of which there is no real understanding. Calculations[22] show convincingly that comparison of transitions involving the same L and comparable momenta are rather insensitive to the radius of the state.

A table of effective pion couplings for baryon resonances has been given in ref. 23.

It should be stressed that when the final baryon has spin greater than $\frac{1}{2}$, other couplings can be formed. Although these more complicated couplings contain only higher multipoles (two more units of angular momentum), it is not inconceivable that they could be important for transitions involving large momentum. In general, there are $s + \frac{1}{2}$ terms which can contribute, where s is the lesser of the baryon spins. The general case has been treated (for $s = \frac{3}{2}$) in refs. 14 and 23.

9.4 KINEMATIC STRUCTURE OF VERTEX FUNCTIONS

Consider a "current" $\mathscr{J}(x)$ having known properties under Lorentz transformations and for which the vertex connecting particles of spin s and s' and momenta p and p' is

$$\langle p's'\lambda'|\, \mathscr{J}(0)\, |ps\lambda\rangle \tag{9.77}$$

Here λ and λ' are helicity labels. When the spins are large it is usually tedious to imitate the analysis of the nucleon form factor given in Sec. (9.2). For many purposes it is useful to relate (9.77) to vertices in the brick-wall frame, which is especially simple for kinematical reasons. In that frame it is straightforward to derive a multipole expansion which is useful for low momentum transfer. This should be contrasted with the special frame used in the preceding section.

The basic tool is the Lorentz transformation law for helicity states

$$U(\Lambda)\,|\mathbf{p}s\lambda\rangle = \sum_{\lambda'} |\Lambda\mathbf{p}s\lambda'\rangle D^{s}_{\lambda'\lambda}(R) \tag{9.78}$$

where R is the Wigner rotation, and the known behavior

$$\mathscr{J}^{\Lambda}(0) \equiv U(\Lambda)\mathscr{J}(0)U^{-1}(\Lambda) \tag{9.79}$$

Equation (9.78) allows us to represent the vertices of $\mathscr{J}$ in one frame in terms of matrix elements of $\mathscr{J}^{\Lambda}$ in some other frame:

$$\begin{aligned}\langle p's'\mu'|\, \mathscr{J}\, |ps\mu\rangle &= \langle p's'\mu'|\, U^{-1}(\Lambda)\mathscr{J}^{\Lambda}(0)U(\Lambda)\, |ps\mu\rangle \\ &= \sum_{\lambda\lambda'} D^{s'*}_{\lambda'\mu'}(R)\langle \Lambda\mathbf{p}'s'\lambda'|\, \mathscr{J}^{\Lambda}(0)\, |\Lambda\mathbf{p}s\lambda\rangle D^{s}_{\lambda\mu}(R) \qquad (9.80)\end{aligned}$$

In order to illustrate the method we consider the neutral pion vertex. Since j_π is a pseudoscalar, we have $j_\pi^{\Lambda}(0) = j_\pi(0)$. Transforming to the brick-wall frame (Fig. 9.3) the vertex has the form

$$\langle p's'\mu'|\, j_\pi\, |ps\mu\rangle = \sum_{\lambda\lambda'} D^{s'*}_{\lambda'\mu'}(R')\Pi_{s'\lambda';s\lambda} D^{s}_{\lambda\mu}(R) \qquad (9.81)$$

where the brick-wall frame vertices have the form

$$\Pi_{s'\lambda';s\lambda} = \langle p0s'\lambda'|\, e^{i\pi J_y} j_\pi\, |p0s\lambda\rangle \qquad (9.82)$$

on choosing the z axis along $\mathbf{p}$.

Next consider the symmetry properties of $\Pi_{s'\lambda';s\lambda}$. Since j_π is a scalar and cannot change J_z we expect $\lambda + \lambda'$ to vanish and further that (if $s' \geqslant s$, say) $|\lambda'|$ cannot exceed s. A formal proof is easily given, since J_z changes sign under rotation by $e^{i\pi J}y$:

$$\begin{aligned}\lambda\Pi_{s'\lambda';s\lambda} &= \langle p0s'\lambda'|\, e^{i\pi J_y} j_\pi J\ \ |p0s\lambda\rangle \\ &= -\langle p0s'\lambda'|\, J_z e^{i\pi J_y} j_\pi\, |p0s\lambda\rangle \\ &= -\lambda'\Pi_{s'\lambda';s\lambda} \qquad (9.83)\end{aligned}$$

Hence $\lambda + \lambda'$ vanishes if $\Pi_{s'\lambda';s\lambda}$ is to be non-zero.

$$\Pi_{s'\lambda';s\lambda} = \delta_{\lambda'-\lambda}\Pi_{s'-\lambda;s\lambda}, \qquad |\lambda| \leq s, s' \qquad (9.84)$$

In order to find the implications of parity conservation we consider the operator Y which induces reflections in the x-z plane. The pion current transforms under Y as $Yj_\pi(0)Y^{-1} = -j_\pi(0)$, so

$$\begin{aligned}\Pi_{s'\lambda';s\lambda} &= \langle p0s'\lambda'|\, e^{i\pi J_y} Y^{-1}(Yj_\pi Y^{-1})Y\, |p0s\lambda\rangle \\ &= -\eta^*_{P'}\eta_P(-1)^{s+s'-\lambda-\lambda'}\langle p0s' - \lambda'|\, e^{i\pi J_y} j_\pi\, |p0s - \lambda\rangle \qquad (9.85)\end{aligned}$$

The relative parity factor $\eta^*_{p'}\eta_p$ may be written as $(-1)^\pi$ where $(-1)^\pi$ is a (conventionally real) relative parity factor. Since $\lambda + \lambda'$ vanishes we obtain

$$\Pi_{s'\lambda';s\lambda} = (-1)^{s+s'+\pi+1}\Pi_{s'-\lambda';s-\lambda} \qquad (9.86)$$

The phase factor in this equation is clearly related to the product of normalities (cf. Eq. (9.43) and the following discussion).

Finally one may obtain reality conditions by invoking time reversal invariance. Since the neutral pion current obeys $Tj_\pi T^{-1} = -j_\pi(0)$ we find (cf. Eq. (7.41))

$$\begin{aligned}\Pi^*_{s'\lambda';s\lambda} &= \langle p0s'\lambda'|\, T^{-1}Te^{i\pi J_y}j_\pi T^{-1}T\, |p0s\lambda\rangle \\ &= -\langle p0s'\lambda'|\, e^{i\pi J_y}j_\pi\, |p0s\lambda\rangle \eta^*_{T'}\eta_T \\ &= -\eta^*_{T'}\eta_T \Pi_{s'\lambda';s\lambda} \qquad (9.87)\end{aligned}$$

When the external particles are the same (i.e. in the same internal symmetry multiplet) the pion vertex is purely imaginary. For distinct particles we may choose $\eta_{T'} = \eta_T = 1$ and obtain a uniform result for all vertices.

The general matrix $\Pi_{s'\lambda';s\lambda}$ contains $(2s + 1)(2s' + 1)$ entries for given (s, s'). However the three conditions of Eqs. (9.84), (9.86), (9.87) reduce the number of independent components to $s + \frac{1}{2}$, in agreement with Sec. (9.3).

We now study the multipole expansions obtained by expressing (9.83) in terms of rest states.

$$\begin{aligned}\Pi_{s'\lambda';s\lambda} &= \langle s'\lambda'|\, e^{i\zeta' K_z}e^{i\pi J_y}j_\pi e^{-i\zeta K_z}\, |s\lambda\rangle \\ &= (-1)^{s'-\lambda'}\langle s' - \lambda'|\, j_\pi e^{-i(\zeta+\zeta')K_z}\, |s\lambda\rangle \qquad (9.88)\end{aligned}$$

The boost operator K_z has been discussed previously. The second relation was obtained by using the rotational properties of K_z, the rest state $|s\lambda\rangle$ and noting that j_z commutes with K_z. The parameters ζ, ζ' are given by $\tanh \zeta = p/m$, $\tanh \zeta' = p/m'$. Writing $\nu = \zeta + \zeta'$, we now make the expansion

$$j_\pi e^{-i\nu K_z} = \sum_{n=0}^{\infty} \frac{(-i\nu)^n}{n!} j_\pi(K_z)^n \qquad (9.89)$$

and express each term as a sum of spherical tensors of rank n.

$$j_\pi(K_z)^n = \sum_{J=0}^{n} T^{(n)}_{J,0} \qquad (9.90)$$

The operator $T^{(n)}_{J,0}$ commutes with J_z. If n is even (odd), all odd (even) terms in (9.90) vanish.

To compute $T^{(n)}_{J,0}$ we recall the standard definition of a spherical tensor of rank J

$$RT_{JM}R^{-1} = \sum_{M'} T_{JM'}D^{(J)}_{M'M}(R) \qquad (9.91)$$

Thus if we subject Eq. (9.90) to a rotation we can pick off the $T^{(n)}_{J,0}$ using the orthogonality relation for the rotation matrices

$$\int dR\, D^{j'*}_{m_1'm_2'}(R)\, D^{j}_{m_1m_2}(R) = \frac{8\pi^2}{2j+1}\,\delta_{jj'}\delta_{m_1m_1'}\delta_{m_2m_2'} \tag{9.92}$$

where $dR = \sin\beta\, d\alpha\, d\beta\, d\gamma$, $0 \leqslant (\alpha, \gamma) \leqslant 2\pi$, $0 \leqslant \beta \leqslant \pi$. The result is

$$T^{(n)}_{J,M} = \frac{2J+1}{8\pi^2}\int dR\, D^{J*}_{M,0}(R)(Rj_\pi K_z^n R^{-1}) \tag{9.93}$$

Putting this result in (9.90) gives for (9.89) the result

$$\begin{aligned} j_\pi e^{-i\nu K_z} &= \sum_{J=0}^{\infty} T_{J,0} \\ T_{J,0} &= \sum_{n=J}^{\infty} \frac{2J+1}{8\pi^2}\frac{(-i\nu)^n}{n!}\int dR\, D^{J*}_{00}(R)(Rj_\pi K_z^n R^{-1}) \end{aligned} \tag{9.94}$$

Hence the vertex is now expressed in terms of rest states

$$\Pi_{s'\lambda';s\lambda} = (-1)^{s'-\lambda'}\sum_{J=0}^{\infty}\langle s' - \lambda'|\, T_{J,0}\,|s\lambda\rangle \tag{9.95}$$

and the Wigner-Eckart theorem may be employed to express (9.95) in terms of reduced matrix elements using the 3-j symbol[24]

$$\Pi_{s'\lambda';s\lambda} = (-1)^{2s'}\sum_{J=0}^{\infty}\begin{pmatrix} s' & J & s \\ \lambda' & 0 & \lambda \end{pmatrix}\langle s'\|\, T_J\,\|s\rangle \tag{9.96}$$

The orthogonality relations obeyed by the 3-j symbols allows one to invert Eq. (9.96):

$$\langle s'\|\, T_J\,\|s\rangle = (-1)^{2s'}(2J+1)\sum_{\lambda}\begin{pmatrix} s' & J & s \\ -\lambda & 0 & \lambda \end{pmatrix}\Pi_{s'-\lambda;s\lambda} \tag{9.97}$$

If we note the symmetry relation

$$\begin{pmatrix} s' & J & s \\ -\lambda' & -\mu & -\lambda \end{pmatrix} = (-1)^{s'+s+J}\begin{pmatrix} s' & J & s \\ \lambda' & \mu & \lambda \end{pmatrix} \tag{9.98}$$

then the parity condition (9.86) says that the only terms contributing to (9.96) obey $(-1)^J = (-1)^{\pi+1}$. Recalling that $(-1)^n = (-1)^J$ in Eq. (9.94) we see that the reduced matrix elements $\langle s'\|\, T_J\,\|s\rangle$ are even or odd functions of p for odd or even relative parities of the external particles. Since $\nu \to p(m+m')/mm'$ as $p \to 0$, the dependence on p as $p \to 0$ is p^J.

For further illustrations, and for a detailed treatment of the general electromagnetic vertex, the reader should consult ref. 1.

REFERENCES

1. L. Durand III, P. C. DeCelles, and R. B. Marr, *Phys. Rev.* **126,** 1882 (1962).
2. D. R. Yennie, M. M. Levy, and D. G. Ravenhall, *Rev. Mod. Phys.* **29,** 144 (1957).
3. G. Salzman, *Phys. Rev.* **99,** 973 (1955).
4. S. D. Drell and F. Zachariasen, "Electromagnetic Structure of Nucleons" (Oxford University Press, Oxford, 1960).
5. F. J. Ernst, R. G. Sachs, and K. C. Wali, *Phys. Rev.* **119,** 1105 (1960).
6. M. Rosenbluth, *Phys. Rev.* **79,** 615 (1950).
7. M. E. Rose, "Relativistic Electron Theory" (John Wiley and Sons, Inc., New York, 1961) p. 72.
8. "Nucleon Structure," ed. R. Hofstadter and L. I. Schiff (Stanford University Press, Stanford, Calif., 1964).
9. R. Hofstadter, "Nuclear and Nucleon Structure" (W. A. Benjamin, Inc., New York, 1963).
10. T. A. Griffy and L. I. Schiff, in "High Energy Physics, Vol. I," ed. E. H. S. Burhop (Academic Press, New York, 1967).
11. S. Weinberg, *Phys. Rev.* **133B,** 1318 (1964).
12. J. D. Jackson, *Nuovo Cimento* **34,** 1644 (1964).
13. P. Auvil and J. J. Brehm, *Phys. Rev.* **140B,** 135 (1965), *ibid,* **145,** 1152 (1966).
14. J. G. Rushbrooke, *Phys. Rev.* **143,** 1345 (1966).
15. D. Brudnoy, *Phys. Rev.* **145,** 1229 (1966).
16. P. Carruthers, *Phys. Rev.* **152,** 1345 (1966).
17. M. D. Scadron, *Phys. Rev.* **165,** 1640 (1968).
18. S. S. Schweber, H. A. Bethe, and F. de Hoffmann, "Mesons and Fields, Vol. I" (Row-Peterson, Evanston, Ill., 1955).
19. M. E. Rose, "Elementary Theory of Angular Momentum" (John Wiley and Sons, Inc., New York, 1957).
20. J. M. Blatt and V. F. Weisskopf, "Theoretical Nuclear Physics" (John Wiley and Sons, Inc., New York, 1952) p. 358.
21. L. D. Roper, University of California Report No. UCRL 14193 (unpublished).
22. M. Goldberg, J. Leitner, R. Musto, and L. O'Raifeartaigh, *Nuovo Cimento* **45,** 169 (1966).
23. P. Carruthers and J. Shapiro, *Phys. Rev.* **159,** 1456 (1967).
24. A. R. Edmonds, "Angular Momentum in Quantum Mechanics" (Princeton University Press, Princeton, New Jersey, 1957).

10

INVARIANT AMPLITUDES AND KINEMATICAL PROPERTIES OF TWO-BODY REACTIONS

10.1 INTRODUCTION

The description of spin states by the helicity formalism[1] permits a simple and general description of the general two-body reaction $a + b \to c + d$. However for many common reactions involving particles of relatively low spin it is often convenient to use the so-called "invariant amplitudes." The latter possess simple crossing and analytic properties, and arise naturally if one evaluates the amplitude of interest in low orders of perturbation theory. The helicity amplitudes belong to a particular reference frame (c.m. frame) and have more complicated crossing properties.

Although it is possible to give a general treatment of the invariant amplitude problem,[2,3,4] the ideas are best brought out in specific examples of practical interest. When photons are involved, special care is required to maintain gauge invariance.[5] In the present chapter we treat several common reactions involving massive particles of spin 0, $\frac{1}{2}$, and 1. We relate helicity amplitudes to invariant amplitudes in various channels of interest and in addition give the relevant pole terms involving particles of low spin and mass. We also give the crossing relations and partial wave decompositions. These examples are prototypes for many related reactions which differ only by changes of parity, isospin and other internal quantum numbers.

In Sec. 10.2 we review the concept of "normality" and note its useful role in the construction of the "parity-conserving helicity amplitudes." Section 10.3 is concerned with reactions of the type: spin 1 + spin 0 $\to$ spin 0 + spin 0 for various parities and isospins of the spin 1 particle. In Sec. 10.4 reactions of the type: spin 1 + spin 0 $\to$ spin 1 + spin 0 are discussed.

Section 10.5 deals with the important reaction $N\pi \to N\pi$ and the crossed process $N\bar{N} \to \pi\pi$. Finally references are given to the literature for several other important reactions.

10.2 PARITY-CONSERVING HELICITY AMPLITUDES

In Chapter 9 it was noted that the systematic structure of three particle vertices depends on the relative normality of the fields. The normal parity Bosons and Fermions have spin parity belonging to the sequences

$$\begin{gathered} 0^+, 1^-, 2^+, 3^-, \ldots \\ \tfrac{1}{2}^+, \tfrac{3}{2}^-, \tfrac{5}{2}^+, \tfrac{7}{2}^-, \ldots \end{gathered} \tag{10.1}$$

while the corresponding "abnormal parity" states have J^P given by

$$\begin{gathered} 0^-, 1^+, 2^-, 3^+, \ldots \\ \tfrac{1}{2}^-, \tfrac{3}{2}^+, \tfrac{5}{2}^-, \tfrac{7}{2}^+, \ldots \end{gathered} \tag{10.2}$$

This nomenclature is also useful in elucidating the dependence of scattering amplitudes on the relative parities and spins of the participating particles. Equations (10.1)–(10.2) are summarized by $\nu = \pm 1$ in

$$P = \nu(-1)^{J-v} \tag{10.3}$$

where $v = 0$ for Bosons and $v = \frac{1}{2}$ for Fermions.

Two particle states of definite J^P may be labeled by J, ν as well. Under parity the two particle c.m. helicity state transforms as

$$P\,|JM\lambda_a\lambda_b\rangle = \eta_a\eta_b(-1)^{J-s_a-s_b}\,|JM - \lambda_a - \lambda_b\rangle \tag{10.4}$$

where (η_a, s_a) are the intrinsic parity and spin of particle a. The states of normality ν are linear combinations of the usual helicity states and obey

$$P\,|JM\lambda_a\lambda_b\rangle_\nu = \nu(-1)^{J-v}\,|JM\lambda_a\lambda_b\rangle_\nu \tag{10.5}$$

The states of normality ± 1 are given by

$$\sqrt{2}\,|JM\lambda_a\lambda_b\rangle_\pm = |JM\lambda_a\lambda_b\rangle \pm \eta_a\eta_b(-1)^{s_a+s_b-v}\,|JM - \lambda_a - \lambda_b\rangle \tag{10.6}$$

Hence amplitudes of definite J^P for $a(\lambda_a) + b(\lambda_b) \to c(\lambda_c) + d(\lambda_d)$ are

$$\begin{aligned} M^{J\pm}_{\lambda_c\lambda_d,\lambda_a\lambda_b} &= {}_\pm\langle JM\lambda_c\lambda_d|\,M\,|JM\lambda_a\lambda_b\rangle_\pm \\ &= \langle JM\lambda_c\lambda_d|\,M\,|JM\lambda_a\lambda_b\rangle \\ &\quad \pm \eta_c\eta_d(-1)^{s_c+s_d-v}\langle JM - \lambda_c - \lambda_d|\,M\,|JM\lambda_a\lambda_b\rangle \end{aligned} \tag{10.7}$$

It is useful to introduce the abbreviation

$$\langle\lambda_c\lambda_d| M^J |\lambda_a\lambda_b\rangle \equiv \langle JM\lambda_c\lambda_d| M |JM\lambda_a\lambda_b\rangle \tag{10.8}$$

Many different normalizations of the amplitude occur in the literature. The convenience of the various choices depends on the problem at hand. We choose to expand the invariant amplitude M defined by

$$\langle cd| (S - 1) |ab\rangle = (2\pi)^4 i\delta(p_c + p_d - p_a - p_b)N_aN_bN_cN_d\langle cd| M |ab\rangle \tag{10.9}$$

where $N = (2\omega)^{-\frac{1}{2}}$ for Bosons and $(m/E)^{\frac{1}{2}}$ for Fermions. We always orient

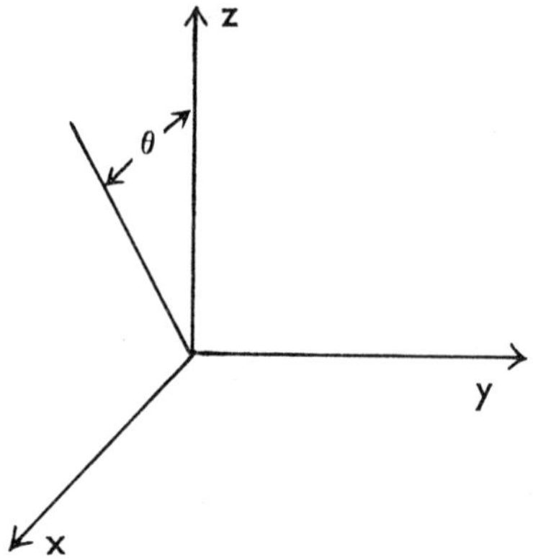

Figure 10.1 The geometry of the scattering process is chosen so that the incident momentum $\mathbf{p}_a$ lies along the z axis and the final momentum $\mathbf{p}_c$ lies in the x-z plane at an angle θ with respect to the z axis.

the z axis along the direction of motion $\mathbf{p}_a$ of particle a, and choose the xz plane to lie in the plane of p_a and p_c; $\hat{p}_c \cdot \hat{p}_a = \cos\theta$ (Fig. 10.1). (For unitarity sums one has to use a more general geometry.) Then the partial wave expansion is

$$\langle cd| M |ab\rangle = \sum_J (2J + 1)d^J_{\lambda\mu}(\theta)\langle\lambda_c\lambda_d| M^J |\lambda_a\lambda_b\rangle$$

$$\langle\lambda_c\lambda_d| M^J |\lambda_a\lambda_b\rangle = \frac{1}{2}\int d\cos\theta\, d^J_{\lambda\mu}(\theta)\langle cd| M |ab\rangle \tag{10.10}$$

$$\lambda = \lambda_a - \lambda_b, \qquad \mu = \lambda_c - \lambda_d$$

Before defining the "parity-conserving helicity amplitudes" we note the dependence of the d^J functions on θ, given by the formula

$$d^J_{\lambda\mu}(\theta) = \left(\frac{(j-\mu)!\,(j+\lambda)!}{(j+\mu)!\,(j-\lambda)!}\right)^{\frac{1}{2}} \frac{\left(\cos\frac{\theta}{2}\right)^{2j}}{(\lambda-\mu)!}\left(\cos\frac{\theta}{2}\right)^{\mu-\lambda}\left(-\sin\frac{\theta}{2}\right)^{\lambda-\mu}$$

$$\cdot\, {}_2F_1\left(\lambda-j, -\mu-j; \lambda-\mu+1; -\tan^2\frac{\theta}{2}\right) \quad (10.11)$$

$${}_2F_1(a, b; c; z) = 1 + \frac{ab}{c}z + \frac{1}{2}\frac{a(a+1)b(b+1)}{c(c+1)}z^2 + \cdots$$

where the latter definition applies if $|z| < 1$ and c is not a negative integer. Here $a = \lambda - j$ and $b = -\mu - j$ are both so that the greatest power of z is equal to the lesser of $|\lambda - j|$, $|\mu + j|$. To remove the singularities at $\theta = 0$, π it is useful to divide by appropriate powers of cos $(\theta/2)$ and sin $(\theta/2)$. In order to have well-behaved functions with simple symmetry properties under $z \to -z$ one defines

$$2e^{J\pm}_{\lambda\mu}(z) \equiv \left(\sqrt{2}\cos\frac{\theta}{2}\right)^{-|\lambda+\mu|}\left(\sqrt{2}\sin\frac{\theta}{2}\right)^{-|\lambda-\mu|} d^J_{\lambda\mu}(\theta)$$

$$\pm (-1)^{\lambda+\lambda_m}\left(\sqrt{2}\sin\frac{\theta}{2}\right)^{-|\lambda+\mu|}\left(\sqrt{2}\cos\frac{\theta}{2}\right)^{-|\lambda-\mu|} d^J_{\lambda-\mu}(\theta) \quad (10.12)$$

where $\lambda_m = \max(|\lambda|, |\mu|)$. We note several useful relations which follow from symmetries of the d functions:

$$\begin{aligned} e^{J\pm}_{-\mu-\lambda}(z) &= e^{J\pm}_{\lambda\mu}(z) \\ e^{J\pm}_{\mu\lambda}(z) &= (-1)^{\lambda-\mu} e^{J\pm}_{\lambda\mu}(z) \\ e^{J\pm}_{\lambda-\mu}(z) &= \pm(-1)^{\lambda+\lambda_m} e^{J\pm}_{\lambda\mu}(z) \\ e^{J\pm}_{\lambda\mu}(-z) &= \pm(-1)^{J-\lambda_m} e^{J\pm}_{\lambda\mu}(z) \end{aligned} \quad (10.13)$$

Various e^J functions and further properties are given in Appendix B.

The "parity-conserving helicity amplitudes" are defined similarly for the scattering amplitude

$$M^{\pm}_{\lambda_c\lambda_d,\lambda_a\lambda_b} \equiv \left(\sqrt{2}\cos\frac{\theta}{2}\right)^{-|\lambda+\mu|}\left(\sqrt{2}\sin\frac{\theta}{2}\right)^{-|\lambda-\mu|} M_{\lambda\mu}$$

$$\pm\, \eta_c\eta_d(-1)^{\lambda+\lambda_m}(-1)^{s_c+s_d-v}\left(\sqrt{2}\sin\frac{\theta}{2}\right)^{-|\lambda+\mu|}\left(\sqrt{2}\cos\frac{\theta}{2}\right)^{-|\lambda-\mu|} M_{\lambda-\mu}$$

$$(10.14)$$

The θ-dependent coefficients, expressed as functions of the crossed channel energy t, have the effect of removing kinematical singularities from $M^{\pm}$.

It is also possible (and often useful) to remove powers of momenta and energy in order that all singularities are "dynamical." These problems have been discussed at length in the literature. The amplitudes (10.14) are useful for continuation by means of complex angular momentum. Equation (10.14) may be written in the form

$$M^{\pm}_{\lambda_c\lambda_d,\lambda_a\lambda_b} = \sum_J (2J+1)[e^{J+}_{\lambda\mu} M^{J\pm}_{\lambda_c\lambda_d,\lambda_a\lambda_b} + e^{J-}_{\lambda\mu} M^{J\mp}_{\lambda_c\lambda_d,\lambda_a\lambda_b}] \qquad (10.15)$$

Equation (10.15) may be inverted using the functions $C^{J\pm}_{\lambda\mu}$ defined by

$$2C^{J\pm}_{\lambda\mu}(z) = \left(\sqrt{2}\cos\frac{\theta}{2}\right)^{|\lambda+\mu|}\left(\sqrt{2}\sin\frac{\theta}{2}\right)^{|\lambda-\mu|} d^J_{\lambda\mu}(\theta)$$
$$\pm (-1)^{\lambda+\lambda_m-1}\left(\sqrt{2}\sin\frac{\theta}{2}\right)^{|\lambda+\mu|}\left(\sqrt{2}\cos\frac{\theta}{2}\right)^{|\lambda-\mu|} d^J_{\lambda-\mu}(\theta) \quad (10.16)$$

The partial wave amplitudes are given by the formula

$$F^{J\pm}_{\lambda_c\lambda_d,\lambda_a\lambda_b} = \tfrac{1}{2}\int_{-1}^{1} d\cos\theta [C^{J+}_{\lambda\mu}(z) M^{\pm}_{\lambda_c\lambda_d,\lambda_a\lambda_b} + C^{J-}_{\lambda\mu}(z) M^{\mp}_{\lambda_c\lambda_d,\lambda_a\lambda_b}] \quad (10.17)$$

The formalism given above is very convenient for the description of Regge poles and anymptotic behavior of the scattering amplitudes. This aspect of the subject is beyond the scope of this book, although we shall give some allied results in the examples discussed in subsequent sections.

10.3 KINEMATICS FOR THE REACTIONS $\omega\pi \rightarrow \pi\pi$ AND $A_1\pi \rightarrow \pi\pi$

The simplest class of reactions involving spin is that in which only one particle has spin. The latter must of course be a Boson. Of this class, spin 1 is the simplest and most interesting case. We give examples of both vector (1^-) and pseudovector (1^+) particles. We choose the spinless particles to be pions. Among the known particles the ω ($I = 0$, $G = -$, $J^P = 1^-$) at 775 MeV and A_1 ($I = 1$, $G = -$, $J^P = 1^+$) at 1070 MeV can couple to three pions. Naturally one cannot perform direct experiments for the reactions $\omega\pi \rightarrow \pi\pi$ and $A_1\pi \rightarrow \pi\pi$. For theoretical calculations it is very useful to represent the many-body resonances, such as ω and A_1, by stable particles. The examples given here are easily adapted to other reactions of the same spin structure.

First consider the reaction $\omega\pi \rightarrow \pi\pi$. For orientation we consider the implication of parity and angular momentum conservation in the c.m. frame (Fig. 10.2). We ask what orbital momenta L of the $\omega\pi$ state lead to a state R of given J^P (Fig. 10.1). For $\omega\pi$ we have $P = (-1)^L$ and $J = L \pm 1,\ L$. Hence states of normal parity $P = (-1)^J$ have $L = J$ while $L = J \pm 1$ states are abnormal. The s wave ($L = 0$) cannot have $J = 0$. Hence the 0^+ state is excluded. For a fixed J the system has L, P given by

$$\begin{aligned} L &= J & P &= (-1)^J \\ L &= J \pm 1 & P &= -(-1)^J \end{aligned} \tag{10.18}$$

Similarly for given J^P the two pion state has $P = (-1)^J$ so only the first set of $\omega\pi$ states in (10.18) occur in $\omega\pi \to \pi\pi$. For each J^P there is only one

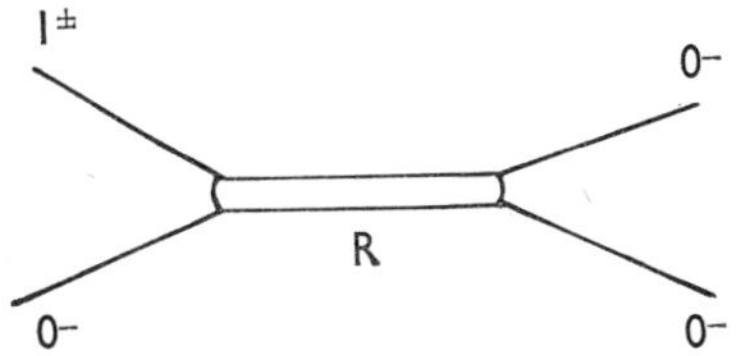

Figure 10.2 This figure illustrates an intermediate state R of definite spin-parity coupling to both 2π and $\omega\pi$ (or $A_1\pi$) states.

independent amplitude and we expect that only one invariant amplitude occurs. We note one further restriction due to Bose symmetry. Since only $I = 1$ occurs in $\omega\pi \to \pi\pi$, total symmetry under permutation of final pions requires odd J.

The s channel is labeled by

$$\omega(p_1, \lambda) + \pi_i(p_2) \to \pi_j(p_3) + \pi_k(p_4) \tag{10.19}$$

and the corresponding S matrix is

$$S = i(2\pi)^4\delta(p_3 + p_4 - p_1 - p_2)e_\mu(p_1\lambda)T^\mu_{ijk}(p_3p_4;\, p_1p_2) \cdot (16E_1\omega_2\omega_3\omega_4)^{-\frac{1}{2}} \tag{10.20}$$

where e_μ is the helicity wave function of the ω meson. T_μ is a pseudovector, and has the form

$$T^{(s)}_{\mu ijk} = i\epsilon_{ijk}\epsilon_{\mu\nu\rho\sigma}p_2^\nu p_3^\rho p_4^\sigma A(s, t, u) \tag{10.21}$$

Clearly only isospin 1 occurs in this reaction. The variables s, t, u are defined by $s = (p_1 + p_2)^2$, $t = (p_1 - p_3)^2$, $u = (p_1 - p_4)^2$, with $s + t + u = m_\omega^2 + 3\mu^2$. Here μ is the pion mass. The form (10.21) is clearly the same as obtained from the effective point coupling

$$\mathscr{L}(x) = g\epsilon_{ijk}\epsilon_{\mu\nu\rho\sigma}\omega^{\mu}\,\partial^{\nu}\pi_i\,\partial^{\rho}\pi_j\,\partial^{\sigma}\pi_k \tag{10.22}$$

The real analytic function $A(s, t, u)$ contains all the dynamical information about the process $\omega\pi \to \pi\pi$. It is a completely symmetric function of its arguments. To prove this we note that the s, t, and u channels describe identical processes. Choosing notation as in our previous discussion of $\pi\pi$ scattering (Sec. 8.2), we define the u channel process and its amplitudes by

$$\begin{aligned} &\omega(p_1, \lambda) + \pi_k(\bar{p}_4) \to \pi_j(p_3) + \pi_i(\bar{p}_2) \\ &T^{(u)}_{\mu kji}(p_3\bar{p}_2; p_1\bar{p}_4) = i\epsilon_{kji}\epsilon_{\mu\nu\rho\sigma}\bar{p}_4^{\nu}p_3^{\rho}\bar{p}_2^{\sigma}A^{u}(s', t', u') \\ &s' = (p_1 + \bar{p}_4)^2 \\ &t' = (p_1 - p_3)^2 \\ &u' = (p_1 - \bar{p}_2)^2 \end{aligned} \tag{10.23}$$

If $\bar{p}_4 \to p_2, \bar{p}_2 \to p_4, k \leftrightarrow i$, (10.23) coincides with (10.19) so that the channel label u occurring in A^u may be suppressed (the same function describes all three channels). With these conventions the crossing condition is

$$T^{(s)}_{\mu ijk}(p_3p_4, p_1p_2) = T^{(u)}_{\mu kji}(p_3 - p_2; p_1 - p_4) \tag{10.24}$$

Noting that when $\bar{p}_2 = -p_2$, $\bar{p}_4 = -p_2(s', t', u')$ have the values (u, t, s); we find, inserting (10.20) and (10.23) into (10.24)

$$A(s, t, u) = A(u, t, s) \tag{10.25}$$

The t channel is treated in the same manner. In this case s-t crossing leads to

$$A(s, t, u) = A(t, s, u) \tag{10.26}$$

The complete symmetry is exhibited by the ρ pole terms (Fig. 10.3) whose vertices are conveniently parametrized by the effective Lagrangian densities

$$\begin{aligned} \mathscr{L}_{\rho\pi\pi} &= f_{\rho\pi\pi}\epsilon_{abc}\rho^{\mu}_{a}\pi_b\,\partial_{\mu}\pi_c \\ \mathscr{L}_{\omega\rho\pi} &= g_{\omega\rho\pi}\epsilon^{\mu\nu\lambda\sigma}\,\partial_{\mu}\omega_{\nu}\,\partial_{\lambda}\boldsymbol{\rho}_{\sigma}\cdot\boldsymbol{\pi} \end{aligned} \tag{10.27}$$

The amplitude A accordingly has pole terms at the ρ mass

$$A_{\text{pole}} = 2f_{\rho\pi\pi}g_{\omega\rho\pi}\left(\frac{1}{s - m_\rho^2} + \frac{1}{t - m_\rho^2} + \frac{1}{u - m_\rho^2}\right) \tag{10.28}$$

Next we describe the partial wave analysis of the s channel reaction in the c.m. frame, with geometry as given by Fig. 10.1. The initial c.m. momentum

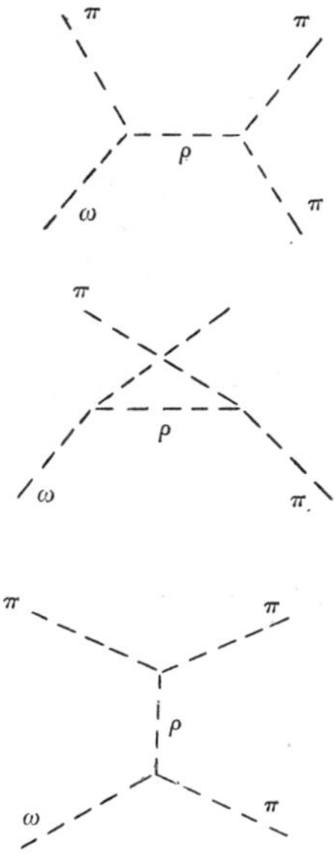

Figure 10.3 The ρ pole contributions (in the s, t, and u channels) to the reaction $\omega\pi \to \pi\pi$ are illustrated.

p and the final momentum p' are given by

$$p^2 = [s - (m_\omega + \mu)^2][s - (m_\omega - \mu)^2]/4s$$
$$p'^2 = (s - 4\mu^2)/4 \tag{10.29}$$

The individual four-momenta are given by

$$\begin{aligned} p_1^\mu &= (E, 0, 0, p) \\ p_2^\mu &= (\omega, 0, 0, -p) \\ p_3^\mu &= (\omega', p' \sin\theta, 0, p' \cos\theta) \\ p_4^\mu &= (\omega', -p' \sin\theta, 0, -p' \cos\theta) \end{aligned} \tag{10.30}$$

Using the wave functions $e^\mu(p_1, \pm 1) = \mp(1/\sqrt{2})(0, 1, \pm i, 0)$, $e^\mu(p_1, 0) = (1/m)(p, 0, 0, E)$ and dropping the isospin factor gives

$$\begin{aligned} M_\lambda \equiv e_\mu(p_1, \lambda)T^\mu &= Wpp' \sin\theta A(s, t, u)/\sqrt{2} \qquad \lambda = \pm 1 \\ &= 0, \qquad \lambda = 0 \end{aligned} \tag{10.31}$$

The vanishing of the longitudinal amplitude is easily understood as a consequence of TP invariance. W is $\sqrt{s}$, the total c.m. energy. $e \cdot T$ is the same as the amplitude M occurring in Eq. (10.9) and has the expansion ($\lambda = \pm 1$)

$$M_{\pm} = \sum_{\text{odd } J} (2J + 1)F_{\pm}^{J} d_{\pm 1,0}^{J}(\theta)$$
$$F_{-}^{J} = \left(\frac{Wpp'[J(J + 1)]^{\frac{1}{2}}}{2\sqrt{2}(2J + 1)}\right)\int_{-1}^{1} dz[P_{J-1}(z) - P_{J+1}(z)]A(s, t, u) \qquad (10.32)$$

The symmetry of A under $z \rightarrow -z$ ($t \leftrightarrow u$) ensures the vanishing of all F_{-}^{J} having integral J. In obtaining (10.32) we have used $d_{\pm 1,0}^{J}(\theta) = \mp \sin \theta P'_J(\cos \theta)/(J(J + 1))^{\frac{1}{2}}$ and some standard identities satisfied by the Legendre functions.

The invariant amplitude A has the simple expansion

$$A(s, t, u) = \sum_{\text{odd } J} G_J(s)P'_J(\cos \theta) \qquad (10.33)$$

From this form it is easy to make a standard Watson-Sommerfeld transform and find the Regge asymptotic formula A $\sim (1 - e^{-i\pi\alpha(s)})t^{\alpha(s)-1}$ as $t \rightarrow \infty$, s fixed. Here the function $\alpha(s)$ is the position of the leading pole of $G_J(s)$ in the angular momentum plane.

Next we study the changes which occur if one changes the parity of the vector meson. We also change the isospin from zero to one in order that our considerations be applicable to the observed A_1 meson. For a 1^+ meson the pattern of (10.18) is interchanged, i.e.

$$L = J, \qquad P = -(-1)^J$$
$$L = J \pm 1, \qquad P = (-1)^J \qquad (10.34)$$

Hence the $\pi\pi$ channel couples to $A\pi$ states having two values of L (except that 0^- is excluded), and we expect two invariant amplitudes. In the present problem all values of isospin are allowed.

We describe the s-channel reaction $A\pi \rightarrow \pi\pi$ by momenta p_i ($i = 1, 2, 3, 4$) and Cartesian charge indices i, j, k, l, which range from 1 to 3:

$$A_i(p_1, \lambda) + \pi_j(p_2) \rightarrow \pi_k(p_3) + \pi_l(p_4) \qquad (10.35)$$

λ is the helicity of the A meson. As usual, the kinematic invariants are defined by $s = (p_1 + p_2)^2$, $t = (p_1 - p_3)^2$, $u = (p_1 - p_4)^2$, with $s + t + u = m_A^2 + 3\mu^2$. Letting e_μ be the helicity wave function of the A meson, the invariant amplitude M_λ may be written as

$$M_\lambda = e_\mu(p_1, \lambda)M^{\mu}_{kl,ij}(p_3p_4, p_1p_2)$$
$$S = i(2\pi)^4\delta(p_1 + p_2 - p_3 - p_4)(iM_\lambda)(16E_1\omega_2\omega_3\omega_4)^{-\frac{1}{2}} \qquad (10.36)$$

The factor i is introduced for notational convenience (to make the invariant amplitudes introduced below real analytic functions of s, t, and u). Before describing the vector structure of M^μ, we take case of the isospin dependence in precise analogy to the treatment of $\pi\pi$ scattering by introducing three (vector) amplitudes A^μ, B^μ, and C^μ (cf. Sec. 8.2)

$$M^\mu_{kl,ij} = \delta_{kl}\delta_{ij}A^\mu + \delta_{ki}\delta_{jl}B^\mu + \delta_{kj}\delta_{li}C^\mu \tag{10.37}$$

The isospin amplitudes are $M_0 = 3A + B + C$, $M_1 = B - C$, and $M_2 = B + C$. For each isospin state there are two invariant functions M_i. To be definite about our notation, we describe the u-channel amplitude by

$$\begin{aligned} A_i(p_1, \lambda) + \pi_l(\bar{p}_4) &\to \pi_k(p_3) + \pi_j(\bar{p}_2) \\ M^{(u)}_\lambda &= e_\mu(p_1, \lambda)M_{kj,il}(p_3\bar{p}_2; p_1\bar{p}_4) \end{aligned} \tag{10.38}$$

and the t-channel reaction by

$$\begin{aligned} A_i(p_1, \lambda) + \pi_k(\bar{p}_3) &\to \pi_j(\bar{p}_2) + \pi_l(p_4) \\ M^{(t)}_\lambda &= e_\mu(p_1, \lambda)M^\mu_{jl,ik}(\bar{p}_2p_4; p_1\bar{p}_3) \end{aligned} \tag{10.39}$$

The crossing relations are

$$\begin{aligned} M^\mu_{kl,ij}(p_3p_4; p_1p_2) &= M^\mu_{kj,il}(p_3 - p_2; p_1 - p_4) \quad (su) \\ M^\mu_{kl,ij}(p_3p_4; p_1p_2) &= M^\mu_{jl,ik}(-p_2p_4; p_1 - p_3) \quad (st) \end{aligned} \tag{10.40}$$

In addition, there is a tu crossing relation which is equivalent to the Bose symmetry requirement in the s channel:

$$M^\mu_{kl,ij}(p_3p_4; p_1p_2) = M^\mu_{lk,ij}(p_4p_3; p_1p_2) \tag{10.41}$$

We have not given the functions M channel labels since all channels are the same. In terms of A, B, and C the crossing relations are

$$\begin{aligned} &\left.\begin{aligned} A_\mu(p_3p_4; p_1p_2) &= C_\mu(p_3 - p_2; p_1 - p_4) \\ B_\mu(p_3p_4; p_1p_2) &= B_\mu(p_3 - p_2; p_1 - p_4) \end{aligned}\right\} su \\ &\left.\begin{aligned} A_\mu(p_3p_4; p_1p_2) &= B_\mu(-p_2p_4; p_1 - p_3) \\ C_\mu(p_3p_4; p_1p_2) &= C_\mu(-p_2p_4; p_1 - p_3) \end{aligned}\right\} st \\ &\left.\begin{aligned} A_\mu(p_3p_4; p_1p_2) &= A_\mu(p_4p_3; p_1p_2) \\ B_\mu(p_3p_4; p_1p_2) &= C_\mu(p_4p_3; p_1p_2) \end{aligned}\right\} tu \end{aligned} \tag{10.42}$$

Each vector amplitude can be expressed in terms of two invariant functions, e.g.

$$\begin{aligned} A^s(p_3p_4;\, p_1p_2) &= (p_2 + p_3)_\mu A(s, t, u) + (p_2 - p_3)_\mu A'(s, t, u) \\ A^u_\mu(p_3\bar{p}_2;\, p_1\bar{p}_4) &= (\bar{p}_4 + p_3)_\mu A(s', t', u') + (\bar{p}_4 - p_3)_\mu A'(s', t', u') \\ A^t_\mu(\bar{p}_2p_4;\, p_4\bar{p}_3) &= (\bar{p}_2 + \bar{p}_3)_\mu A(s'', t'', u'') + (\bar{p}_3 - \bar{p}_2)_\mu A'(s'', t'', u'') \end{aligned} \tag{10.43}$$

where $s' = (p_1 + \bar{p}_4)^2$, $t' = (p_1 - p_3)^2$, $u' = (p_1 - \bar{p}_2)^2$, and $s'' = (p_1 + \bar{p}_3)^2$, $t'' = (p_1 - \bar{p}_2)^2$, $u'' = (p_1 - p_4)^2$. Since the s-, t-, and u-channel reactions are identical, the functions A and A' introduced in Eq. (10.43) do not need channel labels; they are the same functions. Performing the analytic continuation in Eq. (10.42) changes the variables as follows: su crossing, $(s't'u') \to (uts)$; st crossing, $(s''t''u'') \to (tsu)$; and tu crossing, $(stu) \to (sut)$. The invariant functions of Eq. (10.43) are especially suited to st crossing. Noting that p^μ is equivalent to zero because of $e_\mu(p_1, \lambda)$, one easily finds the crossing conditions connecting the invariant amplitudes.

$$\left.\begin{aligned} \begin{pmatrix} A(stu) \\ A'(stu) \end{pmatrix} &= \begin{pmatrix} -1 & 0 \\ 0 & 1 \end{pmatrix} \begin{pmatrix} B(tsu) \\ B'(tsu) \end{pmatrix} \\ \begin{pmatrix} C(stu) \\ C'(stu) \end{pmatrix} &= \begin{pmatrix} -1 & 0 \\ 0 & 1 \end{pmatrix} \begin{pmatrix} C(tsu) \\ C'(tsu) \end{pmatrix} \end{aligned}\right\} st \text{ crossing}$$

$$\left.\begin{aligned} \begin{pmatrix} A(stu) \\ A'(stu) \end{pmatrix} &= \begin{pmatrix} \frac{1}{2} & -\frac{1}{2} \\ -\frac{3}{2} & -\frac{1}{2} \end{pmatrix} \begin{pmatrix} C(uts) \\ C'(uts) \end{pmatrix} \\ \begin{pmatrix} B(stu) \\ B'(stu) \end{pmatrix} &= \begin{pmatrix} 1 & -\frac{1}{2} \\ -\frac{3}{2} & -\frac{1}{2} \end{pmatrix} \begin{pmatrix} B(uts) \\ B'(uts) \end{pmatrix} \end{aligned}\right\} su \text{ crossing} \tag{10.44}$$

$$\left.\begin{aligned} \begin{pmatrix} A(stu) \\ A'(stu) \end{pmatrix} &= \begin{pmatrix} \frac{1}{2} & \frac{1}{2} \\ \frac{3}{2} & -\frac{1}{2} \end{pmatrix} \begin{pmatrix} A(sut) \\ A'(sut) \end{pmatrix} \\ \begin{pmatrix} B(stu) \\ B'(stu) \end{pmatrix} &= \begin{pmatrix} \frac{1}{2} & \frac{1}{2} \\ \frac{3}{2} & -\frac{1}{2} \end{pmatrix} \begin{pmatrix} C(sut) \\ C'(sut) \end{pmatrix} \end{aligned}\right\} tu \text{ crossing}$$

A more useful statement of this result is to note that the isospin amplitudes $3M_I + M_I$, and $M_I - M_I$, are even or odd under $t \leftrightarrow u$ (i.e., $\cos\theta \to -\cos\theta$ in the s channel):

$$\begin{aligned} (3M_I + M_{I'})(stu) &= (-1)^I(3M_I + M_{I'})(sut) \\ (M_I - M_{I'})(stu) &= (-1)^{I+1}(M_I - M_{I'})(sut) \end{aligned} \tag{10.45}$$

These are the combinations which occur in the partial-wave expansion: the exhibited symmetry enforces even (odd) J for even (odd) I. To give the partial-wave expansion of the helicity amplitude M_λ we orient the xz plane in the scattering plane. The initial A momentum defines the z axis; $p_1^\mu = (E, 0, 0, p)$, $p_2^\mu = (\omega, 0, 0, -p)$, $p_3^\mu = (\omega', p' \sin\theta, 0, p' \cos\theta)$. We then obtain for the s-channel amplitudes

$$\begin{aligned} M_{\pm 1}^I &= \pm p' \sin\theta (M^I - M'^I) \\ m_A M_0^I &= p\omega'(3M^I + M'^I) - Ep' \cos\theta (M^I - M'^I) \end{aligned} \tag{10.46}$$

If we write

$$M_\lambda = \sum_J (2J + 1) F_\lambda^{IJ} d_{\lambda 0}^J(\theta) \tag{10.47}$$

we find the partial-wave amplitudes

$$\begin{aligned} F_{\pm 1}^{IJ} &= \frac{P'[J(J+1)]^{\frac{1}{2}}}{2\sqrt{2}(2J+1)} \int_{-1}^{1} dz (M_I' - M_I)(P_{J-1}(z) - P_{J+1}(z)) \\ F_0^{IJ} &= \frac{1}{2m_A}\left[p\omega' \int_{-1}^{1} dz (3M_I + M_I') P_J(z) \right. \\ &\quad \left. + Ep' \int_{-1}^{1} dz (M_I' - M_I) z P_J(z)\right] \end{aligned} \tag{10.48}$$

From these expressions, one sees that the particular linear combinations occurring in (10.45) are quite natural.

Since all three channels are the same, the number of independent invariant functions is not six but two. Motivated by the possible absence of $I = 2$ resonance poles, we write ($I_s = 2$)

$$\begin{aligned} M_2(s, t, u) &\equiv g(t, u) = (B + C)(stu) \\ M_2'(s, t, u) &\equiv g'(t, u) = (B' + C')(stu) \end{aligned} \tag{10.49}$$

We now use the crossing conditions to get a complete set of s-channel invariant amplitudes (A, B, C, A', B', C'): for $I_u = 2$,

$$\begin{aligned} M_2(uts) &= g(t, s) = \tfrac{1}{2}(A + B)(stu) - \tfrac{1}{2}(A' + B')(stu) \\ M_2'(uts) &= g'(t, s) = -\tfrac{3}{2}(A + B)(stu) - \tfrac{1}{2}(A' + B')(stu) \end{aligned} \tag{10.50}$$

and for $I_t = 2$

$$\begin{aligned} M_2(tsu) &= g(s, u) = -(A + C)(stu) \\ M_2'(tsu) &= g'(s, u) = (A' + C')(stu) \end{aligned} \tag{10.51}$$

The s-channel isospin amplitudes may be written explicitly as

$$\begin{aligned} M_2(s,t,u) &= g(t,u) \\ M_1(s,t,u) &= \tfrac{1}{2}[g(t,s) - g'(t,s) + 2g(s,u)] \\ M_0(s,t,u) &= \tfrac{1}{2}[\tfrac{3}{2}(g(t,s) - g'(t,s)) - 3g(s,u) - g(t,u)] \\ M_2'(s,t,u) &= g'(t,u) \\ M_1'(s,t,u) &= -\tfrac{1}{2}[3g(t,s) + g'(t,s) + 2g'(s,u)] \\ M_0'(s,t,u) &= \tfrac{1}{2}\{-\tfrac{3}{2}[3g(t,s) + g'(t,s)] + 3g'(s,u) - g'(t,u)\} \end{aligned} \tag{10.52}$$

The ρ meson plays a prominent role in the reaction $A\pi \to \pi\pi$. It contributes s-, t-, and u-channel poles. To define the pole terms we use the effective interaction Lagrangian (10.27) for $\rho\pi\pi$ and

$$\mathscr{L}_{A\rho\pi} = G_S \epsilon_{abc} \pi_a A_{b\mu} \rho_c^\mu + G_D \epsilon_{abc} \pi_a \,\partial_\mu A_{b\nu}\, \partial^\nu \rho_c^\mu \tag{10.53}$$

for the $A\rho\pi$ coupling. The ρ pole terms can be expressed in the form

$$\begin{aligned} M - M' &= \frac{4f_{\rho\pi\pi} G_S \mathscr{P}^1_{kl,ij}}{s - m_\rho^2} + \frac{f_{\rho\pi\pi}[2G_S + (s-u)G_D]}{t - m_\rho^2}\mathscr{P}^1_{jl,ik} \\ &\quad + \frac{f_{\rho\pi\pi}[2G_S + (s-t)G_D]}{u - m_\rho^2}\mathscr{P}^1_{kj,il} \\ 3M + M' &= -\frac{2f_{\rho\pi\pi} G_D (t-u)}{s - m_\rho^2}\mathscr{P}^1_{kl,ij} - \frac{f_{\rho\pi\pi}[6G_S + (u-s)G_D]}{t - m_\rho^2}\mathscr{P}^1_{jl,ik} \\ &\quad + \frac{f_{\rho\pi\pi}[6G_S + (t-s)G_D]}{u - m_\rho^2}\mathscr{P}^1_{kj,il} \end{aligned} \tag{10.54}$$

where the $\mathscr{P}^1$'s are isospin-1 projection operators in the s, t, and u channels. Using isospin crossing matrices, one can easily compute the isospin amplitudes (for $I_s = 1$, $\mathscr{P}^1_{kl,ij} \to 1$, $\mathscr{P}^1_{jl,ik} \to \tfrac{1}{2}$, $\mathscr{P}^1_{kj,il} \to \tfrac{1}{2}$; for $I_s = 2$, $\mathscr{P}^1_{kl,ij} \to 0$, $\mathscr{P}^1_{jl,ik} \to -\tfrac{1}{2}$, $\mathscr{P}^1_{kj,il} \to \tfrac{1}{2}$).

The scalar (0^+) isoscalar meson ϵ (700) also contributes to the reaction under consideration. In order to interpret this term in conventional language, we define the couplings

$$\begin{aligned} \mathscr{L}_{\varepsilon\pi\pi} &= g_{\varepsilon\pi\pi}\, \epsilon \boldsymbol{\pi} \cdot \boldsymbol{\pi} \\ \mathscr{L}_{A\varepsilon\pi} &= g_{A\varepsilon\pi} \mathbf{A}_\mu \cdot \partial^\mu \boldsymbol{\pi} \epsilon \end{aligned} \tag{10.55}$$

The ϵ pole terms arising from these couplings are:

$$
\begin{aligned}
M - M' &= -6g_{\epsilon\pi\pi}g_{A\varepsilon\pi}\left(\frac{\mathscr{P}^0_{jl,ik}}{t - m_\epsilon^2} - \frac{\mathscr{P}^0_{kj,il}}{u - m_\epsilon^2}\right) \\
3M + M' &= -6g_{\epsilon\pi\pi}g_{A\epsilon\pi}\left(\frac{2\mathscr{P}^0_{kl,ij}}{s - m_\epsilon^2} + \frac{\mathscr{P}^0_{jl,ik}}{t - m_\epsilon^2} + \frac{\mathscr{P}^0_{kj,il}}{u - m_\epsilon^2}\right)
\end{aligned}
\tag{10.56}
$$

where $\mathscr{P}^0_{kl,ij}$ is the $I_s = 0$ projection operator.

In order to see that the pole terms determine the asymptotic behavior of the full amplitude we consider in more detail the $I_s = 1$ s-channel amplitude. This contains poles at s, t, $u = m^2$, which correspond to Regge contributions of the form $t^{\alpha-\Delta}/\sin \pi\alpha(s)$, e.g., for fixed s, large t, etc., with $\alpha \approx 1$. The exponent Δ is determined by comparison with (10.51). For fixed s, large t we find ($I_s = 1$)

$$
\begin{aligned}
(3M_1 + M_1')(s, t, u) &\to t^{\alpha(s)} \\
(M_1 - M_1')(s, t, u) &\to t^{\alpha(s)-1}
\end{aligned}
\tag{10.57}
$$

to be the leading asymptotic behavior.

Taking $s \to \infty$, keeping t fixed near m_ρ^2, gives

$$
\begin{aligned}
(3M + M')(s, t, u) &\to s^{\alpha(t)} \\
(M - M')(s, t, u) &\to s^{\alpha(t)}
\end{aligned}
\tag{10.58}
$$

while for u fixed, $s \to \infty$

$$
\begin{aligned}
(3M + M')(s, t, u) &\to s^{\alpha(u)} \\
(M - M')(s, t, u) &\to s^{\alpha(u)}
\end{aligned}
\tag{10.59}
$$

Explicit models for the reactions $A\pi \to \pi\pi$ and $\omega\pi \to \pi\pi$ are studied in refs. 12 and 13. Other work in this area may be traced from these papers.

10.4 KINEMATICS FOR THE REACTIONS $A_1\pi \to A_1\pi$, $\omega\pi \to \omega\pi$ AND $\rho\pi \to \rho\pi$

The inelastic reactions considered in Sec. (10.3) are particularly simple because (for a given isospin) the nature of the final 2π state excludes one choice of normality. If we consider elastic scattering of vector or pseudovector particles off scalars (or pseudoscalars, as here) more possibilities occur. In the present section we consider the reactions $A_1\pi \to A_1\pi$ and $\omega\pi \to \omega\pi$ in detail. The reaction $\rho\pi \to \rho\pi$ has the isospin structure of the former and the spin-parity structure of the latter. The system of reactions under consideration is of considerable theoretical interest since the mesons in question presumably dominate matrix elements of the electromagnetic and weak currents. Hence the amplitudes are constrained by the algebraic relations of current algebra.

We describe the s-channel reaction $A\pi \to A\pi$ by momenta p^i $(i = 1, 2, 3, 4)$ and Cartesian charge indices i, j, k, l which run from 1 to 3:

$$A_i(p_1, \lambda) + \pi_j(p_2) \to A_k(p_3, \lambda') + \pi_l(p_4) \tag{10.60}$$

λ and λ' denote the initial and final helicity states. As usual the invariants are defined by $s = (p_1 + p_2)^2$, $t = (p_1 - p_3)^2$, and $u = (p_1 - p_4)^2$, with $s + t + u = 2m_A^2 + 2\mu^2$. Letting $e_\mu(p, \lambda)$ be the spin-1 helicity wave function, the invariant amplitude may be written as

$$M_{\lambda'\lambda} = e_\mu^*(p_3, \lambda')T^{\mu\nu}_{kl,ij}(p_3p_4;\, p_1p_2)e_\nu(p_1, \lambda) \tag{10.61}$$

where $M_{\lambda'\lambda}$ is related to the S matrix by

$$S - 1 = (2\pi)^4 i\delta(p_1 + p_2 - p_3 - p_4)M_{\lambda'\lambda}(16E_1E_3\omega_2\omega_4)^{-\frac{1}{2}} \tag{10.62}$$

where E_i and ω_j denote the energies of the A and π mesons. Before describing the tensor structure of $T^{\mu\nu}$ we take care of the isospin dependence in precise analogy to the treatment of the $\pi\pi \to \pi\pi$ and $A\pi \to \pi\pi$ problems by introducing three tensor amplitudes $A^{\mu\nu}$, $B^{\mu\nu}$, and $C^{\mu\nu}$:

$$T^{\mu\nu\ j}_{kl,ij} = A^{\mu\nu}\delta_{kl}\delta_{ij} + B^{\mu\nu}\delta_{ki}\delta_{lj} + C^{\mu\nu}\delta_{kj}\delta_{li} \tag{10.63}$$

The amplitudes having $I_s = 0$, 1, and 2 are $M_0 = 3A + B + C$, $M_1 = B - C$, and $M_2 = B + C$.

A convenient set of invariant amplitudes $T_i(s, t, u)$ is defined by the expansion

$$\begin{aligned} T_{\mu\nu}(p_3p_4;\, p_1p_2) = g_{\mu\nu}T_1(stu) + P_\mu P_\nu T_2(stu) + (P_\mu Q_\nu + Q_\mu P_\nu)T_3(stu) \\ + Q_\mu Q_\nu T_4(stu) \end{aligned}$$

$$\begin{aligned} P_\mu &= \tfrac{1}{2}(p_2 + p_4)_\mu \\ Q_\mu &= \tfrac{1}{2}(p_1 + p_3)_\mu \end{aligned} \tag{10.64}$$

where the various T could be isospin amplitudes, or one of the A, B, C amplitudes.

In order to give a precise and convenient statement of the crossing conditions, we define the u channel by

$$A_i(p_1, \lambda) + \pi_l(\bar{p}_4) \rightarrow A_k(p_3, \lambda') + \pi_j(\bar{p}_2)$$

$$\begin{aligned} M^u_{\lambda'\lambda}(p_3\bar{p}_2; p_1\bar{p}_4) &= e^*_\mu(p_3, \lambda')M^{u\mu\nu}_{kj,il}e_\nu(p_1, \lambda) \\ M^{u\mu\nu}_{kj\,il} &= A^{u\mu\nu}\delta_{kj}\delta_{il} + B^{u\mu\nu}\delta_{ki}\delta_{lj} + C^{u\mu\nu}\delta_{kl}\delta_{ij} \\ T^u_{\mu\nu}(p_3\bar{p}_2; p_1\bar{p}_4) &= g_{\mu\nu}T_1(s't'u') + \bar{P}_\mu\bar{P}_\nu T_2(s't'u') \\ &\quad + (\bar{P}_\mu Q_\nu + Q_\mu\bar{P}_\nu)T_3(s't'u') + Q_\mu Q_\nu T_4(s't'u') \end{aligned} \tag{10.65}$$

where $s' = (p_1 + \bar{p}_4)^2, t' = (p_1 - p_3)^2, u' = (p_1 - \bar{p}_3)^2$, and $\bar{P}_\mu = \frac{1}{2}(\bar{p}_2 + \bar{p}_4)_\mu$. We have omitted channel labels on the T_i because with the preceding conventions the T_i are the same invariant functions as occur in Eq. (10.64). If we continue $T^u_{\mu\nu}(p_3\bar{p}_2; p_1\bar{p}_4)$ to the point $\bar{p}_2 = -p_2, \bar{p}_4 = -p_4$, we obtain the amplitude for the s-channel reaction, the crossing conditions being

$$\begin{aligned} A_{\mu\nu}(p_3p_4; p_1p_2) &= C_{\mu\nu}(p_3 - p_2; p_1 - p_4) \\ B_{\mu\nu}(p_3p_4; p_1p_2) &= B_{\mu\nu}(p_3 - p_2; p_1 - p_4) \end{aligned} \tag{10.66}$$

In terms of invariant amplitudes, Eqs. (10.66) become

$$\begin{aligned} A_i(s, t, u) &= \epsilon_i C_i(u, t, s) \\ B_i(s, t, u) &= \epsilon_i B_i(u, t, s) \\ \epsilon_i &= 1, \qquad i = 1, 2, 4 \\ &= -1, \qquad i = 3 \end{aligned} \tag{10.67}$$

The t-channel reaction is written as

$$A_i(p_1, \lambda) + A_k(\bar{p}_3, \lambda') \rightarrow \pi_j(\bar{p}_2) + \pi_l(p_4)$$

$$\begin{aligned} M^t_{\lambda'\lambda} &= -e_\mu(p_1, \lambda)e^{(2)}_\nu(p_3, \lambda')M^{t\mu\nu}_{jl,ik}(\bar{p}_2p_4; p_1\bar{p}_3) \\ M^{t\mu\nu}_{jl,ik} &= A^{t\mu\nu}\delta_{jl}\delta_{ik} + B^{t\mu\nu}\delta_{ji}\delta_{kl} + C^{t\mu\nu}\delta_{jk}\delta_{il} \\ T^t_{\mu\nu}(\bar{p}_2p_4; p_1\bar{p}_3) &= g_{\mu\nu}T^t_1(s''t''u'') + R_\mu R_\nu T^t_2(s''t''u'') \\ &\quad + (R_\mu S_\nu + S_\mu R_\nu)T^t_3(s''t''u'') + S_\mu S_\nu T^t_4(s''t''u'') \end{aligned} \tag{10.68}$$

where $R_\mu = \frac{1}{2}(p_4 - \bar{p}_2)_\mu$, $S_\nu = \frac{1}{2}(p_1 - \bar{p}_3)_\nu$, $s'' = (p_1 + \bar{p}_3)^2$, $t'' = (p_1 - \bar{p}_2)^2$ and $u'' = (p_1 - p_4)^2$.

The conventions used here were motivated by the following considerations. First of all, the vector wave function with the superscript (2) differs from the usual wave function by a phase factor $(-1)^{1-\lambda}$, and is the wave function for particle number 2 in the convention of Jacob and Wick. The phase factor $(-1)^{\lambda}$ is useful because, when we continue from the t to the s channel, we obtain $e(-p_3, \lambda') = (-1)^{\lambda}e^*(p_3, \lambda')$; thus $-e^{(2)}(\bar{p}_3, \lambda')$ turns into $e^*(p_3, \lambda')$ as desired. By choosing the basis formed from R and S rather than P and Q, we find that the label number on an invariant function does not change on continuation from the t to the s channel (note that R, $S \rightarrow P$, Q when $\bar{p}_2 \rightarrow -p_2$, $\bar{p}_3 \rightarrow -p_3$).

The st crossing relations are

$$\begin{aligned} A_{\mu\nu}(p_3p_4;\, p_1p_2) &= B^t_{\mu\nu}(-p_2p_4;\, p_1 - p_3) \\ C_{\mu\nu}(p_3p_4;\, p_1p_2) &= C^t_{\mu\nu}(-p_2p_4;\, p_1 - p_3) \\ A^t_i(t, s, u) &= B_i(s, t, u) \\ B^t_i(t, s, u) &= A_i(s, t, u) \\ C^t_i(t, s, u) &= C_i(s, t, u) \end{aligned} \tag{10.69}$$

(For notational convenience we do not label the s-channel functions with a superscript s.)

Finally, we have an important relation which follows from Bose symmetry in the t channel. In terms of isospin amplitudes this is

$$T_i^{I_t}(t, s, u) = \epsilon_i(-1)^{I_t}T_i^{I_t}(t, u, s) \tag{10.70}$$

We now indicate how to express the isospin amplitudes in terms of amplitudes having $I = 2$ in the s, t, and u channels. The result is perfectly general but motivated by the simplicity of writing down $I = 2$ amplitudes when reasonable assumptions are made concerning their lack of resonance structure. We define functions f_i^s and f_i^t by

$$\begin{aligned} f_i^s(t, u) &\equiv T_i^{I_s=2}(s, t, u) = B_i(s, t, u) + C_i(s, t, u) \\ f_i^s(t, s) &\equiv T_i^{I_u=2}(u, t, s) = B_i(u, t, s) + C_i(u, t, s) \\ &\qquad = \epsilon_i[B_i(s, t, u) + A_i(s, t, u)] \\ f_i^t(s, u) &\equiv T_i^{I_t=2}(t, s, u) = B_i^t(t, s, u) + C_i^t(t, s, u) \\ &\qquad = A_i(s, t, u) + C_i(s, t, u) \end{aligned} \tag{10.71}$$

It is now a simple algebraic task to find the s-channel isospin amplitudes

$$\begin{aligned} T_i^0(s, t, u) &= \tfrac{1}{2}[3\epsilon_i f_i^s(t, s) + 3f_i^t(s, u) - f_i^s(t, u)] \\ T_i^1(s, t, u) &= \epsilon_i f_i^s(t, s) - f_i^t(s, u) \\ T_i^2(s, t, u) &\equiv f_i^s(t, u) \end{aligned} \tag{10.72}$$

In a similar way the t-channel isospin amplitudes are found to be

$$T_i^0(t, s, u) = \tfrac{1}{2}[3f_i^s(t, u) + 3\epsilon_i f_i^s(t, s) - f_i^t(s, u)]$$
$$T_i^1(t, s, u) = \epsilon_i f_i^s(t, s) - f_i^s(t, u) \qquad (10.73)$$
$$T_i^2(t, s, u) = f_i^t(s, u)$$

Next we consider the partial-wave expansion of the parity-conserving helicity amplitudes. First consider the s channel ($A\pi \to A\pi$). For a given J, states of either parity can appear, except that $J^P = 0^-$ is forbidden. In the center-of-mass (c.m.) frame of this reaction we choose $p_1^\mu = (E, 0, 0, p)$, $p_2^\mu = (\omega, 0, 0, -p)$, $p_3^\mu = (E, p \sin\theta, 0, p\cos\theta)$, $p_4^\mu = (\omega, -p\sin\theta, 0, -p\cos\theta)$. Parity-conserving amplitudes $M_{\lambda'\lambda}^{\nu}$ dominated asymptotically by states of normality $\nu = P(-1)^J$ are defined by (cf. Eq. (10.14))

$$M_{00}^{+} = 2M_{00}$$
$$M_{01}^{+} = 2M_{01}/\sin\theta \qquad (10.74)$$
$$M_{11}^{\pm} = \frac{M_{11}}{1+\cos\theta} \pm \frac{M_{-11}}{1-\cos\theta}$$

These amplitudes are given in terms of the invariant amplitudes T_i by the relations

$$M_{00}^{+} = \frac{2p^2}{m_A^2}[T_1 + \omega(\omega + E)T_2 + E(\omega + E)T_3]$$
$$+ \frac{2\cos\theta}{m_A^2}[-E^2T_1 + p^2E(\omega(T_2 - T_3) + E(T_3 - T_4))]$$
$$+ \frac{p^2E^2}{2m_A^2}(1+\cos\theta)^2[T_2 + T_4 - 2T_3]$$
$$\qquad (10.75)$$
$$\sqrt{2}M_{01}^{+} = \frac{2}{m_A}[ET_1 + \tfrac{1}{2}p^2\omega(T_3 - T_2) + \tfrac{1}{2}p^2E(T_4 - T_3)]$$
$$+ \frac{p^2E}{2m_A}(1+\cos\theta)(2T_3 - T_2 - T_4)$$
$$M_{11}^{+} = -T_1 - \tfrac{1}{4}p^2\cos\theta(2T_3 - T_2 - T_4)$$
$$M_{11}^{-} = \tfrac{1}{4}p^2(2T_3 - T_2 - T_4)$$

Partial-wave amplitudes $F_{\lambda'\lambda}^{J\nu}$ of parity $P = \nu(-1)^J$ are defined in terms of helicity states normalized to unity:

$$\begin{aligned} &F_{00}^{J} \\ &F_{10}^{J+} = (F_{10}^{J} + F_{-10}^{J})/\sqrt{2} \\ &F_{11}^{J\pm} = F_{11}^{J} \pm F_{-11}^{J} \end{aligned} \tag{10.76}$$

The partial-wave expansions are given by Eq. (10.15):

$$\begin{aligned} M_{00} &= \sum_J (2J+1) F_{00}^{J} P_J(z) \\ M_{01}^{+} &= \sqrt{2} \sum_J \frac{2J+1}{[J(J+1)]^{\frac{1}{2}}} F_{10}^{J+} P'_J(z) \\ M_{11}^{\pm} &= \sum_J \frac{2J+1}{J(J+1)} \{F_{11}^{J\pm}[P'_J(z) + zP''_J(z)] - F_{11}^{J\mp} P''_J(z)\} \end{aligned} \tag{10.77}$$

The unitary conditions for our amplitudes are

$$\begin{aligned} \operatorname{Im} F_{00}^{J} &= -\pi\rho(|F_{00}^{J}|^2 + |F_{10}^{J+}|^2) \\ \operatorname{Im} F_{10}^{J+} &= -\pi\rho(F_{11}^{J\pm} F_{10}^{J\pm *} + F_{10}^{J\pm} F_{00}^{J*}) \\ \operatorname{Im} F_{11}^{J\pm} &= -\pi\rho(|F_{11}^{J\pm}|^2 + |F_{10}^{J\pm}|^2) \\ \rho &= p/16\pi^2 W \end{aligned} \tag{10.78}$$

In the t channel, Bose symmetry of the pions dictates that only states of normal parity $(-1)^J$ occur. The c.m. variables are defined by $p_1^{\mu} = (E, 0, 0, p)$, $p_2^{\mu} = (\omega, p' \sin\theta_t, 0, p'\cos\theta_t)$, where $p^2 = \frac{1}{4}(t - 4m_A{}^2)$, $(p')^2 = \frac{1}{4}(t - 4\mu^2)$, and $\cos\theta_t = (s-u)/4pp'$. The parity-conserving amplitudes are

$$\begin{aligned} M_{1-1}^{+} &= 2M_{1-1}^{t}/\sin^2\theta_t, \qquad & M_{11}^{+} &= 2M_{11}^{t} \\ M_{10}^{+} &= 2M_{10}^{t}/\sin\theta_t, \qquad & M_{00}^{+} &= 2M_{00}^{t} \end{aligned} \tag{10.79}$$

A simple calculation yields

$$\begin{aligned} M_{1-1}^{+} &= -(p')^2 T_2 \\ M_{11}^{+} &= -2T_1 + (p')^2 \sin^2\theta_t T_2 \\ M_{10}^{+} &= (\sqrt{2}/m_A)[E(p')^2 \cos\theta_t T_2 - pp'ET_3] \\ m_A^2 M_{00} &= (E^2 + p^2)T_1 - (Ep'\cos\theta_t)^2 T_2 - p^2E^2T_4 \\ &\quad + 2E^2pp'\cos\theta_t T_3 \end{aligned} \tag{10.80}$$

The amplitudes M_{1-1}^{+}, M_{10}^{+}, M_{11}^{+}, and M_{00}^{+} are expected to exhibit Regge behavior $s^{\alpha-2}$, $s^{\alpha-1}$, s^{α}, and s^{α}, respectively, as $s \to \infty$, t fixed. (This follows from the properties of the $e^{J\pm}$ functions.) Thus the invariant amplitudes T_i are bounded by

$$T_1 \sim s^{\alpha(t)}, \qquad T_2 \sim s^{\alpha(t)-2}, \qquad T_3 \sim s^{\alpha(t)-1}, \qquad T_4 \sim s^{\alpha(t)} \tag{10.81}$$

where $\alpha(t)$ is the leading positive normality trajectory, which is here the $\rho - f$ trajectory.

Applying a similar analysis to the s-channel amplitudes, we find that the expected asymptotic behavior is

$$M_{00} \sim t^{\alpha}, \qquad M_{10}^{+} \sim t^{\alpha-1}, \qquad M_{11}^{+} \sim t^{\alpha-1}, \qquad M_{11}^{-} \sim t^{\alpha-2} \tag{10.82}$$

(in the absence of a negative normality trajectory having $\alpha_{-} > \alpha - 1$) in the limit $t \to \infty$, s fixed. Study of Eqs. (10.75) gives the following behavior of the invariant amplitudes:

$$\begin{aligned} T_1 &\sim t^{\alpha-1}, & T_2 - T_3 &\sim t^{\alpha-1} \\ T_3 - T_4 &\sim t^{\alpha-1}, & 2T_3 - T_2 - T_4 &\sim t^{\alpha-2} \end{aligned} \tag{10.83}$$

so that the behavior of the *individual* T_i is not determined. If the T_i were completely independent, T_2, T_3, and T_4 would all go as $t^{\alpha-2}$, which would have undesirable consequences, such as the absence of ρ poles. Thus each of the T_i $(i = 2, 3, 4)$ has t^{α} asymptotic behavior but is correlated with the other T_i in order to satisfy (10.83):

$$\begin{aligned} T_2 &= a(s)t^{\alpha} + b(s)t^{\alpha-1} + \cdots \\ T_3 &= a(s)t^{\alpha} + \tfrac{1}{2}[b(s) + c(s)]t^{\alpha-1} + \cdots \\ T_4 &= a(s)t^{\alpha} + c(s)t^{\alpha-1} + \cdots \end{aligned} \tag{10.84}$$

The invariant amplitudes T_i contain poles at the ρ mass whose residues may be related to $A\rho\pi$, $AA\rho$, and $\rho\pi\pi$ couplings defined in a standard way by effective Lagrangian densities. In addition to the $\rho\pi\pi$ and $A\rho\pi$ couplings defined by Eqs. (10.27) and (10.53) we need the $AA\rho$ coupling:

$$\mathscr{L}_{AA\rho} = \epsilon_{abc}[g_1 A_{\mu}^{a}\, \partial_{\nu} A^{b\mu} \rho^{c\nu} + g_2 A_{\mu}^{a}\, \partial^{\mu} A_{\nu}^{b} \rho^{c\nu} + g_3\, \partial^{\lambda} A_{\mu}^{a}\, \partial^{\mu} A_{\nu}^{b}\, \partial^{\nu} \rho_{\lambda}^{c}] \tag{10.85}$$

These expressions are unique only when all three particles are on the mass shell, which is appropriate for computing residues of pole terms.

In momentum space, the $A\rho\pi$ and $AA\rho$ vertices for $A(p_1) \to \rho(p_2) + \pi(q)$

and $\rho(p_1) \to A(p_2) + A(p_3)$ are described by the tensors

$$\Gamma_{\mu\nu} = G_S g_{\mu\nu} + G_D p_{1\mu} p_{2\nu}$$
$$\Gamma_{\mu\nu\lambda} = g_1 g_{\mu\nu}(p_2 + p_3)_\lambda + g_2[g_{\mu\lambda}(p_2 + p_3)_\nu + g_{\nu\lambda}(p_2 + p_3)_\mu] \quad (10.86)$$
$$+ g_3(p_2 + p_3)_\mu(p_2 + p_3)_\nu(p_2 + p_3)_\lambda$$

The relation of these vertices to current algebra has been studied in ref. 14 and in subsequent work.

The $I_s = 1$ s-channel ρ poles are given by

$$T_1(s, t, u) = 2G_S^2/(s - m_\rho^2)$$
$$T_2(s, t, u) = \frac{2}{s - m_\rho^2}\left\{-\frac{G_S^2}{m_\rho^2} - \frac{G_S G_D}{m_\rho^2}(m_\rho^2 + m_A^2 - m_\pi^2) + G_D^2\left[(m_A^2 - \tfrac{1}{2}t) - \frac{(m_\rho^2 + m_A^2 - m_\pi^2)^2}{4m_\rho^2}\right]\right\}$$
$$T_3(s, t, u) = \frac{2}{s - m_\rho^2}\left\{-\frac{G_S^2}{m_\rho^2} + \frac{G_S G_D}{m_\rho^2}(m_\rho^2 - m_A^2 + m_\pi^2) + G_D^2\left[(m_A^2 - \tfrac{1}{2}t) - \frac{(m_\rho^2 + m_A^2 - m_\pi^2)^2}{4m_\rho^2}\right]\right\} \quad (10.87)$$
$$T_4(s, t, u) = \frac{2}{s - m_\rho^2}\left\{-\frac{G_S^2}{m_\rho^2} + \frac{G_S G_D}{m_\rho^2}(3m_\rho^2 - m_A^2 + m_\pi^2) + G_D^2\left[(m_A^2 - \tfrac{1}{2}t) - \frac{(m_\rho^2 + m_A^2 - m_\pi^2)^2}{4m_\rho^2}\right]\right\}$$

The amplitudes T_2, T_3, and T_4 differ only in the $G_S G_D$ terms. In particular, the t dependence is identical. Further, these amplitudes appear in the partial-wave expansion only in the combinations $T_2 - T_3$, $T_3 - T_4$. The latter combinations are not only independent of t near $s = m_\rho^2$ but also are equal:

$$\text{Res}\,(T_2 - T_3)|_{s=m_\rho^2} = \text{Res}\,(T_3 - T_4)|_{s=m_\rho^2}$$
$$\text{Res}\,(2T_3 - T_2 - T_4)|_{s=m_\rho^2} = 0 \quad (10.88)$$

Equations (10.87) and (10.88) should be compared with the conditions (10.84) for $\alpha \approx 1$. Clearly an amplitude lacking t^α and $t^{\alpha-1}$ terms conspiring in the indicated manner would correspond to a vanishing $A\rho\pi$ coupling.

Calculation of the ρ exchange term leads to a t-channel pole given by $(I_t = 1)$

$$\begin{aligned} T_1(t, s, u) &= -\frac{2f_{\rho\pi\pi}g_1}{t - m_\rho^2}(s - u) \\ T_2(t, s, u) &= 0 \\ T_3(t, s, u) &= -\frac{8f_{\rho\pi\pi}g_2}{t - m_\rho^2} \\ T_4(t, s, u) &= -\frac{8f_{\rho\pi\pi}g_3}{t - m_\rho^2}(s - u) \end{aligned} \tag{10.89}$$

The preceding pole terms are compatible with the asymptotic behavior indicated in (10.81) and (10.83) and may be used to derive the latter (with the exception that $\text{Res}\, T_2|_{t=m_\rho^2} = 0$ does not lead uniquely to the behavior $s^{\alpha-2}$).

Equations (10.87) and (10.89) show that Res T_i at $\alpha_s = 1$ (or $\alpha_t = 1$) are linear in t (or s). In addition, the $I_s = 1$ amplitudes have to satisfy a relation which is a special case of the relation (10.84) (valid for all s), which guarantees the correct asymptotic signature. A slightly more stringent condition is required to prevent parity doubling on the leading trajectory, namely, that $\text{Res}\,(2T_3 - T_2 - T_4)|_{\alpha(s)=N}$ vanish for $I_s = 1$ to order z^N and z^{N-1}. This follows on computing residues of the positive and negative normality partial-wave amplitudes for $F_{11}^{J\pm}$ at $\alpha(s_N) = N$. One finds that for $J = N$, $\text{Res}\,(2T_3 - T_2 - T_4) \sim z^{N-2}$ gives no pole in the odd normality amplitude as expected.

Finally, consider the contribution of the s-channel ρ pole to the parity-conserving helicity amplitudes. The latter have residues easily computed to be

$$\begin{aligned} \text{Res}\, M_{00}^+ &= -4\cos\theta(EG_S + p^2 m_\rho G_D)^2/m_A^2 \\ \text{Res}\, M_{01}^+ &= 2\sqrt{2}(EG_S^2 + p^2 m_\rho G_S G_D)/m_A \\ \text{Res}\, M_{11}^+ &= -2G_S^2 \\ \text{Res}\, M_{11}^- &= 0 \end{aligned} \tag{10.90}$$

Here E and p^2 are to be evaluated for $s = m_\rho^2$. The angular dependence exhibited in Eq. (10.90) ensures the absence of poles in amplitudes having $J \neq 1$.

Next we consider the reactions $\omega\pi \to \omega\pi$ and $\omega\omega \to \pi\pi$, which are quite similar to the reactions just considered. For $\omega\pi$ scattering there is no exotic channel ($I_s = 1$ and $I_t = 0$ only). We describe the s channel by momenta p_i ($i = 1, 2, 3, 4$) and pion Cartesian charge indices i, j:

$$\omega(p_1\lambda) + \pi_i(p_2) \to \omega(p_3\lambda') + \pi_j(p_4) \tag{10.91}$$

The invariant amplitudes are defined in precise analogy to Eqs. (10.64), (10.65) and (10.68). The s-u crossing relation is

$$T_i(s, t, u) = \epsilon_i T_i(u, t, s) \tag{10.92}$$

with ϵ_i defined in Eq. (10.67). The isospin structure of the present problem is trivial. In order to find the relation of the parity conserving helicity amplitudes to the invariant amplitudes one simply changes the sign of the normality on the left hand sides of Eqs. (10.74) and (10.75), and changes $m_A \to m_\omega$. The partial wave expansions are as in (10.77) with all normalities changed. The Regge asymptotic behavior corresponding to the (normal) ρ trajectory is easily found. In the t channel the asymptotic behavior is given in Eq. (10.81).

$$\begin{aligned} M_{00}^- &\sim 0 \\ M_{10}^- &\sim 0 \\ M_{11}^- &\sim t^{\alpha-2} \\ M_{11}^+ &\sim t^{\alpha-1} \end{aligned} \tag{10.93}$$

In (10.93) the zero for M_{00}^- and M_{10}^- means that a single positive-normality trajectory will not contribute to these amplitudes.

The B trajectory ("abnormal") exhibits the same asymptotic behavior in $\omega\pi$ scattering as ρ does in $A\pi$ scattering (with an over-all change of normality):

$$\begin{aligned} M_{00}^- &\sim t^{\alpha_B} \\ M_{01}^- &\sim t^{\alpha_B-1} \\ M_{11}^- &\sim t^{\alpha_B-1} \\ M_{11}^+ &\sim t^{\alpha_B-2} \end{aligned} \tag{10.94}$$

Next consider how the invariant amplitudes T_i behave in the limit $t \to \infty$. This behavior must be correlated if the particle content of the leading trajectory is to agree with that assumed at the outset. First we note that the amplitudes behave as in Eq. (10.83):

$$
\begin{aligned}
T_1 &\sim t^{\alpha_B-1} \\
T_2 - T_3 &\sim t^{\alpha_B-1} \\
T_3 - T_4 &\sim t^{\alpha_B-1} \\
2T_3 - T_2 - T_4 &\sim t^{\alpha_B-2}
\end{aligned}
\tag{10.95}
$$

where the individual T_2, T_3, $T_4 \sim t^{\alpha_B}$.

The proper description of the ρ trajectory requires quite different behavior. To begin, note that $2T_3 - T_2 - T_4 \sim t^{\alpha-1}$ follows from the asymptotic behavior of M_{11}^+. In order that M_{11}^- go as $t^{\alpha-2}$, we see that T_1 must go as t^α and that the t^α, $t^{\alpha-1}$ terms cancel. Thus far we have ($z \to \infty$)

$$
\begin{aligned}
2T_3 - T_2 - T_4 &\sim t^{\alpha-1} \\
T_1 + \tfrac{1}{4}p^2z(2T_3 - T_2 - T_4) &\sim t^{\alpha-2}
\end{aligned}
\tag{10.96}
$$

In order that M_{01}^- go as $t^{\alpha-2}$, we need, in addition, the condition

$$(\omega + \tfrac{1}{2}E)(T_3 - T_2) + \tfrac{1}{2}E(T_4 - T_3) \sim t^{\alpha-2} \tag{10.97}$$

This equation (and (10.96) also implies that $T_2 - T_3$ and $T_4 - T_3$ go as $t^{\alpha-1}$. Finally, consider M_{00}^-. When $z \to \infty$, the z and $(1 + z)^2$ terms go as $t^{\alpha-1}$ by virtue of Eqs. (10.96) and (10.97). In order that the first bracket go as $t^{\alpha-1}$, we need T_2 and T_3 to cancel the t^α part of T_1, which necessitates

$$T_1 + sT_i \sim t^{\alpha-1}, \qquad i = 2, 3, 4 \tag{10.98}$$

In summary, we have the following behavior of the invariant amplitudes giving a leading trajectory having the correct spin-parity structure:

$$
\begin{aligned}
T_i &\sim t^\alpha \qquad (i = 1, 2, 3, 4) \\
T_2 - T_3 &\sim t^{\alpha-1} \\
T_3 - T_4 &\sim t^{\alpha-1} \\
T_1 + \tfrac{1}{4}p^2z(2T_3 - T_2 - T_4) &\sim t^{\alpha-2} \\
(\omega + \tfrac{1}{2}E)(T_3 - T_2) + \tfrac{1}{2}E(T_4 - T_3) &\sim t^{\alpha-2} \\
T_1 + sT_i &\sim t^{\alpha-1} \qquad (i = 2, 3, 4)
\end{aligned}
\tag{10.99}
$$

This intricate set of conditions is satisfied by the s-channel ρ-pole terms arising from the effective Lagrangian density (10.27), from which we find

$$\begin{aligned} T_1 &= \tfrac{1}{4}g^2[s^2 + 2st - 2s(m_\omega^2 + m_\pi^2) + (m_\omega^2 - m_\pi^2)^2]/(s - m_\rho^2) \\ T_1 &= g^2 s p^2 z/(s - m_\rho^2) \\ T_2 &= g^2(m_\omega^2 - \tfrac{1}{2}t)/(s - m_\rho^2) \\ T_3 &= g^2(m_\pi^2 - s - \tfrac{1}{2}t)/(s - m_\rho^2) \\ T_4 &= g^2(2s - m_\omega^2 + 2m_\pi^2 - \tfrac{1}{2}t)/(s - m_\rho^2) \end{aligned} \tag{10.100}$$

In the second line we have simplified T_1 by introducing the c.m. momentum p. It is instructive to check in detail how Eqs. (10.99) are satisfied by the Born terms.

The B-meson poles satisfy (10.95). To normalize the amplitude to a conventional coupling-constant description, we define effective $B\omega\pi$ couplings g_s and g_D in direct analogy to the $A\rho\pi$ couplings:

$$\mathscr{L}_{B\omega\pi} = g_s \boldsymbol{\pi} \cdot \mathbf{B}_\mu \omega^\mu + g_D \boldsymbol{\pi} \cdot \partial_\mu \mathbf{B}_\nu \, \partial^\nu \omega^\mu \tag{10.101}$$

The s-channel poles are then given by

$$\begin{aligned} T_1 &= g_s^2/(s - m_B^2) \\ T_2 &= \left\{ -\frac{g_s^2}{m_B^2} - \frac{g_s g_D}{m_B^2}(s + m_\omega^2 - m_\pi^2) + g_D^2\left[m_\omega^2 - \tfrac{1}{2}t - \frac{(s + m_\omega^2 - m_\pi^2)^2}{4m_B^2} \right] \right\} \\ T_3 &= \left\{ -\frac{g_s^2}{m_B^2} + \frac{g_s g_D}{m_B^2}[2m_B^2 - s - m_\omega^2 + m_\pi^2] + g_D^2[m_\omega^2 - \tfrac{1}{2}t - (s + m_\omega^2 - m_\pi^2)^2/4m_B^2] \right\} \\ T_4 &= \left\{ -\frac{g_s^2}{m_B^2} + \frac{g_s g_D}{m_B^2}[4m_B^2 - m_\omega^2 + m_\pi^2 - s] + g_D^2[m_\omega^2 - \tfrac{1}{2}t - (s + m_\omega^2 - m_\pi^2)^2/4m_B^2] \right\} \end{aligned} \tag{10.102}$$

A number of recent works have dealt with various properties of the reactions under consideration. The kinematical analysis given here follows that of refs. 12, 15, and 16. Further references can be traced from these papers.

The reaction $\rho\pi \to \rho\pi$ shares many features of the reactions $A\pi \to A\pi$ and $\omega\pi \to \omega\pi$. $\rho\pi$ scattering is more complex than the latter two reactions since more particles contribute to its singularity structure. Since the kinematics involve simple modifications of relations given earlier we confine our discussion to the pole terms. The contribution of the A pole in $\rho\pi$ scattering may be found by appropriate label changes from the amplitudes for the ρ pole in $A\pi$ scattering, Eqs. (10.87). Similarly, the ρ pole in the t channel gives a structure similar to that of Eq. (10.89). The ω pole contributions are as in Eq. (10.100) but with modified labels. Only the pion pole has no analogue in the preceding analysis. A simple perturbation calculation gives for the tensor $T_{\mu\nu}$ (defined in analogy to (10.61))

$$T_{\mu\nu} = \frac{8f^2_{\rho\pi\pi}}{m^2_\pi - s} p_{4\mu}p_{2\nu}\mathscr{P}^1_{kl,ij} + \frac{8f^2_{\rho\pi\pi}}{m^2_\pi - u} p_{2\mu}p_{4\nu}\mathscr{P}^1_{kj,il} \qquad (10.103)$$

where $\mathscr{P}^1$ is the $I = 1$ projection operator.

The s channel isospin amplitudes are

$$T^{I_s=0}_{\mu\nu} = -\frac{8f^2_{\rho\pi\pi}}{m^2_\pi - u} p_{2\mu}p_{4\nu}$$

$$T^{I_s=1}_{\mu\nu} = \frac{8f^2_{\rho\pi\pi}}{m^2_\pi - s} p_{4\mu}p_{2\nu} + \frac{4f^2_{\rho\pi\pi}}{m^2_\pi - u} p_{2\mu}p_{4\nu} \qquad (10.104)$$

$$T^{I_s=2}_{\mu\nu} = \frac{4f^2_{\rho\pi\pi}}{m^2_\pi - u} p_{2\mu}p_{4\nu}$$

These relations may be converted to the basis T_i used earlier if desired.

An interesting feature of reactions considered above is the correlation of asymptotic behavior that must exist among the invariant amplitudes for large t (fixed s) in order that the particles on the leading trajectory have the correct spin and parity. To appreciate this situation, one has to note that the detailed form of the correlation of asymptotic behavior depends on the choice of invariant amplitudes. To be definite, we refer to our treatment of $A\pi$ scattering, for which the invariant amplitudes T_i were chosen appropriate to the t channel (i.e. they were independent as $s \to \infty$). The s-channel reaction may be described in terms of the same functions T_i if we choose the s-channel basis to be the appropriate continuation of the t-channel basis; in this case, the T_i cross into themselves. This simple crossing property is attained at the expense of dependency relations among the T_i as $t \to \infty$, s fixed. If we had used another set of amplitudes M_i, *independent* in the s

channel, then the crossing relations (from M to T) would exhibit a complexity comparable to the above "conspiracy." Conversely, if we were to use the M_i to describe the t-channel amplitude, we would have to enforce asymptotic relationships as $s \to \infty$, t fixed. Although a given channel may naturally lead to a given set of amplitudes (for a given type of trajectory) the crossed channel generally prefers a distinct set (or a particular dependency relation). Hence, in practice it seems useful when constructing a set of amplitudes to use a set of independent amplitudes in one channel and to enforce appropriate dependency relations in the crossed channels.

10.5 KINEMATICS FOR THE REACTIONS $N\pi \to N\pi$ AND $N\bar{N} \to \pi\pi$

The study of pion–nucleon scattering has provided many important insights into the nature of strong interactions. In order to describe this reaction we label the relevant quantum numbers in the various channels as indicated by

$$\begin{aligned}
N_\alpha(p, \lambda) + \pi_i(q) &\to N_{\alpha'}(p', \lambda') + \pi_j(q') \quad (s) \\
N_\alpha(p, \lambda) + \pi_j(\bar{q}') &\to N_{\alpha'}(p', \lambda') + \pi_i(\bar{q}) \quad (u) \qquad (10.105) \\
N_\alpha(p, \lambda) + \bar{N}(\bar{p}', \bar{\lambda}') &\to \pi_i(\bar{q}) + \pi_j(q') \qquad (t)
\end{aligned}$$

The subscripts label isospin states while λ, λ', etc. describe helicities.

The invariant transition matrices are defined by

$$\begin{aligned}
S^{(s)} - 1 &= i(2\pi)^4\delta(p + q - p' - q')T^{(s)}_{\lambda'\lambda}(p'\alpha'qj'; p\alpha qi)m/(4EE'\omega\omega')^{\frac{1}{2}} \\
S^{(u)} - 1 &= i(2\pi)^4\delta(p + \bar{q} - p' - \bar{q}')T^{(u)}_{\lambda'\lambda}(p'\alpha'\bar{q}i; p\alpha\bar{q}'j)m/(4EE'\omega\omega')^{\frac{1}{2}}
\end{aligned}$$

(10.106)

$$S^{(t)} = i(2\pi)^4\delta(p + \bar{p}' - \bar{q} - q')T^{(t)}_{\bar{\lambda}'\lambda}(\bar{q}iq'j; p\alpha\bar{p}'\bar{\alpha}')m/(4E\bar{E}'\bar{\omega}\omega')^{\frac{1}{2}}$$

The u channel amplitude is found from $T^{(s)}$ by relabeling. The amplitudes T may be expressed in the form

$$\begin{aligned}
T^{(s)}_{\lambda'\lambda} &= \bar{u}(p', \lambda')\chi^\dagger_{\alpha'}T^{(s)}_{ji}\chi_\alpha u(p, \lambda) \\
T^{(u)}_{\lambda'\lambda} &= \bar{u}(p', \lambda')\chi^\dagger_{\alpha'}T^{(u)}_{ij}\chi_\alpha u(p, \lambda) \qquad (10.107) \\
T^{(t)}_{\bar{\lambda}'\lambda} &= \bar{v}(\bar{p}', \lambda')\chi^{c\dagger}_{\bar{\alpha}'}T^{(t)}_{ij}\chi_\alpha u(p, \lambda)
\end{aligned}$$

The amplitudes $T^{(s)}_{ji}$, $T^{(u)}_{ij}$, $T^{(t)}_{ij}$ are 2×2 matrices in isospin space and 4×4 matrices in spin space. They may be resolved into isospin components using the projection operators of Chapter 8.

$$\begin{aligned} T_{ij}^{(s)} &= \sum_{I_s} \mathscr{P}_{ji}^{I_s} T^{I_s} \\ T_{ij}^{(u)} &= \sum_{I_u} \mathscr{P}_{ij}^{I_u} T^{I_u} \\ T_{ij}^{(t)} &= \sum_{It} \mathscr{P}_{ij}^{I_t} T^{I_t} \end{aligned} \tag{10.108}$$

In practice one often uses the "plus-minus" amplitudes introduced below.

Before proceeding further we consider the *s-t* crossing relation. (The *s-u* crossing relation is simpler and will be left as an exercise.) By using the *LSZ* formalism to "pull in" the fields of the particles to be crossed, we find

$$T_{\lambda'\lambda}^{(s)}(p'\alpha'q'j;\, p\alpha qi) = \int dx\, dy e^{-iq\cdot x} e^{ip'\cdot y} \bar{u}(p'\lambda')\chi_{\alpha}^{\dagger}\, m(x, y)$$

$$T_{\lambda'\lambda}^{(t)}(\bar{q}iq'j;\, p\alpha\bar{p}'\bar{\alpha}') = \int dx\, dy e^{i\bar{q}\cdot x} e^{-i\bar{p}'\cdot x} \bar{v}(\bar{p}', \bar{\lambda}')\chi_{\bar{\alpha}'}^{c\dagger}\, m(x, y) \tag{10.109}$$

$$m(x, y) = K_x \mathscr{K}_y \left(\frac{2\omega' E}{m}\right)^{\frac{1}{2}} \langle \pi_j(q')|\, T(\pi_i(x)\psi(y))\, |N_\alpha(p, \lambda)\rangle$$

where $K = \partial_x^2 + \mu^2$, $\mathscr{K}_y = i\gamma^\mu\, \partial_\mu - m$. If we continue $\bar{q}$, $\bar{p}'$ to the points $\bar{q} = -q$, $\bar{p}' = -p$, the exponentials coincide in (10.109). Moreover, as shown below, $v(-p', \lambda')$ coincides with $u(p', \lambda')$ except for a phase. Hence apart from the possible multivaluedness of the amplitudes, we see that the continued annihilation amplitude coincides with the πN scattering amplitude. Note that it is necessary to make an ordering convention for the $N\bar{N}$ state to avoid sign errors; our convention is $|N\bar{N}\rangle = a\dagger(N)b\dagger(\bar{N})\,|0\rangle$.

In order to relate $v(-p, \lambda)$ to $u(p, \lambda)$ we examine the explicit form of $v(-p, \lambda)$ given by Eq. (3.102). Choosing $\phi = 0$ as in Fig. (10.1) and using the first of Eqs. (3.100) gives

$$v(-p, \lambda) = \begin{pmatrix} -2\lambda \sinh \tfrac{1}{2}\bar{\omega}\chi_\lambda(\hat{p}) \\ \cosh \tfrac{1}{2}\bar{\omega}\chi_\lambda(\hat{p}) \end{pmatrix} \tag{10.110}$$

where $\bar{\omega}$ denotes that E has been continued to $-E$. From the relations

$$\begin{aligned} \cosh \tfrac{1}{2}\omega &= [(E + m)/2m]^{\frac{1}{2}} \\ \sinh \tfrac{1}{2}\omega &= [(E - m)/2m]^{\frac{1}{2}} \end{aligned} \tag{10.111}$$

we see that (10.110) is analytic in E except for cuts joining the branch points at $E = \pm m$ (Fig. 10.4). In the physical region of the annihilation channel E_t

has an infinitesimal positive imaginary part ϵ, while in the s channel $\operatorname{Im} t = -\epsilon$. The appropriate path of continuation is therefore from point I to point II in Fig. (10.4). When $E \to -E$ along this path we find $(E - m)^{\frac{1}{2}} \to i(E + m)^{\frac{1}{2}}$, $(E + m)^{\frac{1}{2}} \to -i(E - m)^{\frac{1}{2}}$. Comparing this result with the last two equations we find

$$v(-p, \lambda) = i(-1)^{\frac{1}{2}+\lambda} u(-p, \lambda) \tag{10.112}$$

Note that the helicity does not change in the process of continuation. Taking

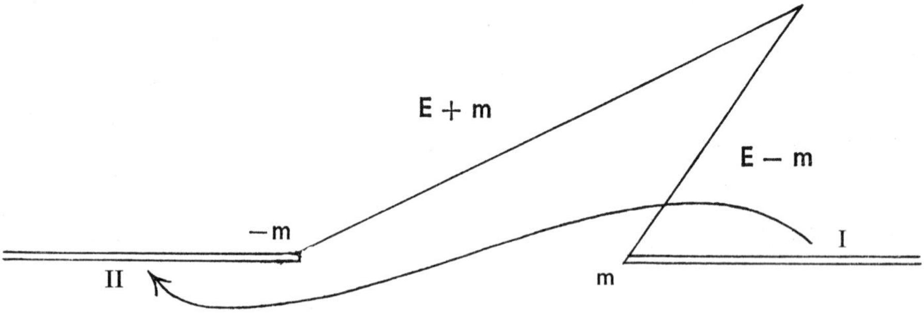

Figure 10.4 The continuation of the square root functions occurring in the nucleon spinor follows a definite path when s-t crossing is performed.

Eq. (7.7) into account gives the s-t crossing relation

$$T^{(t)}_{\lambda'\lambda}(-qi, q'j; p\alpha - p' - \alpha') = (-1)^{\frac{1}{2}+\alpha'} i(-1)^{\frac{1}{2}-\lambda'} T^{(s)}_{\lambda'\lambda}(p'\alpha' q'j; p\alpha qi) \tag{10.113}$$

The s-u crossing conditions are much simpler because only pion crossing occurs:

$$T^{(s)}_{\lambda'\lambda}(p'\alpha' - qi; p\alpha - q'j) = T^{(s)}_{\lambda'\lambda}(p'\alpha' q'j; p\alpha qi) \tag{10.114}$$

(The s and u channels are identical.)

The 4×4 matrices $T^{(s)}_{ji}$, $T^{(u)}_{ij}$, and $T^{(t)}_{ij}$ occurring in (10.107) may be written as

$$\begin{aligned} T^{(s)}_{ji} &= A_{ji}(s, t, u) + \tfrac{1}{2}(q + q') \cdot \gamma B_{ji}(s, t, u) \\ T^{(u)}_{ij} &= A_{ij}(s', t', u') + \tfrac{1}{2}(\bar{q}' + \bar{q}) \cdot \gamma B_{ij}(s', t', u') \\ T^{(t)}_{ij} &= A^{t}_{ij}(s'', t'', u'') + \tfrac{1}{2}(-\bar{q} + q') \cdot \gamma B^{t}_{ij}(s'', t'', u'') \end{aligned} \tag{10.115}$$

We have suppressed the channel labels for $A^{(s)}$, $A^{(u)}$ since the same functions describe both reactions.

The s-u and s-t crossing relations are

$$\left.\begin{aligned} A_{ji}(s, t, u) &= A_{ij}(u, t, s) \\ B_{ji}(s, t, u) &= -A_{ij}(u, t, s) \end{aligned}\right\} su$$
$$\left.\begin{aligned} A^t_{ij}(t, s, u) &= A_{ji}(s, t, u) \\ B^t_{ij}(t, s, u) &= B_{ji}(s, t, u) \end{aligned}\right\} st \qquad (10.116)$$

The second pair of relations states that the t channel amplitudes are obtained by analytic continuation of the s-channel amplitudes. The first pair of equations gives a constraint on the amplitudes A and B, so that we obtain a crossing *symmetry*.

The amplitudes A_{ji} and B_{ji} are the invariant amplitudes but are matrices in isospin space. These may be expanded in isospin amplitudes using the projection operators (8.24–8.25). It is conventional to use another set of amplitudes having simpler crossing properties. Hence we introduce $A^{\pm}$, $B^{\pm}$ by writing

$$A_{ji}(s, t, u) = \delta_{ji}A^+(s, t, u) + \tfrac{1}{2}[\tau_j, \tau_i]A^-(s, t, u) \qquad (10.117)$$

Writing this in terms of isospin projection operators gives in the s channel

$$\begin{aligned} A^+ &= \tfrac{1}{3}A^{\frac{1}{2}} + \tfrac{2}{3}A^{\frac{3}{2}} \\ A^- &= \tfrac{1}{3}A^{\frac{1}{2}} - \tfrac{1}{3}A^{\frac{3}{2}} \end{aligned} \qquad (10.118)$$

The same decomposition holds for the B amplitudes. In terms of the plus-minus amplitudes the s-u crossing relations are

$$\begin{aligned} A^{\pm}(s, t, u) &= \pm A^{\pm}(u, t, s) \\ B^{\pm}(s, t, u) &= \mp B^{\pm}(u, t, s) \end{aligned} \qquad (10.119)$$

If we write A^t_{ij} as

$$A^t_{ij}(t, s, u) = \delta_{ij}A^{t(+)}(t, s, u) + \tfrac{1}{2}[\tau_i, \tau_j]A^t(t, s, u)] \qquad (10.120)$$

the isospin amplitudes are (cf. Eqs. (8.46) and (8.48))

$$\begin{aligned} A^{t(+)} &= A^0/\sqrt{6} \\ A^{t(-)} &= \tfrac{1}{2}A^1 \end{aligned} \qquad (10.121)$$

The B amplitudes obey the same relations as in (10.120) and (10.121). The s-t crossing relations now have the form

$$\begin{aligned} A^{t\pm}(t, s, u) &= \pm A^{\pm}(s, t, u) \\ B^{t\pm}(t, s, u) &= \mp B^{\pm}(s, t, u) \end{aligned} \qquad (10.122)$$

It is interesting to note that the Bose symmetry condition in the annihilation channel ($q'j \leftrightarrow \bar{q}i$) has the same form as the s-u crossing condition (10.119).

In Chapter 8 the partial wave analysis of πN scattering was considered from the point of view of the non-covariant two-component spinors referred to the fixed z axis. The amplitudes f_1 and f_2 considered there may be related to the A, B amplitudes by expanding the relation

$$f = \frac{m}{4\pi W} \bar{u}(p', s')Tu(p, s) \tag{10.123}$$

using the Dirac spinors of Eq. (3.93). The result of this calculation is

$$\begin{aligned} f_1 &= \frac{E + m}{8\pi W} [A + (W - m)B] \\ f_2 &= \frac{E - m}{8\pi W} [-A + (W + m)B] \end{aligned} \tag{10.124}$$

The following kinematic relations for πN scattering are frequently useful.

$$\begin{aligned} p^2 &= [s - (m + \mu)^2][s - (m - \mu)^2]/4s \\ E &= (s + m^2 - \mu^2)/2\sqrt{s} \\ \omega &= (s - m^2 + \mu^2)/2\sqrt{s} \end{aligned} \tag{10.125}$$

The invariant s-channel helicity amplitudes $T_{\lambda'\lambda} = \bar{u}(p', \lambda')Tu(p, \lambda)$ may be computed using explicit spinors of Eq. (3.101) and the form (10.115).

$$\begin{aligned} T_{++} &= T_{--} = \cos\frac{\theta}{2}\left[A + \frac{s - (m^2 + \mu^2)}{2m} B\right] \\ T_{+-} &= -T_{-+} = e^{-i\phi} \sin\frac{\theta}{2}\left[\frac{E}{m} A + \omega B\right] \end{aligned} \tag{10.126}$$

The isospin labels have been suppressed. These amplitudes have the following partial wave expansion:

$$\begin{aligned} T_{++} &= 2\cos\frac{\theta}{2} \sum_J M^J_{++}[P'_{J+\frac{1}{2}}(z) - P'_{J-\frac{1}{2}}(z)] \\ T_{+-} &= 2e^{-i\phi} \sin\frac{\theta}{2} \sum_J M^J_{+-}[P'_{J+\frac{1}{2}}(z) + P'_{J-\frac{1}{2}}(z)] \end{aligned} \tag{10.127}$$

The unitarity condition obeyed by $M^J_{\lambda'\lambda}$ is

$$\operatorname{Im} M^J_{\lambda'\lambda} = \frac{mp}{4\pi W} \sum_{\lambda''} M^J_{\lambda'\lambda''} M^{J*}_{\lambda\lambda''} \tag{10.128}$$

This expression may be diagonalized by using parity eigenstates. We label these by the orbital momentum l introduced in Chapter 8

$$\begin{aligned} |Jl+\rangle &\equiv [|JM+\rangle + |JM-\rangle]/\sqrt{2} \\ |Jl-\rangle &\equiv [|JM+\rangle - |JM-\rangle]/\sqrt{2} \end{aligned} \tag{10.129}$$

On the left hand side the $\pm$ notation corresponds to $J = l \pm \frac{1}{2}$ while on the right hand side the $\pm$ notation refers to nucleon helicity $\pm\frac{1}{2}$. The orbital momentum l is not conserved and is only a conventional way of keeping track of the parity. For nonrelativistic energies l does label the orbital momentum, however. Corresponding to the states (10.129) we introduce amplitudes of definite parity and angular momentum J:

$$\begin{aligned} M_{(l+1)-} &= M^J_{++} + M^J_{+-} \qquad P = (-1)^{J-\frac{1}{2}} \\ M_{l+} &= M^J_{++} - M^J_{+-} \qquad P = (-1)^{J+\frac{1}{2}} \end{aligned} \tag{10.130}$$

The unitary condition (10.128) now allows one to express $M_{l\pm}$ in terms of conventional phase shifts

$$\begin{aligned} M_{l\pm} &= (e^{2i\delta_{l\pm}} - 1)/2i\rho \\ \rho &= mp/4\pi W \end{aligned} \tag{10.131}$$

The $\delta_{l\pm}$ are real in the elastic scattering region. In order to compare the expansion (10.127) with that previously obtained, we define $f_{\lambda'\lambda}$ by

$$f_{\lambda'\lambda} = \frac{m}{4\pi W} T_{\lambda'\lambda} \tag{10.132}$$

The amplitudes $f_{\lambda'\lambda}$ have the expansions

$$\begin{aligned} f_{++} &= \cos\frac{\theta}{2} \sum_J [f_{(J-\frac{1}{2})+} + f_{(J+\frac{1}{2})-}][P'_{J+\frac{1}{2}}(z) - P'_{J-\frac{1}{2}}(z)] \\ f_{+-} &= e^{-i\phi} \sin\frac{\theta}{2} \sum_J [f_{(J-\frac{1}{2}+)} - f_{(J+\frac{1}{2})+}][P'_{J+\frac{1}{2}}(z) + P'_{J-\frac{1}{2}}(z)] \end{aligned} \tag{10.133}$$

Comparison with the expansions of f_1 and f_2 shows the relation

$$\begin{aligned} f_{++} &= \cos\frac{\theta}{2}(f_1 + f_2) \\ f_{+-} &= e^{-i\phi} \sin\frac{\theta}{2}(f_1 - f_2) \end{aligned} \tag{10.134}$$

The relation between the two expansions is further elucidated by writing out the 2×2 matrix of helicity amplitudes

$$\mathbf{F} = \begin{pmatrix} f_{++} & f_{+-} \\ f_{-+} & f_{--} \end{pmatrix} = \begin{pmatrix} (f_1 + f_2)\cos\dfrac{\theta}{2} & e^{-i\phi}\sin\dfrac{\theta}{2}(f_1 - f_2) \\ -e^{i\phi}\sin\dfrac{\theta}{2}(f_1 - f_2) & (f_1 + f_2)\cos\dfrac{\theta}{2} \end{pmatrix} \tag{10.135}$$

We note that $\mathbf{F}$ may be written in the form

$$\mathbf{F} = (f_1 + f_2)\cos\frac{\theta}{2} + i\hat{n}\cdot\boldsymbol{\sigma}\sin\frac{\theta}{2}(f_1 - f_2) \tag{10.136}$$

where $\hat{n}$ is defined in Eq. (3.98). It is now straightforward to relate (10.135) to the 2×2 matrix $\mathbf{f} = f_1 + \boldsymbol{\sigma}\cdot\hat{p}'\,\boldsymbol{\sigma}\cdot\hat{p}f_2$. Let the z axis lie along $\mathbf{p}$ as usual so that $\chi_\lambda(\hat{p})$ are the $\chi(p)$, $\chi(n)$ of Eq. (7.8). Then we write

$$f_{\lambda'\lambda} = \chi^\dagger_{\lambda'}(\hat{p})\mathbf{F}\chi_\lambda(\hat{p}) \tag{10.137}$$

Since $\chi_{\lambda'}(\hat{p}') = e^{-i\hat{n}\sigma\cdot\theta/2}\chi_{\lambda'}(\hat{p})$ we write (10.137) as

$$\begin{aligned} f_{\lambda'\lambda} &= \chi_{\lambda'}(\hat{p}')(e^{-i\hat{n}\cdot\sigma\theta/2}\mathbf{F})\chi_\lambda(\hat{p}) \\ f_{\lambda'\lambda} &= \chi^\dagger_{\lambda'}(\hat{p}')\mathbf{f}\chi_\lambda(\hat{p}) \end{aligned} \tag{10.138}$$

verifying one's intuitive expectations.

Next we consider $N\bar{N} \to \pi\pi$. Instead of using directly the third of Eqs. (10.107) we define a similar amplitude in which the N wave function is that of "particle number two" in the conventions of ref. 1. The connection is

$$\begin{aligned} v_2(-\mathbf{p}, \lambda) &= (-1)^{\frac{1}{2}-\lambda}\exp(-i\pi\sigma_y/2)v(\mathbf{p}, \lambda), \quad i.e. \\ v_2(-\mathbf{p}, \lambda) &= \begin{pmatrix} \sinh\frac{1}{2}\omega\chi^*_\lambda(\hat{p}) \\ -2\lambda\cosh\frac{1}{2}\omega\chi^*_\lambda(\hat{p}) \end{pmatrix} \end{aligned} \tag{10.139}$$

The helicity amplitudes are then given by

$$\begin{aligned} F_{++} &= F_{--} = \frac{p}{m}A - q\cos\theta_t B \\ F_{+-} &= -F_{-+} = -\frac{Eq}{m}\sin\theta_t B \end{aligned} \tag{10.140}$$

where the momenta and scattering angle are given by

$$p^2 = \frac{t - 4m^2}{4}, \qquad q^2 = \frac{t - 4\mu^2}{4}$$
$$\cos\theta_t = \frac{s - u}{4pq} \tag{10.141}$$

The partial wave expansions are

$$F_{++} = \sum_J (2J + 1) P_J(\cos\theta_t) F^J_{++}(t)$$
$$F_{+-} = \sum_J \frac{(2J + 1)}{\sqrt{J(J + 1)}} \sin\theta_t P'_J(\cos\theta_t) F^J_{+-}(t) \tag{10.142}$$

The differential cross section is given by

$$d\sigma/d\Omega = \frac{p}{q} \left| \frac{m}{8\pi E} F_{\lambda'\lambda} \right|^2 \tag{10.143}$$

Further useful information may be found in ref. 16.

Finally consider the pole structure of A and B due to the nucleon exchange, ρ exchange and ε exchange. The effective meson nucleon couplings are

$$\mathscr{L}_{NN\pi} = i g_{\pi NN} \bar{N} \gamma_5 \boldsymbol{\tau} N \cdot \boldsymbol{\pi}$$
$$\mathscr{L}_{NN\rho} = f_{\rho NN} \bar{N} \gamma_\mu \frac{\boldsymbol{\tau}}{2} N \cdot \boldsymbol{\rho}^\mu + \frac{g_{\rho NN}}{4m} \bar{N} \sigma_{\mu\nu} \boldsymbol{\tau} N \cdot (\partial^\mu \boldsymbol{\rho}^\nu - \partial^\nu \boldsymbol{\rho}^\mu) \tag{10.144}$$
$$\mathscr{L}_{NN\varepsilon} = g_{\varepsilon NN} \varepsilon \bar{N} N$$

while the $\rho\pi\pi$ and $\varepsilon\pi\pi$ couplings are given in Eqs. (10.27) and (10.55). The pole terms shown in Fig. (10.5) then give the following contributions to the amplitudes $A^{\pm}$ and $B^{\pm}$:

$$A^+ = g_{\varepsilon\pi\pi} g_{\varepsilon NN}/(m_\varepsilon^2 - \mathrm{t})$$
$$A^- = \left(\frac{f_{\rho\pi\pi} g_{\rho NN}}{2m}\right)\left(\frac{s - u}{m_\rho^2 - \mathrm{t}}\right)$$
$$B^+ = g_{\pi NN}^2 \left(\frac{1}{m^2 - \bar{s}} - \frac{1}{m^2 - u}\right) \tag{10.145}$$
$$B^- = \frac{f_{\rho\pi\pi} f_{\rho NN}}{m_\rho^2 - t} - \frac{2 f_{\rho\pi\pi} g_{\rho NN}}{m_\rho^2 - t} + g_{\pi NN}^2 \left(\frac{1}{m^2 - \bar{s}} + \frac{1}{m^2 - u}\right)$$

It we assume that the invariant amplitudes have Regge behavior the structure

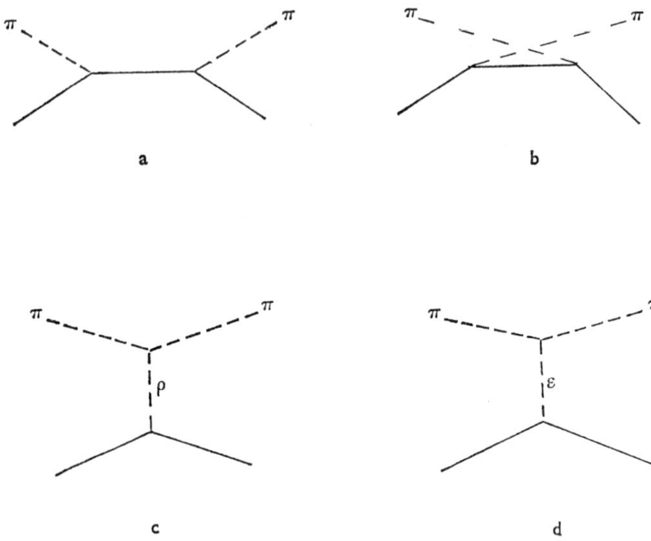

Figure 10.5 Some pole terms in πN scattering due to lighter "particles" are indicated (see Eq. (10.145)).

of the pole terms can be used to determine that behavior. All the particles considered in (10.145) are "normal." We see that as $s \to \infty$ a normal meson exchange gives the behavior $A \sim s^{\alpha}$, $B \sim s^{\alpha-1}$. The nucleon exchange gives $B \sim t^{\alpha-\frac{1}{2}}$ as $t \to \infty$. These relations agree with those obtained by a more general analysis.

10.6 FURTHER REFERENCES FOR KINEMATICS OF TWO BODY REACTIONS

The number of papers and books dealing with the kinematics of scattering processes is so great that any partial list is bound to be inadequate. Since any list is better than none at all, we give a few additional references dealing with several interesting reactions. There exist several modern textbooks containing such information.[18–21] Alternative approaches to the invariant amplitude problem have been studied in refs. 22–25.

The analytic properties of helicity amplitudes have been extensively studied in recent years from many points of view. In addition to refs. 7, 8, 9 we cite refs. 26 and 27, from which earlier contributions may be traced.

The crossing properties of helicity amplitudes are very important. References 28–30 are basic contributions to this subject. A useful treatment of three particle kinematics is given in ref. 31.

Next we turn to specific processes. Nucleon–nucleon scattering has been treated thoroughly in ref. 5 of Chapter 8. Standard treatments of pion photoproduction from nucleons are given in refs. 32–33. This process and several other reactions involving photons are treated systematically in ref. 5, with attention given to the construction of invariant amplitudes free of kinematic singularities. The kinematics of Compton scattering of photons from pions and nucleons is given in ref. 5. An extensive review of the properties of the matrix elements $\langle \pi N | J_\mu | N \rangle$ and $\langle \pi N | J^5_\mu | N \rangle$, where J_μ is the hadron electric current and J^5_μ the axial vector current, has been given in ref. 34. Electroproduction of nucleon resonances of any spin has been studied in ref. 35.

A quite different systematic analysis of the spin structure of scattering amplitudes has been developed in refs. 36 and 37. Many other pertinent references may be found in contemporary research papers.

PROBLEMS

1. Prove the crossing conditions stated in Eqs. (10.24), (10.40), (10.66), and (10.69).
2. Verify the expression (10.75) for the $A\pi \to A\pi$ parity conserving helicity amplitudes.
3. Prove that the $\lambda = 0$ amplitude for $\omega\pi \to \pi\pi$ vanishes.
4. Derive the pole terms for the reactions discussed in the text [cf. Eqs. (10.28), (10.54), (10.56), (10.87), (10.89), (10.100), (10.102), (10.103), and (10.145)] using the effective couplings of Eqs. (10.27), (10.53), (10.55), (10.85), (10.101), and (10.144).
5. If we change the isospin of the A meson to zero (call the resulting meson H) without changing the other quantum numbers, we obtain a reaction intermediate in complexity between $A\pi \to \pi\pi$ and $\omega\pi \to \pi\pi$. The reaction

$$H(p_1, \lambda) + \pi_j(p_2) \to \pi_k(p_3) + \pi_l(p_4)$$

has $I = 1$ in all channels. Show that the invariant helicity amplitude is

$$M_{\lambda klj} = i\epsilon_{jkl}e_\mu(p_1, \lambda)M^\mu(p_3p_4; p_1p_2)$$
$$M_\mu = (p_2 + p_3)_\mu M + (p_2 - p_3)_\mu M'$$

and that the relations for s-u and s-t crossing are

$$M^{\mu}_{kjl}(p_3 - p_2; p_1 - p_4) = M^{\mu}_{klj}(p_3p_4; p_1p_2)$$

$$M^{\mu}_{jlk}(-p_2p_4; p_1 - p_3) = M^{\mu}_{klj}(p_3p_4; p_1p_2)$$

The invariant amplitudes satisfy the crossing relations

$$\begin{pmatrix} M(stu) \\ M'(stu) \end{pmatrix} = \begin{pmatrix} -\frac{1}{2} & \frac{1}{2} \\ \frac{3}{2} & \frac{1}{2} \end{pmatrix} \begin{pmatrix} M(uts) \\ M'(uts) \end{pmatrix}$$

$$\begin{pmatrix} M(stu) \\ M'(stu) \end{pmatrix} = \begin{pmatrix} 1 & 0 \\ 0 & -1 \end{pmatrix} \begin{pmatrix} M(tsu) \\ M'(tsu) \end{pmatrix}$$

$$\begin{pmatrix} M(stu) \\ M'(stu) \end{pmatrix} = \begin{pmatrix} -\frac{1}{2} & -\frac{1}{2} \\ -\frac{3}{2} & \frac{1}{2} \end{pmatrix} \begin{pmatrix} M(sut) \\ M'(sut) \end{pmatrix}$$

From the last of these prove the Bose Symmetry condition

$$M(stu) - M'(stu) = -[M(sut) - M'(sut)]$$

$$3M(stu) + M'(stu) = 3M(sut) + M'(sut)$$

exactly as for the $I = 1$ $A\pi \to \pi\pi$ amplitudes.

In order to define the ρ pole terms we define the effective Lagrangian

$$\mathscr{L}H\rho\pi = G_S'\boldsymbol{\pi} \cdot \boldsymbol{\rho}^{\mu}H_{\mu} + G_D'\boldsymbol{\pi}\, \partial^{\nu}\boldsymbol{\rho}^{\mu}\, \partial_{\mu}H_{\nu}$$

Show that the resulting Born terms are

$$M - M' = \frac{-2f_{\rho\pi\pi}G_S'}{s - m_\rho^2} + \tfrac{1}{2}f_{\rho\pi\pi}\frac{[2G_S' + (s-t)G_D']}{u - m_\rho^2} + \tfrac{1}{2}f_{\rho\pi\pi}\frac{[2G_S' + (s-u)G_D']}{t - m_\rho^2}$$

$$3M + M' = \frac{f_{\rho\pi\pi}(t-u)G_D'}{s - m_\rho^2} + \tfrac{1}{2}f_{\rho\pi\pi}\frac{[6G_S' - (s-t)G_D']}{u - m_\rho^2} + \tfrac{1}{2}f_{\rho\pi\pi}\frac{[-6G_S' + (s-u)6_D']}{t - m_\rho^2}$$

The structure of these amplitudes resembles the $A\pi \to \pi\pi$ poles. Derive the asymptotic behavior of the foregoing amplitudes.

REFERENCES

1. M. Jacob and G. C. Wick, *Ann. Phys. (N.Y.)* **7,** 404 (1959).
2. D. Williams, Lawrence Radiation Laboratory Report UCRL 11113 (1963).
3. K. Hepp, *Helv. Phys. Acta.* **36,** 355 (1963); *ibid* **37,** 55 (1964).
4. A. Hearn, *Nuovo Cimento* **21,** 333 (1961).
5. W. A. Bardeen and Wu-Ki Tung, *Phys. Rev.* **173,** 1423 (1968).
6. M. Gell-Mann, M. L. Goldberger, F. E. Low, E. Marx, and F. Zachariasen, *Phys. Rev.* **133B,** 145 (1964).
7. Y. Hara, *Phys. Rev.* **136B,** 507 (1964).
8. L. L. Wang, *Phys. Rev.* **142,** 1187 (1965).
9. P. D. B. Collins and E. J. Squires, "Regge Poles in Particle Physics" (Springer-Verlag, Berlin, 1968).

10. M. Ademollo, H. R. Rubenstein, G. Veneziano and M. A. Virasoro, *Phys. Rev.* **176,** 1904 (1968).
11. P. Carruthers and F. Cooper, *Phys. Rev.* **D1,** 1223 (1970).
12. A. Capella, B. Diu, J. M. Kaplan and D. Schiff, *Nuovo Cimento* **64A,** 361 (1969).
13. P. Carruthers and E. Lasley, *Phys. Rev.* **D1,** 1204 (1970).
14. E. Abers and V. Teplitz, *Phys. Rev.* **D1,** 624 (1970).
15. G. F. Chew, M. L. Goldberger, F. E. Low, and Y. Nambu, *Phys. Rev.* **106,** 1337 (1957).
16. W. R. Frazer and J. R. Fulco, *Phys. Rev.* **117,** 1603 (1960).
17. V. Singh, *Phys. Rev.* **129,** 1889 (1963).
18. A. O. Barut, "Theory of the Scattering Matrix" (The Macmillan Co., London, 1967).
19. H. Burkhardt, "Dispersion Relation Dynamics" (North-Holland Publ. Co., Amsterdam, 1969).
20. R. J. Eden, "High Energy Collisions of Elementary Particles" (Cambridge Univ. Press, Cambridge, England, 1967).
21. H. Pilkuhn, "The Interactions of Hadrons (North-Holland Publ. Co., Amsterdam, 1967).
22. M. S. Marinov, *Ann. Phys. (N.Y.)* **49,** 357 (1968).
23. V. de Alfaro, S. Fubini, G. Furlan and C. Rossetti, *Ann. Phys. (N.Y.)* **44,** 165 (1967).
24. C. Rebbi, *Ann. Phys. (N.Y.)* **49,** 106 (1968).
25. M. D. Scadron and H. F. Jones, *Phys. Rev.* **173,** 1734 (1968).
26. E. Leader, *Phys. Rev.* **166,** 1599 (1968).
27. J. D. Jackson and G. E. Hite, *Phys. Rev.* **169,** 1248 (1968).
28. T. L. Trueman and G. C. Wick, *Ann. Phys. (N.Y.)* **26,** 322 (1964).
29. I. Muzinich, *J. Math. Phys.* **5,** 1481 (1964).
30. G. Cohen-Tannoudji, A. Morel and H. Navelet, *Ann. Phys. (N.Y.)* **46,** 239 (1968).
31. G. C. Wick, *Ann. Phys. (N.Y.)* **18,** 65 (1962).
32. G. F. Chew, M. L. Goldberger, F. E. Low, and Y. Nambu, *Phys. Rev.* **106,** 1345 (1957).
33. J. S. Ball, *Phys. Rev.* **124,** 2014 (1961).
34. S. Adler, *Ann. Phys. (N.Y.)* **50,** 189 (1968).
35. J. D. Bjorken and J. D. Walecka, *Ann. Phys. (N.Y.)* **38,** 35 (1966).
36. P. L. Csonka, M. J. Moravscik, and M. D. Scadron, *Phys. Rev.* **143,** 1324 (1966).
37. M. J. Moravscik, in "Recent Developments in Particle Physics," ed. M. J. Moravscik (Gordon and Breach, New York, 1966) p. 197.

Appendix A

NOTATION AND OTHER CONVENTIONS

The world of theoretical physics is divided into tribes according to the trivial but easily recognized superficial characteristics of metric, gamma matrices, etc. The conventions used in this book are essentially the same as those of the widely used textbooks by Bjorken and Drell.[1] Another widely used set of conventions is associated with the name "Pauli metric." The reader is advised to consult the "translation dictionary" of Adler and Dashen.[2] This dictionary permits a member of one tribe to check papers from another tribe for sign errors.

We employ a metric tensor $g_{\mu\nu}$ whose non-vanishing elements are $g_{00} = -g_{11} = -g_{22} = -g_{33} = 1$. A superscript μ on a four-vector or tensor is called contra-variant while a subscript is a covariant index. The space components of a contravariant four-vector are identified with a physical 3-vector. For example,* the four-vector $(t, \mathbf{x})$ is identified as x^μ whereas the covariant counterpart x_μ has components $(t, -\mathbf{x})$. (A brief but full discussion of tensors in special relativity may be found in the Appendix of a book by Rindler.[3]) The scalar product of two four-vectors V and W is written $V \cdot W$ and is

$$V \cdot W = g_{\mu\nu} V^\mu W^\nu = V_\mu W^\mu = V^\mu W_\mu = V^0 W^0 - \mathbf{V} \cdot \mathbf{W} \tag{A.1}$$

A four-vector V is space-like, time-like or lightlike according to whether the conditions $V^2 < 0$, $V^2 > 0$, $V^2 = 0$ are satisfied. The four-dimensional

* As usual the velocity of light c and Planck's constant (rather, $h/2\pi = \hbar$) are set equal to unity.

gradient ∂_μ is

$$\partial_\mu = \frac{\partial}{\partial x^\mu} = \left(\frac{\partial}{\partial t}, \nabla\right) \tag{A.2}$$

The operator $\partial_\mu \partial^\mu$ is denoted as ∂^2, which is the same as the symbol $\Box$ of ref. 1. The Levi-Civita symbol $\epsilon_{\mu\nu\rho\sigma}$ is chosen to satisfy ϵ_{0123} $(= -\varepsilon^{0123}) = +1$.

The gamma matrices γ^μ satisfy the anti-commutation rules

$$\{\gamma^\mu, \gamma^\nu\} \equiv \gamma^\mu\gamma^\nu + \gamma^\nu\gamma^\mu = 2g^{\mu\nu}I \tag{A.3}$$

and the Hermiticity conditions

$$\gamma^{0\dagger} = \gamma^0, \qquad \gamma^{i\dagger} = -\gamma^i = \gamma^0\gamma^i\gamma^0 \tag{A.4}$$

where $i = 1, 2, 3$. We have used two explicit sets of gamma matrices. The usual "low-energy" set is given by

$$\gamma^0 = \begin{pmatrix} 1 & 0 \\ 0 & -1 \end{pmatrix}$$

$$\gamma^i = \begin{pmatrix} 0 & \sigma_i \\ -\sigma_i & 0 \end{pmatrix} \tag{A.5}$$

$$\gamma_5 \equiv i\gamma^0\gamma^1\gamma^2\gamma^3 = -i\gamma_0\gamma_1\gamma_2\gamma_3 = \begin{pmatrix} 0 & 1 \\ 1 & 0 \end{pmatrix}$$

Here "unity" means the 2×2 unit matrix and σ_i the Pauli spin matrices. The σ_i occur so frequently that we do not raise or lower indices but always write σ_1, σ_2, σ_3:

$$\sigma_1 = \begin{pmatrix} 0 & 1 \\ 1 & 0 \end{pmatrix}, \qquad \sigma_2 = \begin{pmatrix} 0 & -i \\ i & 0 \end{pmatrix}, \qquad \sigma_3 = \begin{pmatrix} 1 & 0 \\ 0 & -1 \end{pmatrix} \tag{A.6}$$

The "high-energy set" of gamma matrices was employed in Sec. 3.4 and is given by

$$\gamma^0 = \begin{pmatrix} 0 & 1 \\ 1 & 0 \end{pmatrix} \quad \gamma^i = \begin{pmatrix} 0 & -\sigma_i \\ \sigma_i & 0 \end{pmatrix} \quad \gamma_5 = \begin{pmatrix} 1 & 0 \\ 0 & -1 \end{pmatrix} \tag{A.7}$$

In this case γ_5 is diagonal and $(1 \pm \gamma_5)/2$ projects into "left- and right-handed" subspaces.

From the gamma matrices we can construct sixteen quantities Γ_A which span the 4×4 spin space of the Dirac theory. These are conventionally grouped as follows:

$$\begin{aligned}
\Gamma_S &= I \\
\Gamma_P &= i\gamma_5 \\
\Gamma_{V\mu} &= \gamma_\mu \\
\Gamma_{A\mu} &= \gamma_\mu\gamma_5 \\
\Gamma_{T\mu\nu} &= \sigma_{\mu\nu}; \qquad \sigma_{\mu\nu} \equiv \frac{i}{2}[\gamma_\mu, \gamma_\nu]
\end{aligned} \tag{A.8}$$

The matrices $(\Gamma_A)_{\alpha\beta}$ act as Clebsch-Gordan coefficients in constructing tensor quantities from $\bar{\psi}_\alpha\psi_\beta$. Thus one can construct the so-called bilinear covariants from Dirac spinors $\bar{\psi}_1$ and ψ_2:

$$\begin{aligned}
S_{21} &= \bar{\psi}_2\psi_1 \\
P_{21} &= \bar{\psi}_2 i\gamma_5\psi_1 \\
V_{\mu 21} &= \bar{\psi}_2\gamma_\mu\psi_1 \\
A_{\mu 21} &= \bar{\psi}_2\gamma_\mu\gamma_5\psi_1 \\
T_{\mu\nu 21} &= \bar{\psi}_2\sigma_{\mu\nu}\psi_1
\end{aligned} \tag{A.9}$$

where the listed quantities are respectively scalar, pseudoscalar, vector, axial vector and antisymmetric tensor densities under homogeneous Lorentz transformations, including space reflections. When $\psi_1 = \psi_2$ the quantities in (A.9) are Hermitian because the Γ_A of (A.8) were chosen to satisfy $\Gamma_A^\dagger = \gamma_0\Gamma_A\gamma_0$.

It is frequently useful to know how the densities (A.9) transform under space inversion, time reversal and charge conjugation. These operations satisfy

$$\begin{aligned}
P\psi(x)P^{-1} &= \eta_P P_0\psi(x_P) \\
T\psi(x)T^{-1} &= \eta_T T_0\psi(x_T) \\
C\psi(x)C^{-1} &= \eta_C C_0\bar{\psi}^T(x)
\end{aligned} \tag{A.10}$$

where P_0, T_0, C_0 are 4×4 matrices and $\bar{\psi}^T$ denotes the transpose of $\bar{\psi}$. In Chapter 3 we used the explicit forms $P_0 = \gamma_0$, $T_0 = \gamma_3\gamma_1$, $C_0 = i\gamma_0\gamma_2$. These are especially useful solutions of the more general relations

$$\begin{aligned}
P_0\gamma_\mu P_0^{-1} &= g_{\mu\mu}\gamma_\mu \\
T_0\gamma_\mu^* T_0^{-1} &= g_{\mu\mu}\gamma_\mu \\
C_0\gamma_\mu^T C_0^{-1} &= -\gamma_\mu
\end{aligned} \tag{A.11}$$

Under P, C, and T the densities (A.9) transform as indicated in Table A.1. When internal symmetries are considered similar quantities occur with isospin matrices or SU(3) matrices standing inside the inner products in (A.9).

Table A.1 The behavior of the bilinear covariants under T, C, and P transformations is indicated. We have omitted the arbitrary phase factor which can contribute when $\psi_1 \neq \psi_2$.

	P	C	T	CPT
$S_{21}(\mathbf{x}, t)$	$S_{21}(-\mathbf{x}, t)$	$S_{12}(\mathbf{x}, t)$	$S_{21}(\mathbf{x}, -t)$	$S_{12}(-\mathbf{x}, -t)$
$P_{21}(\mathbf{x}, t)$	$-P_{21}(-\mathbf{x}, t)$	$P_{12}(\mathbf{x}, t)$	$-P_{21}(\mathbf{x}, -t)$	$P_{12}(-\mathbf{x}, -t)$
$V_{\mu 21}(\mathbf{x}, t)$	$g_{\mu\mu} V_{\mu 21}(-\mathbf{x}, t)$	$-V_{\mu 12}(\mathbf{x}, t)$	$g_{\mu\mu} V_{\mu 21}(\mathbf{x}, -t)$	$-V_{\mu 12}(-\mathbf{x}, -t)$
$A_{\mu 21}(\mathbf{x}, t)$	$-g_{\mu\mu} A_{\mu 21}(-\mathbf{x}, t)$	$A_{\mu 12}(\mathbf{x}, t)$	$g_{\mu\mu} A_{\mu 21}(\mathbf{x}, -t)$	$-A_{\mu 12}(-\mathbf{x}, -t)$
$T_{\mu\nu 21}(\mathbf{x}, t)$	$g_{\mu\mu} g_{\nu\nu} T_{\mu\nu 21}(-\mathbf{x}, t)$	$-T_{\mu\nu 12}(\mathbf{x}, t)$	$-g_{\mu\mu} g_{\nu\nu} T_{\mu\nu 21}(\mathbf{x}, -t_{\mu\nu})$	$T_{\mu\nu 12}(-\mathbf{x}, -t)$

REFERENCES

1. J. D. Bjorken and S. D. Drell, "Relativistic Quantum Mechanics" and "Relativistic Quantum Fields" (McGraw-Hill Publishing Co., New York, 1965).
2. S. L. Adler and R. F. Dashen, "Current Algebras" (W. A. Benjamin, Inc., New York, 1968).
3. W. Rindler, "Spectial Relativity" (Oliver and Boyd, Edinburgh, 1960).

Appendix B

PROPERTIES OF THE FUNCTIONS $d^j_{m'm}(\theta)$

The representation matrices of the rotation group $SU(2)$ occur frequently in applications. In this appendix we give a skeletal outline of their properties.[1] First, we write the rotation operator in terms of the Euler angles α, β, γ:

$$R(\alpha, \beta, \gamma) = e^{-i\alpha J_z} e^{-i\beta J_y} e^{-i\gamma J_z} \tag{B.1}$$

The matrices $D^j_{m'm}(\alpha, \beta, \gamma)$ are given by the matrix element

$$\langle jm' | R(\alpha, \beta, \gamma) | jm \rangle = e^{-im'\alpha} d^j_{m'm}(\beta) e^{-im\gamma} \tag{B.2}$$

where the functions $d^j_{m'm}$ are given by

$$d^j_{m'm}(\theta) = \langle jm' | e^{-i\theta J_y} | jm \rangle \tag{B.3}$$

In Eq. (10.11) an explicit formula was given for the more general case in which $2j$ is not necessarily an integer. For $2j =$ integer the functions $d^j_{m'm}$ are polynomials of order $2j$ in $\cos \theta/2$, $\sin \theta/2$. In Table B.1 we give the values

Table B.1 The functions $d^j_{m'm}(\theta)$ are given for $j = \frac{1}{2}, 1$

m' \ m	$\frac{1}{2}$	$-\frac{1}{2}$
$\frac{1}{2}$	$\cos\frac{\theta}{2}$	$-\sin\frac{\theta}{2}$
$-\frac{1}{2}$	$\sin\frac{\theta}{2}$	$\cos\frac{\theta}{2}$

m' \ m	1	0	-1
1	$\cos^2\frac{\theta}{2}$	$-\frac{\sin\theta}{\sqrt{2}}$	$\sin^2\frac{\theta}{2}$
0	$-\frac{\sin\theta}{\sqrt{2}}$	$\cos\theta$	$-\frac{\sin\theta}{\sqrt{2}}$
-1	$\sin^2\frac{\theta}{2}$	$\frac{\sin\theta}{\sqrt{2}}$	$\cos^2\frac{\theta}{2}$

for $j = \frac{1}{2}$ and 1.

The following symmetry relations are frequently useful:

$$d^j_{m'm}(\theta) = (-1)^{j+m'} d^j_{m',-m}(\pi - \theta)$$
$$d^j_{m'm}(\theta) = d^j_{-m,-m'}(\theta) = (-1)^{m'-m} d^j_{mm'}(\theta) \tag{B.4}$$

Recursion formulas as well as other useful information may be found in Edmonds[2] and Jacob and Wick.[3]

For integral $j\ (= l)$ the following relations are often useful:

$$d^l_{m0}(\theta) = (-1)^m d^l_{0m}(\theta) = \left(\frac{4\pi}{2l + l}\right)^{\frac{1}{2}} P_{lm}(\theta)$$
$$d^l_{00}(\theta) = P_l(\cos\theta) \tag{B.5}$$
$$d^l_{10}(\theta) = -[l(l + 1)]^{-\frac{1}{2}} \sin\theta P'_l(\cos\theta)$$

For half-integral $j = l + \frac{1}{2}$ one has analogous formulas:

$$d^j_{\frac{1}{2},\frac{1}{2}} = (l + 1)^{-1} \cos\frac{\theta}{2}(P'_{l+1} - P'_l)$$
$$d^j_{-\frac{1}{2},\frac{1}{2}} = (l + 1)^{-1} \sin\frac{\theta}{2}(P'_{l+1} + P'_l)$$
$$d^j_{\frac{1}{2},\frac{3}{2}} = (l + 1)^{-1} \sin\frac{\theta}{2}\left\{\left[\frac{l}{l + 2}\right]^{\frac{1}{2}} P'_{l+1} + \left[\frac{l + 2}{l}\right]^{\frac{1}{2}} P'_l\right\} \tag{B.6}$$
$$d^j_{-\frac{1}{2},\frac{3}{2}} = (l + 1)^{-1} \cos\frac{\theta}{2}\left\{-\left[\frac{l}{l + 2}\right]^{\frac{1}{2}} P'_{l+1} + \left[\frac{l + 2}{l}\right]^{\frac{1}{2}} P'_l\right\}$$

REFERENCES

1. M. E. Rose, "Elementary Theory of Angular Momentum" (John Wiley and Sons, Inc., New York, 1957).
2. A. R. Edmonds, "Angular Momentum in Quantum Mechanics" (Princeton University Press, Princeton, N.J., 1957).
3. M. Jacob and G. C. Wick, *Ann. Phys.* (*N.Y.*) **7**, 404 (1959).

INDEX